Comparative endocrinology of prolactin

Comparative endocrinology of prolactin

D. M. Ensor

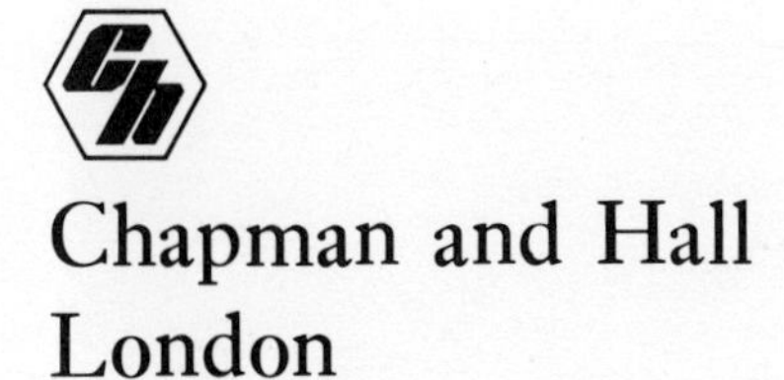

Chapman and Hall
London

A Halsted Press Book
John Wiley & Sons, New York

First published 1978
by Chapman and Hall Ltd
11 New Fetter Lane, London EC4P 4EE

Filmset in Great Britain by
Northumberland Press Ltd, Gateshead, Tyne and Wear
and printed at the University Printing House, Cambridge

ISBN 0 412 12720 2

Distributed in the U.S.A. by Halsted Press
a division of John Wiley & Sons, Inc., New York

Library of Congress Cataloging in Publication Data

Ensor, D. M.
Comparative endocrinology of prolaction.
Bibliography: p.
Includes index.
1. Lactogenic hormones. 2. Endocrinology, Comparative.
I. Title. [DNLM: 1. Physiology, Comparative.
2. Prolactin—Metabolism. WK515 E59c]
QP572.L3E67 596'.01'6 78-7778
ISBN 0-470-26394-6

Contents

Preface

Throughout the last decade our knowledge of prolactin physiology has expanded considerably. This increase in information has now reached such a level that the task of attempting to give a detailed account in a single volume is fraught with difficulty. This preface is thus something of an apologia to those colleagues and friends whose work has made this book possible, and to those authors whose valuable work has been overlooked or not given its true place in the text.

Initially this book was envisaged as primarily dealing with prolactin physiology in the lower vertebrates; however, it rapidly became apparent that to leave out any discussion of mammalian and human physiology would limit the value of the text, hence the inclusion of the final three chapters.

To acknowledge in full the help and advice of colleagues would require a further volume. Yet, without the help of many friends this work would never have been completed. I would particularly like to thank the other workers in our laboratory, Dr. K. J. Brewer, Ms. C. M. Beynon, Mr. D. J. Flint, Mr. K. Howard, Ms. L. Shipp and Mr. G. Wilson, without whose efforts and help none of this would have been possible. The editorial help of Dr. John Ball, who struggled with my idiosyncrasies, is very gratefully acknowledged, not only for his direct assistance in the preparation of the manuscript, but for his encouragement and advice throughout our acquaintance. Throughout the preparation of this manuscript, my wife has shown endless forebearance and her tolerance has contributed greatly to the final product. Finally, but not least, I would like to thank Miss D. S. Paterson and Miss Anita Callaghan, who typed the manuscript and never failed to translate my hieroglyphics.

1 Introduction

The importance of prolactin in the endocrine physiology of most vertebrates was hardly suspected until the early 1960's. Since that time our knowledge about this remarkably versatile pituitary hormone has grown enormously (see reviews by Mazzi, 1969; Bern and Nicoll, 1972; Nicoll, 1974), and a large part of this book has been occasioned by this great body of new information.

1.1 Structure and evolution of prolactin

A hormone similar to mammalian prolactin is now known to occur in all vertebrate groups, with the possible exception of the cyclostomes.

1.1.1 *Teleost fishes*

There can now be little doubt that a prolactin is secreted by the teleost pituitary, although its evident dissimilarities from tetrapod prolactin has led to the suggestion that it might better be termed 'paralactin' (Ball, 1965), a neologism which has not, however, found wide favour. According to Ball (1969*a*), many misconceptions concerning the biological properties of teleost prolactin stem originally from misinterpretation of the work of Noble *et al.* (1936, 1938) and Leblond and Noble (1937), and for a long time the mistaken idea prevailed that teleost prolactin controlled parental behaviour, and that it could stimulate proliferation of the lining of the pigeon crop-sac, the classical biological property of 'true' prolactins. However, more critical work by Chadwick (1966) and Nicoll and Bern (1968) has shown that teleost prolactin does not display full and typical crop-sac activity, though able to produce a partial response. This partial response is qualitatively different from the inflammatory response often produced by the injection of inert carrier substances (Chadwick, 1966), and it may involve a hydration of the tissues (Ensor, Ball and Ingleton, unpublished observation) rather than the characteristic increase in cell division seen in the classical 'prolactin response'. Nevertheless, although the full range of biological activities associated with tetrapod prolactin may not be present in the teleosts, there is now considerable evidence that the teleost pituitary secretes a prolactin-like hormone.

Details of the actions of prolactin in fish are discussed in Chapter 2. For the present, we may note that in bioassays in lower vertebrates, both ovine prolactin and teleost pituitary extracts produce parallel linear responses, indicating the presence of similar biologically active sequences in both the ovine and the teleost prolactin molecules (Grant and Grant, 1958; Grant and Pickford, 1959; Ensor and Ball, 1968). Furthermore, Emmart (1969) was able to localise prolactin-containing cells in the teleostean rostral pars distalis, using fluorescent antibodies to ovine prolactin. More recently McKeown and van Overbeeke (1971) have developed a radioimmunoassay based on Emmart's original antibodies, and have demonstrated that antiserum to ovine prolactin will cross-react with a hormone (presumably teleost prolactin) in teleost plasma. Nonetheless, mammalian and teleostean prolactins are by no mean identical; for example, Blüm (1973) has purified from two teleost species a prolactin which does not cross-react with antibodies to ovine prolactin.

Our knowledge of the structure of teleostean prolactin and its relation to tetrapod hormones has very recently been transformed, with the publication by Farmer *et al.* (1977) of a detailed account of their extensive work in isolating and characterising prolactin from the pituitary of *Tilapia mossambica* (recently re-named *Sarotherodon mossambicus*). Here, for the first time, we have detailed information about a non-mammalian prolactin. *Tilapia* prolactin has a molecular weight of 19 400, rather less than the molecular weight of ovine prolactin (22 700, Li *et al.*, 1970; Li, 1976 from Farmer *et al.*) and *Tilapia* growth hormone (22 200, Farmer *et al.*, 1976*b*). The molecule has a single NH_2-terminal residue (valine) and a single COOH-terminal residue (half-cystine). Most interestingly, *Tilapia* prolactin contains only four half-cystine residues, indicating that it has only two disulphide bridges, a characteristic of all known vertebrate growth hormones, but a contrast to mammalian prolactins which have three disulphide bridges (see p. 3). It is probable that *Tilapia* prolactin lacks the NH_2-terminal disulphide bridge found in the ovine hormone (Fig. 1.1). In its amino-acid composition, *Tilapia* prolactin resembles ovine growth hormone and *Tilapia* growth hormone rather than mammalian prolactins, although in their physico-chemical properties, as in their cells of origin, growth hormone and prolactin are certainly distinct hormones in *Tilapia*. *Tilapia* prolactin is considerably more potent than ovine prolactin in two teleost bioassays: sodium retention in seawater-adapted *Tilapia*, and reduction of water-permeability in the urinary bladder of *Gillichthys* (see

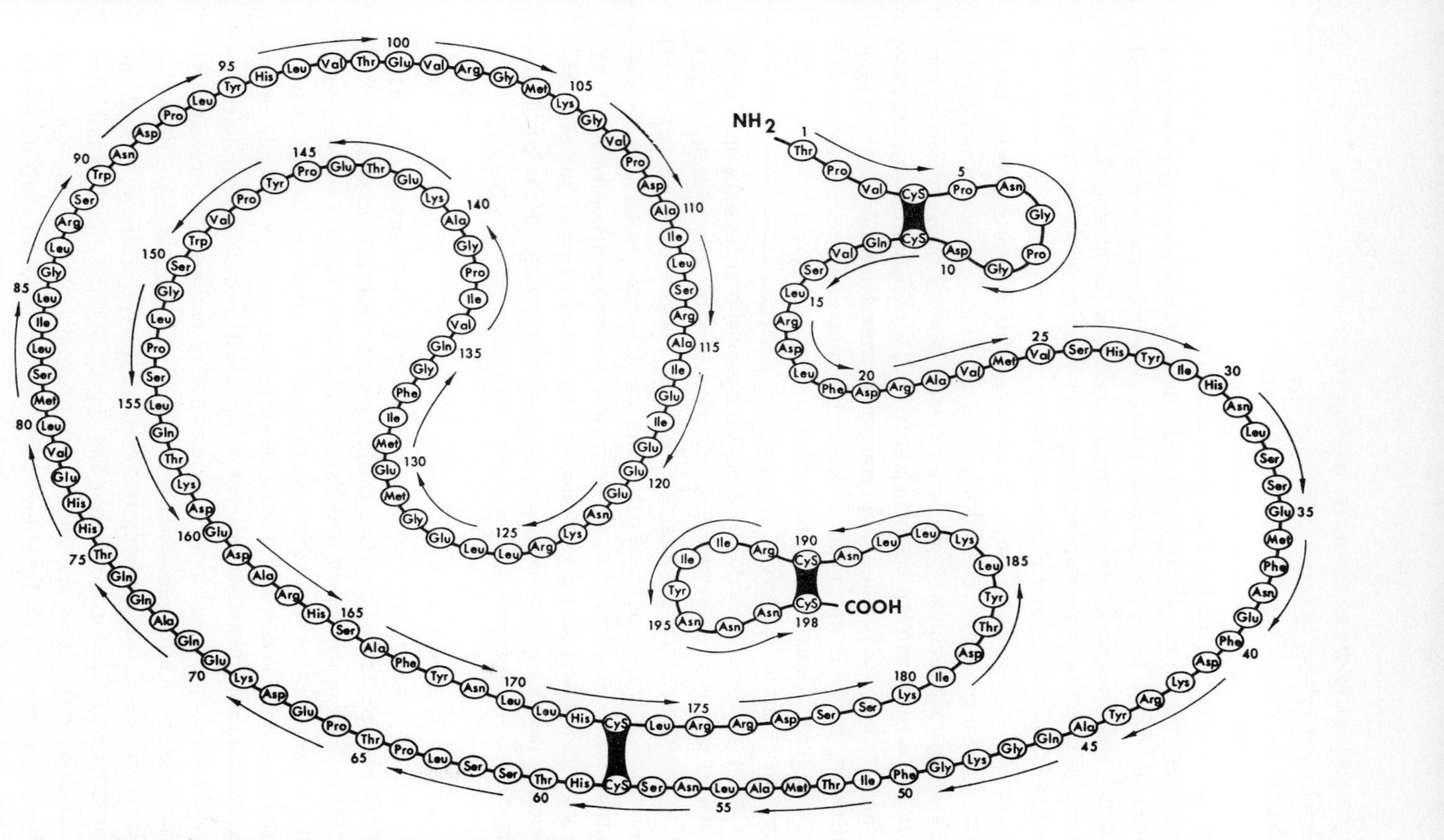

Fig. 1.1 The amino acid sequence of ovine prolactin (from Li, 1972).

Chapter 2); but, as foreshadowed in the earlier work on teleosts discussed above, *Tilapia* prolactin has no activity on the mammary gland or on the pigeon crop-sac. Possibly the acquisition of these 'classical' prolactin activities and of the third disulphide loop coincided in the tetrapod molecule. Given its resemblance to ovine growth hormone in amino-acid content and disulphide bridge configuration, it is fascinating that *Tilapia* prolactin exhibits equivocal but suggestive activity in the rat tibia assay for growth hormone, and that it cross-reacts in two growth hormone (rat and turtle) radioimmunoassays. In its circular dichroism spectrum, however, *Tilapia* prolactin resembles porcine prolactin, but it did not cross-react in agar diffusion with antisera to ovine, porcine or rat prolactin, nor with rat or turtle growth hormone antisera. A specific rabbit antiserum to *Tilapia* prolactin gave a precipitin reaction against *Tilapia* prolactin in agar diffusion, but did not form a precipitin line against *Tilapia* serum, *Tilapia* growth hormone, ovine prolactin, or pituitary extracts of perch, sturgeon or shark (Farmer *et al.*, 1977).

One of the most valuable outcomes of these elaborate studies on *Tilapia* prolactin and growth hormone is that they support the hypothesis that these two vertebrate hormones originated from a common ancestral molecule, while also confirming that they are well-differentiated in teleosts. Furthermore, it seems that while growth hormone structure has been strongly conserved during evolution (Farmer *et al.*, 1976*a*), the prolactin molecule has been more labile, presumably a correlate of the enormous range of effects reported for prolactin throughout the vertebrates.

1.1.2 *Non-teleostean fishes*

The *elasmobranch* pituitary contains a fish-type prolactin, able to stimulate the amphibian water-drive but lacking pigeon crop and mammotrophic activities (Ball, 1969; Ensor and Ball, 1972; Nicoll, 1974). *Cyclostomes*, in contrast, may not secrete a distinct prolactin: water-drive and pigeon crop activities cannot be detected in cyclostome pituitary extracts (Nicoll, 1974), and immunohistochemical studies have furnished no evidence that the cyclostome adenohypophysis contains a hormone resembling the prolactins of either teleosts or tetrapods (Aler *et al.*, 1971). The *ganoid fishes*, primitive relatives of the teleosts, apparently secrete a fish-type prolactin, related immunochemically to ovine prolactin (Aler, 1971), and active on the amphibian water-drive but not on the pigeon crop (Nicoll, 1974). In the *lungfishes* (Dipnoi), we meet for the first time in the evolutionary series a prolactin with pigeon crop activity; however, it possesses only partial mammotrophic activity, which apparently emerged in its full and typical form only later, at the time of the evolution of the amphibians from the extinct relatives of the lungfishes (Nicoll, 1974). The related *coelacanth* (*Latimeria chalumnae*) contains in its pituitary, cells which resemble typical lactotrophs (van Kemenade and Kremers, 1975), but beyond this there is no definite evidence that this interesting 'living fossil' secretes a prolactin.

1.1.3 *Tetrapods*

A number of biological tests have been used to assess the properties of tetrapod prolactins. These include stimulation of water-drive in the red eft (the terrestrial stage of the red-spotted newt), generally termed the amphibian water-drive (Grant and Grant, 1958); growth in anuran tadpoles (Licht and Nicoll, 1971); proliferation of the lining of the crop-sac of Columbiformes, the pigeon crop action (Chadwick, 1966); and the classical mammotrophic and luteotrophic actions of the hormone in mammals. Most of the tetrapod prolactins which have been tested appear to possess this entire range of activities, and all those tested are active on osmoregulation in teleosts (Bern and Nicoll, 1969). However, reptilian and avian preparations display only questionable luteotrophic effects in mammals, and amphibian material has not yet been examined for luteotrophic activity (Nicoll, 1974).

In an attempt to generalise the available data, it could be said that as the vertebrates evolved the potential biological activities of the prolactin molecule were increased. Thus, fish prolactin can produce effects only in fishes and amphibians, whereas most mammalian prolactins are active over a wide range of vertebrate groups. However, all the vertebrates alive today and available to the investigator are the ends of branches away from the main trunk of the evolutionary 'tree', and we should not forget that a hormonal molecule could incur loss of properties as well as gains in the course of evolution.

The purification and characterisation of nonmammalian prolactins still remains a considerable problem. Nicoll and Licht (1971) have compared the electrophoretic mobilities of prolactin and growth hormone from a wide range of tetrapod species. As can be seen from Fig. 1.2 there is a considerable variation in the Rf of the various prolactins, which may reflect changes in biological activities.

In the Mammalia, prolactin-like activity has been reported for a number of hormones. It would appear that all the species that have been studied possess a prolactin *per se*. In addition two other hormones, human growth hormone (hGH) and human placental lactogen (hPL, also known as human chorionic somatomammotrophin or hCS) possess lactotrophic and other activities normally associated with prolactin. There is also evidence that

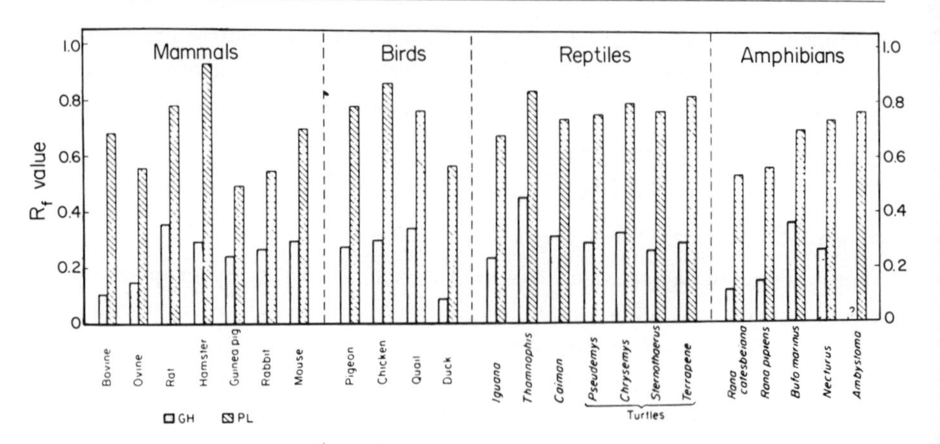

Fig. 1.2 Relative electrophoretic mobilities (R_f) of tetrapod prolactins (hatched bars) and growth hormones (stippled bars). Prolactin values are from Nicoll and Nichols (1971). R_f values represent the ratio between the distance moved from the origin by each hormone and the total length of the column from origin to ion front. (From Nicoll and Licht, 1974.)

the placenta in other mammals may secrete a prolactin-like hormone, with luteotrophic functions (see Chapter 6).

The three lactotrophic hormones that have been fully characterised (ovine prolactin, hGH and hPL) display similar physico-chemical properties; for example, their molecular weights agree closely, being 23 300, 21 500 and 21 600 respectively. The molecular structure of ovine prolactin is shown in Figure 1.1. hGH differs mainly in possessing two rather than three cysteine disulphide bridges (Fig. 1.3), the looped configuration of the prolactin amido-terminal being absent from the hGH molecule (Li and Dixon, 1971). Li and his associates (see Li, 1972) have made detailed studies of the relationship between hGH and hPL. The chemical structure of both is very similar (see Fig. 1.3 and 1.4). Both hormones are equipotent in stimulating the pigeon crop, but hGH is five times as potent as hPL in promoting growth of the immature rat tibia. Although hPL cross-reacts with antibodies to hGH in Ouchterlony plate tests (Friesen, 1965), it does not form a continuous and identical precipitin line. This indicates differences in antigenic site-distribution, and has been confirmed by Li (1972), who found that the precipitin curves for the two hormones when reacted with antibodies to hGH are very different.

It can be seen that rather than thinking of a single prolactin throughout the vertebrates, it is more realistic and more useful to think of a 'family' of hormones, comprising teleost prolactins, 'classical' tetrapod prolactins, hGH, hPL, and no doubt other members, as yet hardly known, in various non-teleostean fishes and in the placentae of many mammals.

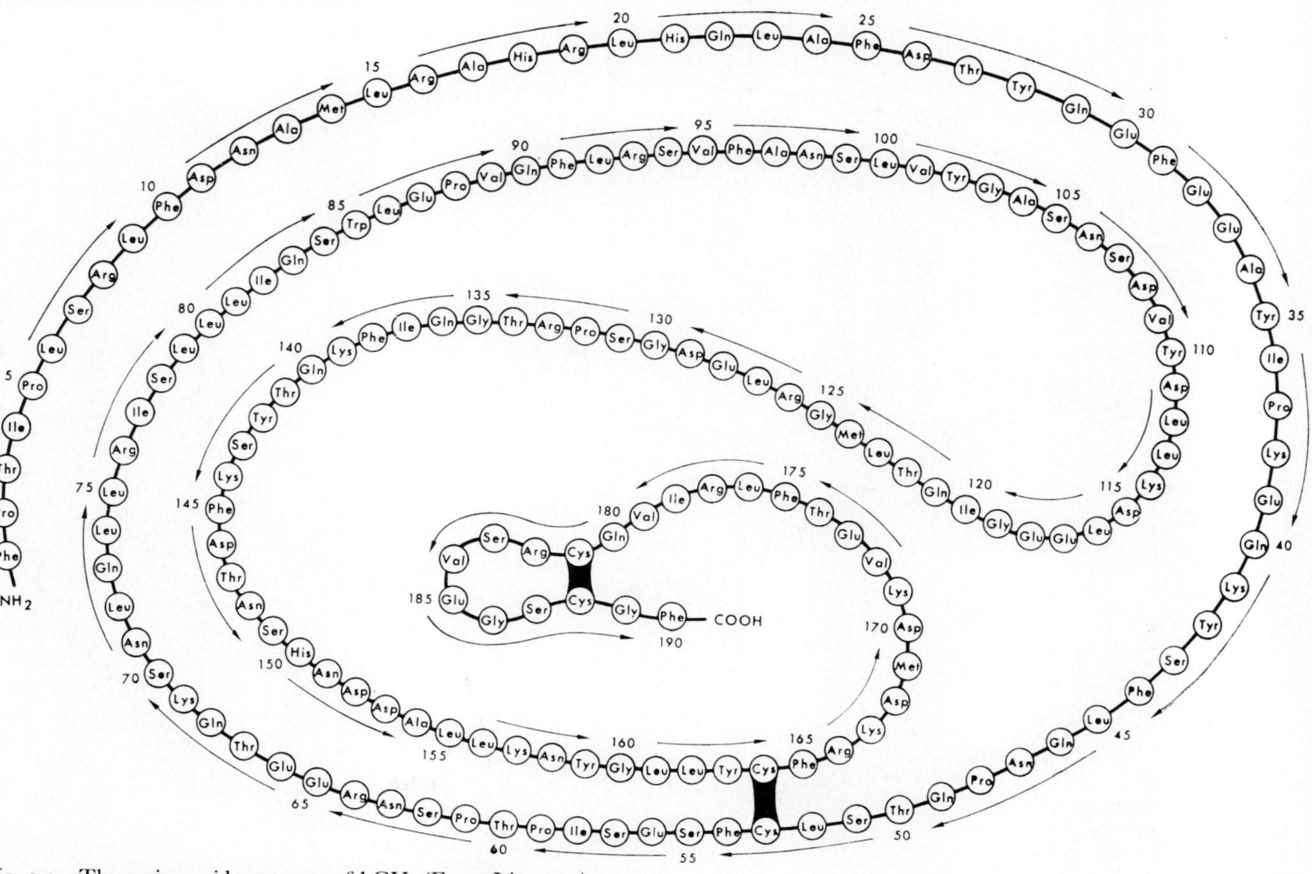

Fig. 1.3 The amino acid sequence of hGH. (From Li, 1972.)

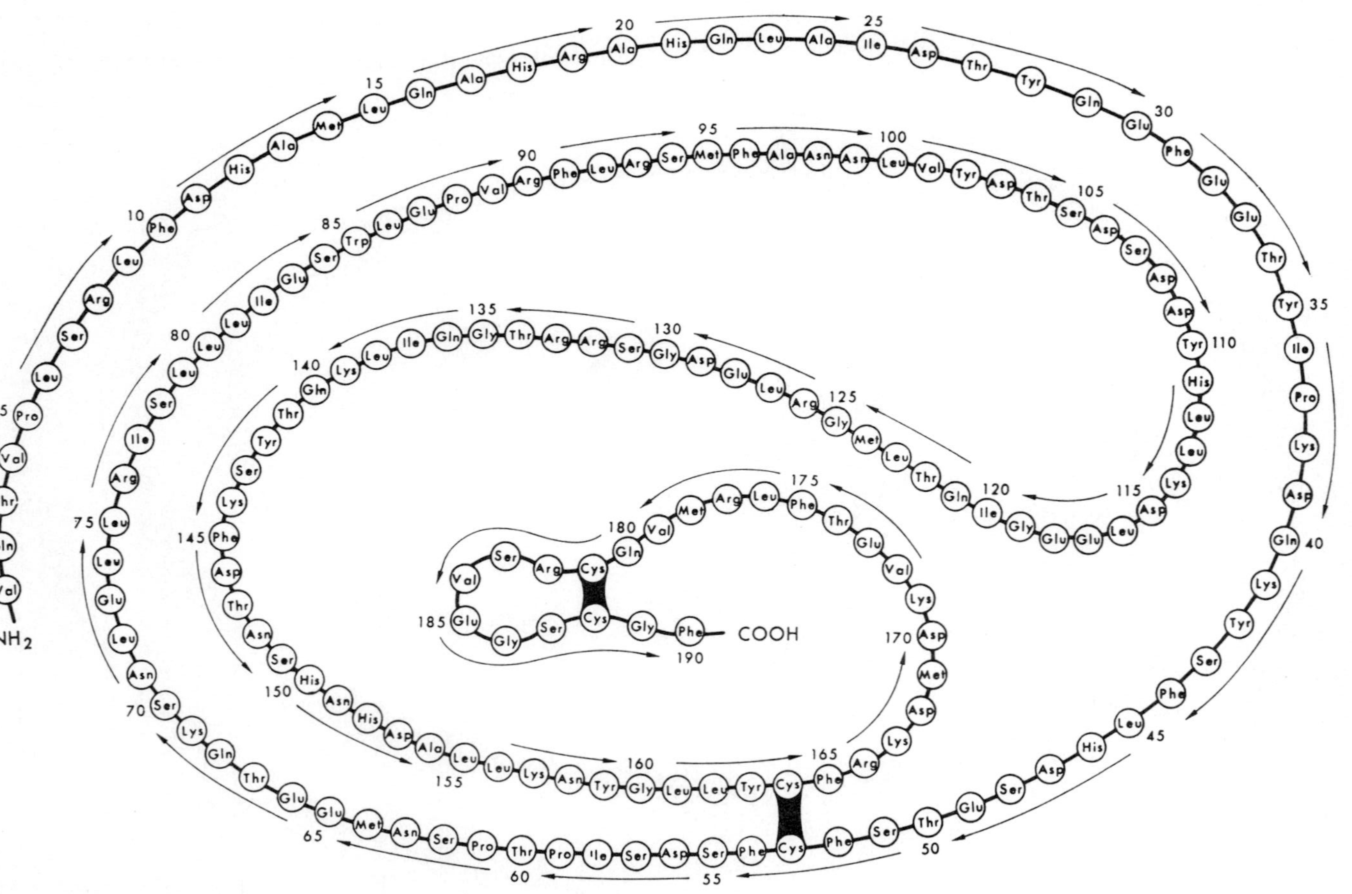

Fig. 1.4 The amino acid sequence of hPL. (From Li, 1972.)

1.2 Cellular origin of prolactin

For a detailed discussion of pituitary cytology beyond the scope of this work, the reader is referred to the recent comprehensive book by Holmes and Ball (1974).

Cells in the pituitary pars distalis fall into two major groups, the so-called acidophils and basophils, depending on their cytological properties, more particularly on the staining reactions of their stored secretory granules. It is now generally agreed on the basis of a great deal of evidence that prolactin in all vertebrates is secreted by an acidophilic cell-type.

Initially, all the acidophils were grouped together as *alpha cells*, but two or three distinct pars distalis acidophil cell-types are now recognised. One of these displays marked variations in activity during the mammalian reproductive cycle (Pearse, 1951; Purves, 1966) and on being found to stain preferentially with azocarmine it was termed the carminophil (Dawson and Friedgood, 1938). The other major acidophil is stable during the mammalian reproductive cycle, and stains preferentially with orange G: thus it was termed the orangeophil and is considered to secrete growth hormone (Purves, 1966). Although originally (Friedgood and Dawson, 1940) considered as the source of luteinising hormone (LH), the carminophils have subsequently been identified as lactotrophs, secreting prolactin (Dawson, 1954; Purves, 1966). This identification was temporarily challenged when Herlant and Racadot (1957) showed that the carminophils contained glycoprotein, and for a while these authorities reverted to the idea that these cells secrete LH. However, the glycoprotein content is very low, and Herlant and Racadot subsequently accepted that the carminophils produce prolactin (Racadot, 1963).

The introduction of Herlant's tetrachrome stain (Herlant, 1960) clarified the picture considerably, and revealed that while the growth hormone cells do indeed stain preferentially with orange G, the lactotrophs are most easily recognised by their strong affinity for erythrosin; thus the lactotrophs are now most usually termed *erythrosinophils* (Holmes and Ball, 1974). Unfortunately the Herlant tetrachrome is by no means easy or certain in its application, and the definitive identification of the lactotrophs in hitherto unknown pituitaries requires a good deal of experimental work to confirm the nature of the cells taking erythrosin (Holmes and Ball, 1974).

More recently, immunofluorescent techniques have been applied to the pituitary of a variety of species (Emmart, 1969; McKeown and van Overbeeke, 1971; Vellano *et al.*, 1970), and have usually confirmed earlier work and permitted a direct and positive demonstration of the prolactin cells. Nevertheless, immunofluorescence histochemical methods are fraught with difficulties, especially with regard to the specificity of the antibody-antigen reactions on which they are based, and this specificity should always be checked rigorously (see Moriarty, 1973).

Combinations of the various histochemical, ultrastructural and experimental techniques now available have led to the identification of the prolactin-secreting cells in members of all the vertebrate groups except cyclostomes. The specific properties of the lactotrophs, and their distribution within the pars distalis, will be discussed later in the accounts of the various groups.

1.3 Hypothalamic control of prolactin secretion

It is difficult at this point to generalise about the control of prolactin, a difficulty that will be appreciated on reading the detailed discussions of this matter in later chapters. In most vertebrates the route of hypothalamic control is the hypothalamo-hypophysial portal system. Teleost fishes are virtually unique in that their pars distalis cells are innervated by the axons of hypothalamic neurosecretory neurons (Holmes and Ball, 1974), and in these fishes the prolactin cells are controlled by direct hypothalamic innervation (Nagahama *et al.*, 1975*a*; Batten and Ball, 1976). The prolactin cells in the few teleost species so far studied are under a primarily inhibitory hypothalamic control (Ball *et al.*, 1972), the major controlling factor apparently being, dopamine released from the aminergic fibres which innervate the prolactin cells (Wigham *et al.*, 1975; Wigham and Ball, 1976; Batten and Ball, 1976; Nagahama *et al.*, 1975*b*). A primarily inhibitory hypothalamic control has also been demonstrated in amphibians and mammals (Holmes and Ball, 1974). In contrast, the evidence for birds, and possibly for reptiles, indicates that two opposing factors, inhibitory and stimulatory, might be produced by the hypothalamus, the balance between the two differing in different species: details of this complex situation will be found in Chapters 4 and 5. Apart from specific hypothalamic control, other factors such as plasma osmolarity, TRH (thyrotrophin releasing hormone), cyclic AMP, thyroid hormone and gonadal steroids, can modulate the secretion of prolactin (McLeod, 1975). The role of these non-hypothalamic factors will be discussed in more detail in later sections.

1.4 Classification of the effects of prolactin

Bern and Nicoll (1968), in their excellent review of prolactin physiology, stated that something of the order of 90 distinct effects have now been documented for this hormone, a number far in excess of the reported actions of all other adenohypophysial hormones combined (Nicoll, 1974). These effects can be grouped according to the classification given by Bern and Nicoll (1968) and summarised in Tables 1.1 and 1.1a. As will be apparent from the Tables, although no unifying principle of hormone action is readily detectable, the various reported effects can be arranged under two broad headings, reproduction and osmoregulation.

Table 1.1 Actions of prolactin related to reproduction

	Teleosts
1	Skin mucus secretion (Blüm and Fiedler, 1964)
2	Reduction of toxic effects of oestrogen
3	Growth and secretion of seminal vesicles (Sundararaj and Nayyer, 1969)
4	Parental behaviour
5	Maintenance of brood pouch in male seahorse (Billard, 1969)
6	Gonadotrophic (Blüm and Weber, 1968)
	Amphibia
7	Water drive (Chadwick, 1941)
8	Secretion of oviducal jelly (de Allende, 1947)
9	Anti-spermatogenic
10	Stimulation of cloacal gland development (Kikuyama, 1976)
	Reptiles
11	Anti-gonadotrophic (Callard and Ziegler, 1970)

Table 1.1a

	Birds
1	Production of crop 'milk' (Chadwick and Jordan, 1971)
2	Formation of brood patch (Jones, 1969)
3	Anti-gonadal (Jones, 1969)
4	Premigratory restlessness (Meier, 1969)
5	Parental behaviour (Arimatsu, 1971)
6	Synergism with steroids on female reproductive tract
7	Suppression of sexual phase of reproductive cycle
	Mammals
8	Mammary development and lactation
9	Preputial gland size and activity (Farnsworth, 1920)
10	Syngergism with androgens on male sex accessory glands
11	Luteotrophic in post-partum maters
12	Luteolytic in rats
13	Increased testes cholesterol
14	Increased androgen binding
15	Stimulation of β-glucuronidase activity in rodent testes (Evans, 1967)
16	Parental behaviour

1.4.1 *Effects related to reproduction*

This category (Table 1.1) includes those actions for which prolactin historically became well-known, the control of lactation in mammals

(Cowie and Tindall, 1971), the control of 'pigeon milk' production by the pigeon crop-sac (Bates, 1962), and the control of parental or 'broody' behaviour (Riddle, 1963). However, as can be seen from Table 1.1, more recent work on a wider range of vertebrate species has greatly extended our knowledge of prolactin as a reproductive hormone.

Throughout the vertebrates a pattern does seem to be discernible, with prolactin affecting parameters associated with care of the young ('parenthood'), such as discus 'milk' production, pigeon crop-sac proliferation, mammalian lactation, and parental behaviour in general. On the other hand, prolactin can also have direct effects on the actual production of the young, acting on the gonads themselves and on accessory sex structures. An important point which will recur throughout this book is the close relationship or synergism between prolactin and the hormonal steroids. Nicoll and Bern (1972) give a list of 30 responses to prolactin which exhibit this relationship.

1.4.2 *Effects related to ion and water balance*

These effects are summarised in Tables 1.2 and 1.2a. Here once again, we see a wide range of responses, which will be discussed in more detail later. Again, it is possible to detect a pattern in these responses. If we consider the vertebrates as falling into two groups, the aquatic and the terrestrial, including the amphibians within the primarily aquatic group, then it can be seen that in aquatic vertebrates prolactin acts primarily as an ion-retaining and diuretic hormone, whereas in terrestrial vertebrates it acts as an ion-excreting and antidiuretic hormone. This statement is obviously a gross simplification of an extremely elaborate and varied situation, but it may be helpful in understanding the complexities of prolactin physiology discussed in the following chapters.

Two 'common denominators' of prolactin action have been proposed by various authors. These are, firstly the stimulation of integumentary tissues (Nicoll and Bern, 1972); and secondly, the possible role of prolactin as a permissive or protective hormone, enhancing the effects of other hormones and protecting the animal from a wide range of environmental stresses. The arguments which have led to this second concept have arisen from attempts to link together the major effects of prolactin on reproduction and osmoregulation. Two illustrations will suffice. Prolactin stimulates lactation, which is a dehydratory stress for the female mammal, and prolactin has also been shown to act as an antidiuretic hormone in mammals (Ensor *et al.*, 1972*a*; Manku *et al.*, 1975). Similarly, birds brooding their clutch may have only limited access to water, and they have high circulating levels of prolactin (Nicoll, 1974) which not only promotes brooding but which

Table 1.2 Actions of prolactin effecting osmoregulatory ability

	Cyclostomes
1	Electrolyte metabolism in hagfish
	Teleosts
2	Survival of hypophysectomised euryhaline freshwater species
3	Restoration of water turnover in hypophysectomised *Fundulus kansae*
4	Restoration of plasma Na^{+} and Ca^{++} in hypophysectomised eels when given cortisol
5	Skin, buccal and gill mucus secretion
6	Reduced gill Na^{+} efflux (Ensor and Ball, 1968)
7	Reduced gill permeability to water
8	Inhibition of gill Na^{+}/K^{+} ATPase
9	Renotropic (Leatherland and Lam, 1969)
10	Increased urinary water elimination and decreased salt excretion
11	Stimulation of renal Na^{+}/K^{+} ATPase
12	Decreased water absorption and increased Na^{+} absorption in flounder bladder
13	Decreased salt and water absorption from eel gut

Table 1.2a Actions of prolactin effecting osmoregulatory ability

	Amphibians
1	Skin and electrolyte changes associated with water drive (Grant and Grant, 1958)
2	Sodium and water transport across toad bladder (Dalton and Smart, 1969)
3	Restoration of plasma Na^{+} in hypophysectomised newts (Crim, 1972)
4	Possible hypercalcemia in toads
5	Sodium transport across isolated frog skin (Howard and Ensor, 1976)
6	Sodium balance in hypect frogs (Howard and Ensor, 1977*a*)
	Reptiles
7	Restoration of plasma Na^{+} levels in hypophysectomised lizard
8	Diuretic effect in aquatic chelonian (Brewer and Ensor, 1976)
9	Na^{+} balance in tortoise (Brewer and Ensor, in press)
	Birds
10	Stimulation of nasal salt secretion (Peaker, Phillips and Wright, 1970)
11	Increased water intake (Ensor and Phillips, 1972)
12	Increased cloacal reabsorption (Ensor and Phillips, personal communication)
	Mammals
13	Lactation
14	Gut uptake in rats (Mainoya and Bern, 1972)
15	Increased Na^{+} retention at renal level (Lockett and Nail, 1965)
16	ADH-like activity (Ensor, Edmondson and Phillips, 1972; Manku *et al.*, 1976)
17	Ca^{++} balance (Horrobin, pers. comm.)

also may protect the bird from dehydration by increasing cloacal water-transfer (Ensor, 1975) and by promoting nasal salt gland secretion in marine species (Peaker *et al.*, 1970)

Although most of the statements made above are hypothetical, and may well require modification in the light of new discoveries, it does appear at present that most of the responses discussed in detail in the rest of this book can be placed into one of the two major categories. It is hoped that this brief introductory discussion may supply the reader unfamiliar with the vast literature on prolactin with a framework in which to hold the profusion of facts that he will meet in the following chapters.

2 Prolactin in fishes

I Teleosts

2.1 Prolactin and osmoregulation in teleosts

Possibly one of the most surprising properties of prolactin is its ability to maintain hypophysectomised teleosts alive in freshwater. In this context Pickford (1973) has described prolactin as 'the ... improbable mammalian hormone'. It is possible that this osmoregulatory activity in fish is 'primitive', since it is possessed by all the prolactin molecules studied so far (Nicoll and Bern, 1968, 1972; Ball, 1969).

The interest in an osmoregulatory role for prolactin stems originally from the work of Burden (1956), who showed that hypophysectomy was followed by the osmotic failure in freshwater of the normally euryhaline killifish, *Fundulus heteroclitus*. Further work by Pickford and Phillips (1959) showed that injections of ovine prolactin would permit survival of the hypophysectomised fish in freshwater.

Since this early work, prolactin has been shown to prevent osmotic failure in a range of teleost species: *Poecilia latipinna* (Ball and Olivereau, 1964), *Xiphophorus maculatus* (Schreibman and Kallman, 1966), *Gambusia* sp. (Chambolle, 1966, 1969), *Tilapia mossambica* (Dharmamba *et al.*, 1967), *Oryzias latipes* (Utida *et al.*, 1971), *Ictalurus* (Chidambaram *et al.*, 1972), *Mugil cephalus* (Sage, 1973) and *Salmo trutta* (Oduleye, 1976). These are species with an absolute requirement for prolactin for survival in freshwater. In contrast, there exists a second group of euryhaline fish, exemplified by *Fundulus kansae* (Stanley and Fleming, 1967), *Anguilla anguilla* (Maetz, Mayer and Chartier-Baraduc, 1967) and *Platichthys flesus* (MacFarlane, 1971), which can survive indefinitely following removal of the pituitary in freshwater, but which nevertheless show an impairment of osmoregulatory functions which can be corrected in part by injections of prolactin.

In many teleosts, studies of the erythrosinophils (lactotrophs) in relation to environmental salinity have shown these cells to be more active in freshwater than in seawater. Usually no other pituitary cell-type shows this correlation, evidence that the endogenous teleost prolactin has the osmoregulatory functions suggested by the effects of injected mammalian prolactin (Ball and Baker, 1969; Olivereau and Ball, 1970; Leatherland and McKeown, 1973, 1974; Holmes and Ball, 1974; Nagahama *et al.*, 1975).

As Bern (1975) states in his review of prolactin and osmoregulation, 'it is becoming increasingly obvious that no accurate generalisation regarding prolactin-dependency for freshwater survival will apply to the Teleostei as a group'. Recent work by Griffiths (1974), reviewing the state of knowledge for the genus *Fundulus*, has suggested that the importance of the pituitary is related to the extremity of environmental conditions faced by the fish. However, the flounder *Platichthys flesus* in many parts of its range exhibits a tidal migration from full seawater to freshwater and back in a 24-hour period (O'Hara, personal communication), and yet can survive in freshwater after pituitary ablation (MacFarlane, 1971). This indicates that Griffiths' suggestion, though valid for *Fundulus* spp., cannot necessarily be generalised to other teleosts.

Interesting recent work by a number of authors has suggested that prolactin may also play a part in the survival of hatchling fish, which for a number of reasons may be under severe osmotic stress (Potts and Parry, 1960). Based largely on changes in pituitary cytology, studies by Nozabi, Tatsumi and Ichikawa (1974) have suggested that prolactin may protect hatchling rainbow trout from failure in freshwater. Other work by Ichikawa *et al.* (1973) has suggested a similar situation in the hatchling guppy. In this case the secretion of prolactin appears to be activated by exposure to freshwater, prolactin administration having little effect on pre-hatchling guppies.

2.1.1 *Prolactin and plasma electrolyte levels*

In those species which fail rapidly following hypophysectomy, the operation appears to cause a marked drop in plasma sodium and chloride levels, changes which in many cases have been shown to be independent of any alteration in water permeability (Ball, 1969). For example, hypophysectomised *Fundulus heteroclitus* transferred to freshwater exhibit over several days a 40–60% drop in serum osmotic pressure, sodium and chloride (Burden, 1956; Pickford *et al.*, 1966). In hypophysectomised *P. latipinna* (see Table 2.1) the fall is even more rapid and results in a 25% drop in plasma sodium in freshwater. This change was not paralleled by a change in plasma potassium levels, suggesting that the effect specifically involved sodium-conserving mechanisms rather than simple haemodilution (Ball and Ensor, 1965). The changes exhibited in both these species can be corrected

Table 2.1 Plasma sodium and potassium concentration in adult female *Poecilia latipinna* (Means ± S.E.)

Treatment	*No. of fish*	*Sodium (mEq/l)*	*Potassium (mEq/l)*
1. Intact, 6 weeks in DSW	10	163·3 ± 0·45	—
2 Intact, reared in FW	9	151·3 ± 1·57	—
3 Hypect 2 weeks, in DSW	7	160·3 ± 0·82	—
Group A. Fish killed 18–24 h after transfer from DSW to FW			
4 Intact	9	129·2 ± 2·73	—
5 Hypect	5	88·1 ± 2·96	—
Group B. Fish operated 3 days after transfer from DSW to FW killed after 4 days			
6 Intact controls at 3 days	8	148·7 ± 1·25	9·4 ± 0·27
7 Sham hypect, inj. CMC	7	144·7 ± 0·72	9·5 ± 0·30
8 Hypect, inj. CMC	10	111·2 ± 1·34	9·8 ± 0·92
9 Hypect, inj. prolactin	10	141·5 ± 1·81	9·8 ± 0·66

DSW = dilute sea water (12 parts /1000); FW = Liverpool tap water; Hypect = hypophysectomised; CMC = 1% sodium carboxymethylcellulose in 0·6% NaCl solution.

'*t*' test for sodium values: 1 and 2, $P<0{\cdot}001$; 1 and 3, $P<0{\cdot}01>0{\cdot}001$; 1 and 4, $P<0{\cdot}001$; 2 and 4, $P<0{\cdot}001$; 4 and 5, $P<0{\cdot}001$; 2 and 6, $P=0{\cdot}3$; 6 and 7, $P<0{\cdot}05>0{\cdot}02$; 7 and 8, $P<0{\cdot}001$; 7 and 9, $P<0{\cdot}001$; 8 and 9, $P<0{\cdot}001$.

The differences between the potassium values in Group B were not significant.

by injecting ovine prolactin. Comparable results have been obtained for *Tilapia mossambica* (Dharmamba and Maetz, 1972) and *Oryzias latipes* (Utida *et al.*, 1971).

In *P. latipinna* the 'set-point' for prolactin release appears to be very sensitive. Thus hypophysectomised fish kept in a dilute seawater of 170 mEq/l Na^{+}, maintain normal plasma sodium levels, about 170 mEq/l. However, if the external concentration of sodium is dropped to 162 mEq/l then in the absence of the pituitary plasma sodium gradually falls to equilibrate with the environment, and this fall is opposed by prolactin treatment. At external sodium levels of 170 mEq/l and above, prolactin therapy has no effect on plasma sodium levels (Ball and Ensor, 1969). In contrast, Dharmamba *et al.* (1973) reported that injection of prolactin into *Tilapia mossambica* in seawater causes inhibition of sodium efflux, and Olivereau and Lemoine (1973) found that prolactin causes hypernatremia in the seawater-adapted eel, *Anguilla anguilla*. Thus it is not possible to generalise from results obtained with prolactin in one species, a point that is illustrated repeatedly in the literature dealing with osmoregulation in teleosts (see Ensor and Ball, 1972; Nicoll, 1974).

2.1.2 *Specificity of prolactin effects*

Little work has been done to demonstrate that actions reported above are specific to prolactin. It is theoretically possible that ovine or other mammalian prolactins could 'mimic' some other teleost pituitary hormone such as growth hormone or ACTH, or a combination of endogenous hormones. However, there is sufficient work on cyprinodonts to suggest that the sodium-retaining activity is possessed only by prolactin.

Figure 2.1 shows the effects of various pituitary hormones on plasma sodium levels in hypophysectomised *P. latipinna* in freshwater. Prolactin alone causes an elevation of plasma sodium levels, and this effect appears

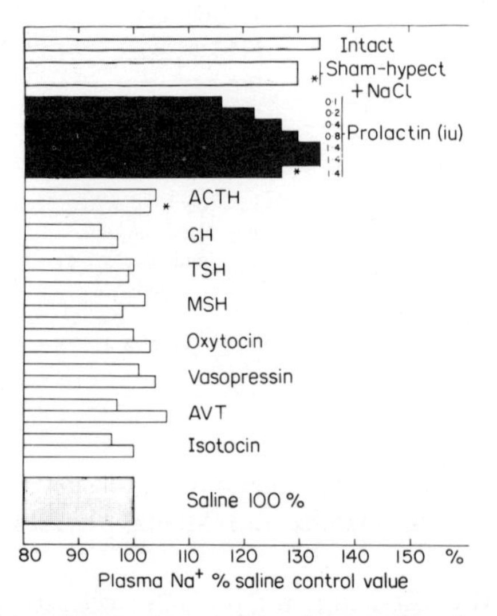

Fig. 2.1 The specificity of prolactin action on plasma sodium levels in *Poecilia latipinna* (Hypect = Hypophysectomised).

to be dose-dependent (Ball and Ensor, 1967). This latter aspect is confirmed by Figs 2.2 and 2.3 which illustrate the dose-response relationship between ovine prolactin and extracts of frog and teleost pituitaries (Ensor and Ball, 1968). As can be seen, the response to ovine prolactin is paralleled by some entity in the three pituitary extracts, indicating that the teleost pituitary does possess a hormone with essential similarities to both ovine and frog prolactins.

Other authors have demonstrated the specificity of mammalian prolactin in maintaining hypophysectomised fish in freshwater, notably Pickford *et al.* (1965) for *F. heteroclitus* and Schreibman and Kallman (1966) for *Xiphophorus maculatus*.

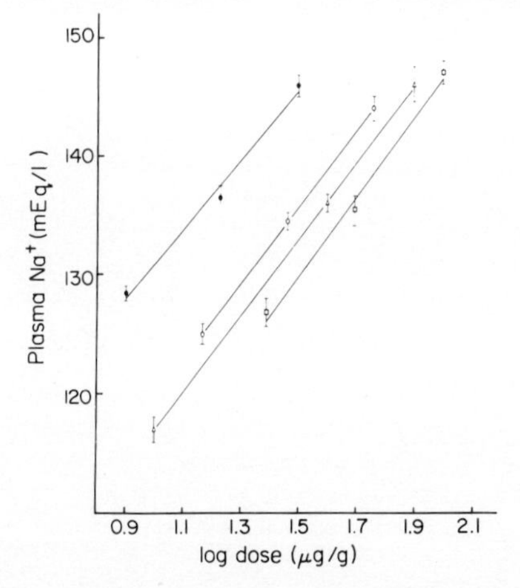

Fig. 2.2 The responses of plasma sodium to prolactin and pituitary homogenates in hypophysectomised *Poecilia latipinna* 12 hours after transfer from dilute seawater to freshwater.
△ Ovine prolactin
□ Female frog (*Rana temporaria*)
● Female *P. latipinna* after 72 hours in freshwater
○ Female *P. latipinna* adapted to dilute seawater.

2.1.3 *Prolactin and whole-body sodium fluxes*

There are five possible sites at which prolactin may act to produce alterations in plasma sodium levels: the branchial epithelium, the kidney, the urinary bladder, the gut and the skin. These sites will each be dealt with in more detail later, but it will be useful first to consider the overall effect of prolactin on whole-body (global) sodium fluxes.

Poecilia latipinna, like other teleosts (Maetz, 1971), adapts to dilution of the external medium by markedly reducing its whole-body sodium turnover rate (Table 2.1). It proved difficult in this fish to measure flux rates after hypophysectomy in freshwater, because of the very short postoperative survival in this medium (Ball and Olivereau, 1964). However, it has been possible to investigate pituitary influences on fluxes in freshwater by hypophysectomising fish in dilute seawater and then transferring them to freshwater and measuring fluxes 4–5 hours after the transfer. The results of such an experiment are shown in Table 2.2. As can be seen, hypophysectomised fish under these conditions are in net sodium imbalance, because sodium efflux remains high. This is not a result of postoperative stress, since it is not seen in the sham-hypophysectomised fish, which like intact fish rapidly reduce their sodium efflux on entering freshwater. Both ovine prolactin and pituitary autografts promote this normal reduction of sodium efflux in hypophysectomised fish (Ensor and Ball,

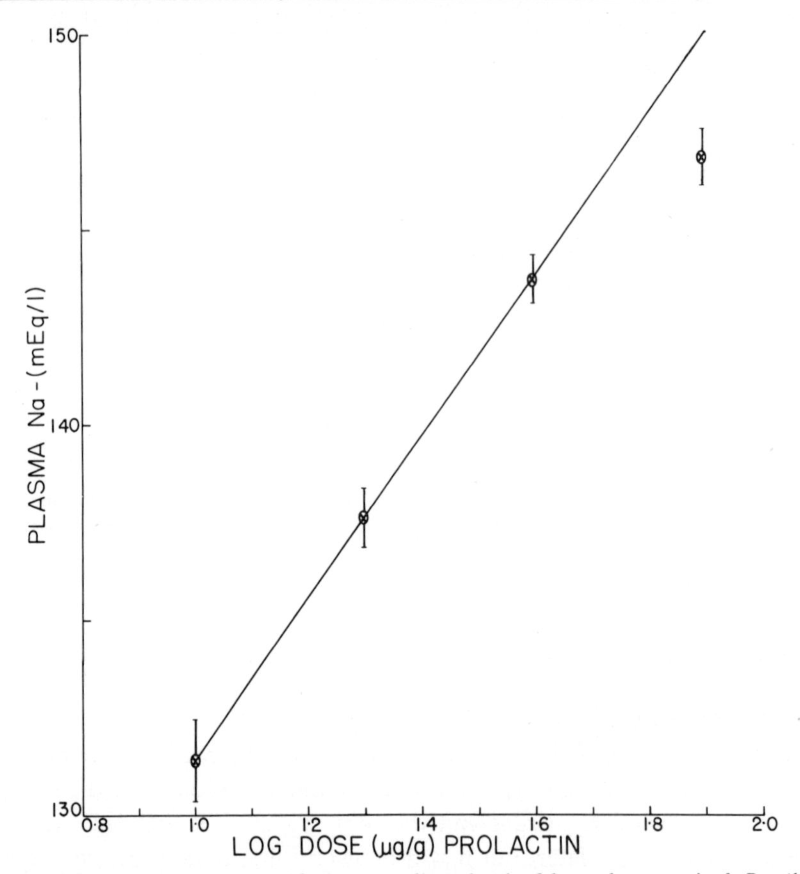

Fig. 2.3 Dose response curve of plasma sodium level of hypophysectomised *Poecilia latipinna* to ovine prolactin.

1972). More will be said of the role of the ectopic pituitary transplant at a later stage.

In dilute seawater, although having no marked effect on plasma sodium levels, hypophysectomy results in a significant increase in total sodium turnover (see Table 2.3) and an expansion of the apparent sodium space. Again, these changes can be corrected by injections of ovine prolactin or by ectopic pituitary autografts (Ensor and Ball, 1972).

Total sodium fluxes have also been measured in *Tilapia mossambica* (Dharmamba, 1970), *Fundulus kansae* (Fleming and Ball, 1972), *Fundulus heteroclitus* (Maetz, *et al.*, 1967), *Anguilla anguilla* (Maetz, Mayer and Chartier-Baraduc, 1967; Mayer, 1970) and *Platichthys flesus* (MacFarlane, 1971; MacFarlane and Maetz, 1974). The results once again underline the problems of generalisation between species barriers. *Tilapia*, like *Poecilia*, showed an increased sodium efflux after hypophysectomy in freshwater

Table 2.2 *Poecilia latipinna:* whole-body sodium fluxes in freshwater, 4 hours after transfer from one-third seawater

Group	*Number of fish*	*Influx, μEq/g/h*	*Outflux, μEq/h/h*	*Net flux μEq/g/h*
Intact	7	0·6 ±0·11	1·6 ±0·16	−1·0 ±0·16
Sham hypect, 7 days+saline	6	0·5 ±0·11	1·7[ab] ±0·24	−1·2[eg] ±0·22
Hypect, 7 days+saline	7	0·3 ±0·17	11·8[acd] ±3·75	−11·5[efh] ±3·92
+cortisol 0·5 μg/g	7	0·3 ±0·28	9·5[b] ±0·86	−9·2[g] ±3·92
+prolactin 0·8 IU/g	8	0·3 ±0·09	2·0[c] ±0·24	−1·7[f] ±0·24
+ectopic pituitary autograft, 14 days	4	0·4 ±0·01	1·7[d] ±0·02	−1·3[h] ±0·02

Fish given a single injection of saline and/or ^{22}Na at transfer. Values paired by superscript differ significantly (P=0·05 or less).

Table 2.3 *Poecilia latipinna:* whole-body sodium turnover rate, plasma sodium levels, and sodium space in one-third seawater (Na^+ 156·7 mEq/l)

Group	*Na^+ turnover rate, % exch. Na^+/h*	*Plasma Na^+ mEq/l*	*Na^+ space, ml/100g*
Intact	15·1 ±1·00[ab] (6)	157·9 ±4·55 (12)	32·7 ±0·45[f] (5)
Hypect+NaCl	28·6 ±2·19[ace] (8)	155·5 ±1·76 (15)	52·3 ±3·52[fg] (6)
+ACTH, 0·75 IU	32·0 ±5·17[b] (9)	156·8 ±1·73 (15)	48·4 ±3·56 (5)
+prolactin, 0·8 IU	12·7 ±1·11[cd] (6)	155·2 ±1·63 (14)	40·4 ±1·91[g] (8)
+ectopic pit. autograft, 18 weeks	18·2 ±1·11[de] (4)	156·6 ±4·93 (6)	

Fish hypophysectomised, rested for 1 week, and injected on alternate days for a further 2 weeks. Sodium outflux measured 4–5 hours after the final injection. Intact controls were injected with 0·6% saline 4 hours before starting flux measurements. All fish were injected with ^{22}Na at this time. Values paired by superscript differ significantly (P=0·05 or less). Numbers of fish in brackets.

(Dharmamba and Maetz, 1972). This, however, was coupled with a decreased sodium influx not observed in *Poecilia*. Prolactin was able to correct both these changes. In contrast to *Poecilia*, prolactin also caused a reduction in sodium efflux in seawater *Tilapia*, resulting in hypernatremia. The results for *Anguilla* and *F. heteroclitus* resemble those for *Poecilia*, in that no alteration was observed in the active uptake component. In teleosts in general, the total sodium turnover is inversely related to the external salinity.

In *F. kansae*, as in *Tilapia*, prolactin can stimulate active uptake of sodium from a sodium-enriched tap water by hypophysectomised fish (Fleming and Ball, 1972). However, unlike other teleosts, *F. kansae* shows no change in sodium efflux following hypophysectomy or prolactin treatment, although prolactin can cause sodium retention in intact *Fundulus kansae* in distilled water. Since sodium influx must obviously be minimal in distilled water, it seems that in these circumstances prolactin can limit sodium efflux, though ineffective in tap water. In *F. kansae*, as in *Tilapia*, exogenous prolactin appears to have a deleterious effect in seawater, causing increased plasma sodium levels (Fleming and Ball, 1972; Potts and Fleming, 1971). However, *Anguilla* and *F. heteroclitus* resemble *Poecilia* in that prolactin is without effect in seawater (Maetz *et al.*, 1967; Maetz, Mayer and Chartier-Baraduc, 1967). MacFarlane (1971) has also shown that prolactin causes a 50% reduction in whole-body sodium turnover in intact seawater-adapted *Platichthys*, and a similar effect was reported by Utida *et al.* (1971) for *Oryzias latipes*. In the case of *Platichthys*, hypophysectomy also reduced sodium turnover, possibly largely a consequence of the removal of ACTH.

2.1.4 *Effects on branchial epithelium*

The possible effects of prolactin on the gills can be divided into two main groups. In essence, teleosts in freshwater can minimise sodium loss across the gills either by increasing active uptake or by reducing the passive efflux. Active uptake appears to be correlated primarily with the so-called 'chloride cells' (ionocytes) situated on the gill lamellae (see Fig. 2.4). The control of passive efflux is apparently more complex. Two possible mechanisms have been suggested, either the production of a mucus sheath surrounding the gill and possibly cross-linked by calcium ions (Oduleye, 1972), or a reduction in blood flow through the gill lamellae (Rankin and Maetz, 1971).

The main evidence for a direct action of prolactin on the gills comes from two sources: measurements of whole-body sodium fluxes with concomitant cannulation of the urinary papilla; and observations of changes in chloride

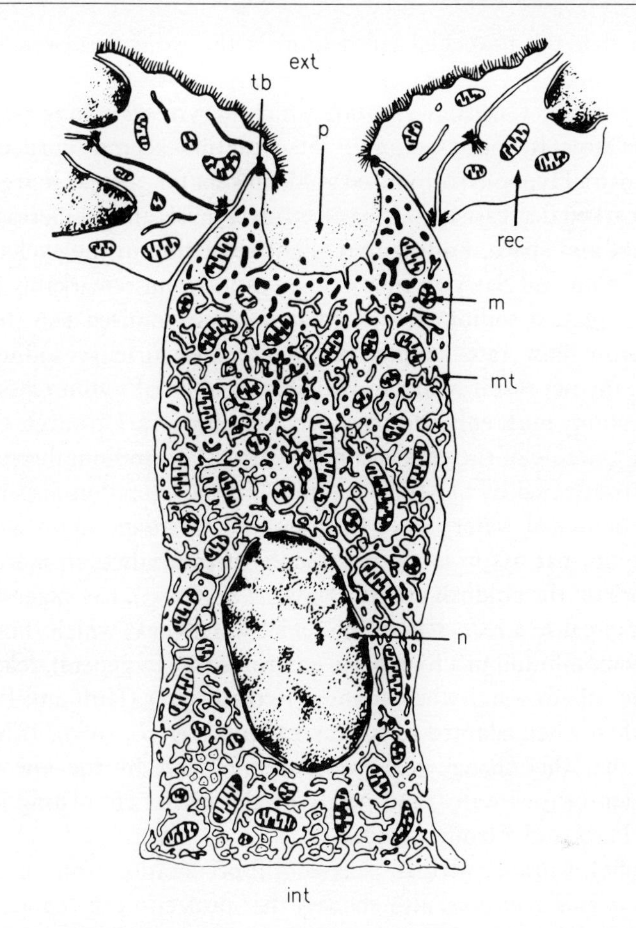

Fig. 2.4 A typical 'chloride cell'. The pit (p) characterises the cell in seawater adapted fish. Numerous mitochondria(m) and microtubules(m.t.) are also typical. (Enlarged 6000×.) n, nucleus; tb, terminal bar; rec, flat respiratory epithelial cell. (From Maetz, 1971.)

cells following hypophysectomy and prolactin replacement therapy. *Fundulus heteroclitus* resembles *Poecilia latipinna* in suffering a marked drop in plasma electrolyte levels in freshwater after hypophysectomy (Pickford *et al.*, 1966, 1970), and the flux responses on transfer from seawater to freshwater are also very similar (Maetz *et al.*, 1967). Renal sodium loss, measured by cannulation of the urinary papilla, accounted for only 17% of the total body sodium efflux (Maetz *et al.*, 1967). It is therefore reasonable to assume that the main site of adjustment in this species is extra-renal. Consideration of relative surface areas leads to the conclusion that the branchial epithelium is the main extra-renal site of sodium exchange.

In the goldfish, *Carassius auratus*, which survives for long periods in freshwater after hypophysectomy, the situation seems more complex (Chavin, 1956). Hypophysectomised goldfish kept for 3 weeks in freshwater display a marked decrease in plasma electrolyte levels and an increase in the apparent sodium space, leading to a slight reduction in the exchangeable sodium (Lahlou and Sawyer, 1969). The tissues seem remarkably tolerant of very low plasma sodium levels. In hypophysectomised fish there is a reduced urine flow rate coupled with increased urinary sodium concentration, the net result being no real change in renal sodium loss. Thus, hypophysectomy must enhance extra-renal sodium loss. However, evidence concerning the role of the gills is contradictory. Gill sodium fluxes appear only slightly affected by hypophysectomy, but the operation does result in decreased branchial water fluxes (Lahlou and Giordan, 1970) which are restored to normal by prolactin treatment. In contradiction, more recent *in vitro* work on the goldfish gill by Ogawa *et al.* (1973), has suggested that prolactin can cause a *reduction* in branchial water flow, which thus would prevent haemodilution in a hypotonic environment. In general, teleosts are more permeable to water when adapted to freshwater (Duff and Fleming, 1972*a*, *b*) than when adapted to seawater (Motais *et al.*, 1969). It has been suggested that this change may be caused in part by the low calcium concentration in freshwater, and partly by the higher circulating levels of prolactin (Potts and Fleming, 1970).

Lam (1969), working with an isolated gill preparation from the stickleback *Gasterosteus aculeatus*, also showed that prolactin can reduce the net osmotic influx of water. More recently, Ogawa (1974, 1975) has confirmed these results for the Japanese eel *Anguilla japonica* and the rainbow trout *Salmo gairdnerii*.

There is an obvious major contradiction here (see Table 2.4). Further-

Table 2.4 The effect of prolactin on branchial water permeability in various teleost species

Stimulatory	*Inhibitory*
Carassius auratus	*Gasterosteus aculeatus*
Fundulus kansae	*Anguilla anguilla*
	Anguilla japonica
	Salmo gairdnerii
	Carassius auratus

more, a number of authors have shown that elevation of freshwater calcium levels produces an effect similar to prolactin in reducing branchial water permeability (Potts and Fleming, 1970; Cuthbert and Maetz, 1972; Bornacin *et al.*, 1972; Ogawa, 1974; Oduleye, 1976). One physico-chemical effect of calcium may be to increase the binding between polar organic molecules in the branchial epithelial membrane, resulting in an overall decreased permeability (Oduleye, 1976). It may also be that calcium ions interact with the mucous sheath produced by prolactin in such a way as to increase its stability and so producing an unstirred layer around the gill (Oduleye and Potts, personal communication). If this idea is correct then it is possible that interactions between varying levels of calcium and varying doses of prolactin could produce different results, although whether such interactions could change the *direction* of the permeability response is open to question.

Whether prolactin stimulates the production of an impermeable mucus layer around the gill, or whether, alternatively, it stimulates the ionocytes or 'chloride cells', is still debatable. Although present in freshwater, the ionocytes are generally more numerous in seawater-adapted fish (Blanc-Livni and Abraham, 1970; Olivereau, 1970; Utida *et al.*, 1971). Evidence is accumulating which suggests that the ionocytes are the main site of the active outward ion transport in seawater fish (Maetz, 1971) and that most of the Na^+/K^+-ATPase activity of the gills is concentrated in these cells (Mizuhira *et al.*, 1970). In seawater the teleost gill has a high level of Na^+/K^+-ATPase, associated with its high rate of outward sodium pumping, and in freshwater the level of this enzyme is much lower (Maetz, 1970). The elevated concentration of the enzyme in the seawater-adapted eel, *Anguilla anguilla*, is promoted by ACTH and cortisol (Mayer, 1970; Milne *et al.*, 1971). When the freshwater-adapted killifish, *Fundulus heteroclitus*, is hypophysectomised, the levels of branchial Na^+/K^+-ATPase greatly increase towards seawater levels; prolactin injections reverse this response to hypophysectomy (Pickford *et al.*, 1971), and may have the same effect in *Anguilla japonica* (Kamiya, 1972). However, in both *Fundulus heteroclitus* and *Anguilla japonica*, the main action of prolactin in freshwater appears to be to reduce the outflux of sodium, so promoting freshwater tolerance. This effect may not primarily involve the ionocytes: it has been suggested (Kerstetter *et al.*, 1970) that while the ionocytes are the site of active electrolyte transport movements, passive electrolyte movements – such as, presumably, sodium outflux in freshwater – take place at other sites in the branchial epithelium. A great deal of evidence links the ionocytes with sodium excretion in seawater, and indicates that the seawater activity of these cells is promoted by ACTH and cortisol. Could the markedly reduced sodium turnover-rate in freshwater be the result of suppression of the ionocytes by high prolactin levels in this medium? Recent work indicates that the situation is too complex for this to be so. In some teleosts, the ionocytes are numerous and active in freshwater, in certain species, indeed more active than in seawater (Mattheij and Stroband, 1971). In work on the stenohaline freshwater cichlid *Cichlasoma biocellatum*, the ionocytes were found to become regressed in seawater, in which condition they could be stimulated by prolactin. The activation of the ionocytes in freshwater and, further, in distilled water, closely paralleled the increased secretory activity of the prolactin cells in the same individuals (Mattheij and Stroband, 1971). From these findings, and in the light of other evidence, the suggestion has been made that there are two types of ionocytes in the teleost gill: one type (the only one in the stenohaline *Cichlasoma*) responsible for electrolyte absorption in hypotonic media and under prolactin control, the other type being responsible for electrolyte excretion in hypertonic media and, as we have seen, probably under ACTH control. This concept is certainly speculative, but certainly it is difficult to explain the strong stimulation of ionocytes in *Anguilla anguilla* in both deionised water and seawater (Olivereau, 1970, 1971) without assuming that these cells, whether of one or two types, are probably involved in active electrolyte transport in either direction, according to the external medium. The *Cichlasoma* work suggests that prolactin might stimulate the ionocytes in hypotonic media, but against this must be set the fact that deionised water induces regression, not hyperactivity, of the prolactin cells in *Anguilla* (Olivereau, 1971; Olivereau and Ball, 1970).

The effects of prolactin on the mucous cells of the gills are better documented. Hypophysectomy strongly reduces the number of mucous cells in *F. heteroclitus* (Burden, 1956). In intact *Anoptichthys jordani*, prolactin injections stimulate the branchial mucous cells, and there exists a fairly close correlation between the number and activity of these cells and the activity of the pituitary lactotrophs (Mattheij and Sprangers, 1969). This is also true for the stickleback, *Gasterosteus aculeatus* (Leatherland and Lam, 1969) and the eel (Olivereau and Olivereau, 1971). Contrary to this a number of investigators, including Ball (1969) and Ball and Ensor (1969) working with *P. latipinna*, have failed to show a correlation between alterations in sodium turnover and mucous cell activity following prolactin treatment and hypophysectomy.

2.1.5 *Effects on integumentary mucous cells*

Though relatively smaller in area than the branchial epithelium, the skin presents another boundary or interface for ionic exchanges, and limitation of ionic efflux by this route is presumably of some importance to freshwater fish. As for the gills, one way in which skin permeability can be reduced is by the increased secretion of integumentary mucus. The effects of prolactin on integumentary mucous cells are particularly well documented in the Cichlidae: prolactin has been shown to stimulate the number and activity of these cells in *Symphysodon aequifasciata* (Blüm and Fiedler, 1964, 1965; Blüm, 1973) and *Cichlasoma biocellatum* (Mattheij and Stroband, 1971). In non-cichlids, hypophysectomy has been shown to decrease the numbers of mucous cells in *Betta splendens* (Schreibman and Kallman, 1965), and in *Carassius auratus* (Ogawa and Johansen, 1967) in which prolactin injections opposed this degenerative change.

Further work in Olivereau's laboratory has centred on the measurement of sialic acid (N-acetylneuraminic acid) (Olivereau and Lemoine, 1971). This compound is a component of mucus, and its levels in the skin reflect the amount of mucus being secreted. The findings confirmed that skin mucification plays a large part in osmoregulation, and showed that the levels of sialic acid fell following hypophysectomy but could be restored to normal by prolactin treatment. What is more, ectopic pituitary autotransplants, which are known to secrete large amounts of prolactin, also maintained sialic acid levels in hypophysectomised eels. Epidermal mucus secretion appears, from these findings, to be stimulated by prolactin in the eel as in other teleosts. However, Olivereau and Lemoine (1972) have more recently questioned the role of endogenous prolactin in stimulating epidermal mucus secretion in the eel, since they found during the transfer of eels to freshwater that there was very little increase in cutaneous sialic acid, although the pituitary prolactin cells became strongly hyperactive.

Blüm and Fiedler (1965) have shown that prolactin stimulates proliferation of the epidermal mucous cells of adult *Symphysodon discus*, a cichlid in which the mucus ('discus milk') serves as a nutrient for young hatchlings. This suggests that even at this level of evolution the two major attributes of 'parental care' and 'osmoregulation' are expressed in the prolactin molecule.

2.1.6 *Effects on kidney function*

Renal effects of prolactin have been reported for a variety of teleosts. In *Fundulus kansae* (Stanley and Fleming, 1966, 1967) and *Carassius auratus* (Lahlou and Sawyer, 1969; Lahlou and Giordan, 1970) the pattern of response is essentially similar. Following removal of the pituitary there is a rise in urinary sodium concentration associated with a reduced urine flow, the overall result being that the net urinary sodium loss does not differ markedly from the intact controls. Prolactin given to hypophysectomised goldfish increases urine flow to normal while reducing its sodium concentration, the net result being an elevated loss of sodium *via* the urine. As in the case of its effects on the goldfish gill, it is difficult to relate the effects of prolactin in the goldfish kidney to any presumed physiological role of the hormone.

Structural responses in the teleost kidney have been associated with prolactin. In *Gasterosteus aculeatus* the structure of the glomerular/capsular apparatus varies with the salinity of the external media. On transer from seawater to freshwater there is an expansion of the glomerular tuft and

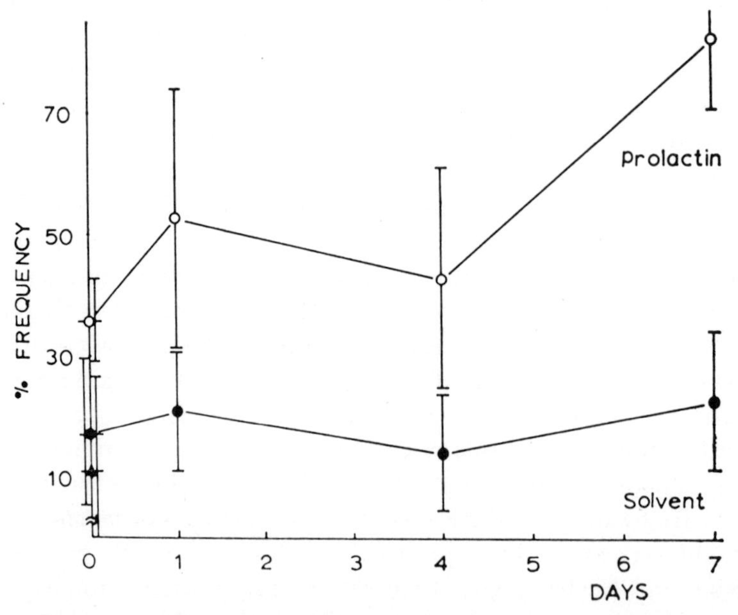

Fig. 2.5 Prolactin and glomerular ratio (tuft diameter/capsule diameter) of sticklebacks *G. aculaetus* form *trachurus*, in seawater and after transfer to freshwater. Abscissa represents days in freshwater, each point is mean ±95% confidence interval. (From Leatherland and Lam, 1969).

a closing of the intracapsular space. These changes have been correlated with a possible expansion of the glomerular filtration surface, and an assumed increase in urine production. The seawater condition of large intracapsular spaces is claimed to be associated with inactivity of the glomerulus (Leatherland and Lam, 1969) (see Fig. 2.5). These workers also found that injection of prolactin into prolactin-deficient seawater *Gasterosteus* increased the size of the glomerular tufts but had no effect on the size of the Bowman's capsule; the hormone also increased the number of capsules which showed no intracapsular space. Thus, prolactin may increase the effective glomeruler filtration rate, at any rate in *Gasterosteus*. To generalise from this, if prolactin maintains or reduces the urine-to-plasma sodium ratio (Lam, 1972), as in *Carassius* (Lahlou and Sawyer, 1969; Lahlou and Giordan, 1970), and if at the same time prolactin increases the glomerular filtration rate and urine flow, then it would follow that there must concomitantly be an increase in the tubular reabsorption of sodium presumably stimulated by prolactin. There is evidence in support of this hypothesis. Thus prolactin increases Na^+/K^+-ATPase activity in the kidney of hypophysectomised *Fundulus heteroclitus*, although it does not completely restore the high enzyme levels found in intact fish (Pickford *et al.*, 1970). Furthermore, using a freeze-etching electron microscopic technique, Wendelaar Bonga and Veenhuis (1974) demonstrated that prolactin induced accumulation of particles on the basal membrane of the renal tubule cells of *Gasterosteus*; since these particles are believed to be associated with transmembrane transport processes, the observations imply that prolactin may directly stimulate reabsorption of substances (including sodium) from the kidney filtrate.

It appears thus far, then, that prolactin may perhaps have two distinct renal actions: stimulation of renal tubular reabsorption, and stimulation of glomerular filtration rate. However, from work on *Fundulus kansae*, the existence of the latter response as an active process has been questioned, the argument being that prolactin primarily acts to increase extra-renal water permeability, and that this leads to an 'obligatory' rise in urine flow (Stanley and Fleming, 1967). However, the fact that evidence from *Carassius* (Lahlou and Giordan, 1970) indicates that the action of prolactin on urine flow is independent of its action on branchial water permeability implies a direct and independent influence of prolactin on glomerular filtration rate. The *in vitro* work by Lam (1969, 1972) showing that prolactin decreased the water permeability of the gill of the marine *Gasterosteus aculeatus* (a species in which, as argued above, prolactin appears to *increase* glomerular filtration) may also be taken as indicating that the renal and branchial effects of prolactin are independent. Further support for a direct prolactin action on the teleost kidney tubule (not the glomerulus) comes from work on the eel, *Anguilla anguilla*, in which prolactin produced morphological stimulatory effects on the renal tubules in intact fish, and to a lesser degree in hypophysectomised fish; ectopic pituitary autotransplants, known to secrete prolactin, maintained the renal tubule in a condition intermediate between that in intact and that in hypophysectomised eels (Olivereau and Lemoine, 1968, 1969; Olivereau, 1970).

2.1.7 *Effects on intestinal transport*

Evidence regarding the effect of prolactin on the teleost intestine is somewhat contradictory. Hirano and Utida (1968) could demonstrate no effect of prolactin on intestinal water absorption in the freshwater eel *Anguilla japonica*. However in seawater *Anguilla japonica*, Utida *et al.* (1972) have shown that prolactin produces a marked depression of water flux and a decrease in the net sodium and chloride transport. It should be noted that as in other teleosts, intestinal water transport is much higher in the eel in seawater than in freshwater, as part of the adaptive osmoregulatory mechanisms (Maetz, 1971; Bern, 1975).

Lemoine and Olivereau (1972) showed that hypophysectomy in the eel caused a fall in the sialic acid level of the intestinal mucus, about 20–30 days after the operation. This fall was totally prevented by replacement therapy with ovine prolactin. Furthermore, prolactin can cause an increase in the size (142–190%) of the mucus-producing cells of the gastric mucosa in the stickleback (Wossughi, 1972), although Olivereau and Lemoine (1972) failed to detect any effects of prolactin on intestinal mucous cells in various marine species, even at high doses. The same workers (Lemoine and Olivereau, 1973) found that gradual transfer of eels from freshwater to seawater resulted in reduced intestinal mucus production as measured by sialic acid levels, which would correlate with all the evidence that prolactin secretion is reduced in seawater. It is difficult to ascribe a role to prolactin in teleost intestinal physiology, particularly bearing in mind that changes in mucus production may well be independent of ion transport levels. Mucus apart, Utida *et al.* (1972) have suggested that prolactin in the freshwater eel may act by suppressing the intestinal chloride pump, thereby reducing intestinal permeability to water. Considering that water and ion transport by the intestine is adaptively higher in marine teleosts than in freshwater, one may suppose that elevated plasma prolactin levels in freshwater teleosts could be physiologically important in 'switching-off' intestinal water transport, unwanted in freshwater.

2.1.8 *Effects on the urinary bladder*

Over the last five years there has been considerable expansion of our knowledge of the role of the urinary bladder in teleost osmoregulation (Hirano *et al.*, 1971; Johnson *et al.*, 1972). The teleost urinary bladder is not homologous with the ectodermal (allantoic) bladder of tetrapods. It is mesodermal, an expansion of the fused mesonephric ducts, and can be regarded as an extension of the renal system involved in the selective absorption of water and ions (Bern, 1975; Forster, 1975). Johnson and his co-workers (Johnson *et al.*, 1974) found that a single injection of prolactin into the seawater-adapted flounder, *Platichthys stellatus*, caused a decrease in the permeability of the bladder to water. This was correlated with an increase in the rate of sodium absorption from the urine by the bladder, and a decrease in urine osmolarity and sodium concentration. They also found that the time course of these events closely paralleled the changes in bladder physiology observed when flounders are transferred from seawater to freshwater. Measurements of flounder pituitary prolactin levels during transfer, using the *Gillichthys* yellowing response, indicated an increase in prolactin production 12 hours after transfer to freshwater. It should, however, be noted that the validity of this *Gillichthys* assay for prolactin is now under question (see Farmer *et al.*, 1977). Hirano (1974, 1975) has confirmed the bladder response to prolactin for *Platichthys* and obtained similar results for another flounder, *Kareius bicoloratus*. He suggests that prolactin acts by reducing water permeability, and increasing the activity of ionic pumps in the bladder wall.

It appears from further work (Johnson, 1973; Doneen and Nagahama, 1973; Doneen and Bern, 1974) that regulation of bladder function in euryhaline teleosts depends on two opposing control mechanisms, prolactin promoting the freshwater state (low water transport, high sodium transport) and cortisol the seawater state (high water transport, low sodium transport). This prolactin/cortisol antagonism is found in other systems, for example the teleost gill. The phenomenon has been used by Doneen (1974) as the basis of an assay for fish prolactin, the cortisol-maintained water permeability in the seawater bladder being inhibited in a dose-dependent manner by prolactin. Nagahama *et al.* (1975) have suggested that the urinary bladder of *Gillichthys* is subdivided cytologically and possibly functionally, having a region of columnar cells structurally similar to the branchial ionocytes, which are activated by transfer from seawater to 5% seawater.

In summary, it can be seen that prolactin has a wide range of actions, which add together in increasing the osmoregulatory ability of teleosts in freshwater. Apart from the effects on glomerular recruitment in the stickleback, these actions may all be basically similar. If the urinary bladder response is indeed mediated by an ionocyte type of cell (Nagahama *et al.*, 1975), then we may postulate that prolactin acts in osmoregulation either on ionocytes or on mucous cells, which vary only in that they are located in different target organs.

2.1.9 *Effect of prolactin on plasma calcium levels*

The fact that hypophysectomy reduces plasma calcium levels in teleosts has been known for a number of years, largely from work on *Anguilla anguilla*

(Fontaine, 1956; Olivereau and Chartier-Baraduc, 1965; Chan and Chester Jones, 1968; Chan *et al.*, 1968).

It seems that calcium is to some extent controlled independently of other plasma electrolytes. In seawater or hard freshwaters, the external calcium levels may be several times higher than the plasma levels, whereas teleosts in soft freshwaters face the problem of calcium loss. Most investigations involving hypophysectomy have used freshwater fish, in which pituitary ablation causes severe hyponatremia and haemodilution, processes which in themselves can markedly alter plasma calcium levels. Thus, evidence for a pituitary mechanism controlling calcium, independent of the control of other ions, has only recently been available.

Olivereau and Lemoine (1973) showed for the eel that the fall in plasma calcium levels following hypophysectomy in freshwater could be corrected by prolactin treatment. However, there were parallel changes in sodium and potassium levels. The same workers (Olivereau and Lemoine, 1973) obtained some evidence for a separate control of calcium in the eel: when male silver eels were gradually transferred from freshwater to seawater, plasma sodium and potassium did not change, but plasma calcium decreased and plasma chloride increased.

Pang (1973) has recently produced firmer evidence for a calcium-regulatory role of prolactin independent of its effects on sodium. Following hypophysectomy of male killifish (*Fundulus heteroclitus*), Pang *et al.* (1971) maintained the fish in an artificial dilute seawater, deficient in calcium but containing normal levels of sodium and other electrolytes. In this medium, the hypophysectomised fish developed tetanic seizures and severe hypocalcaemia, but maintained normal sodium levels. Treatment of such fish with purified mammalian hormones showed that prolactin and ACTH were capable of restoring plasma calcium levels, but that melanophore stimulating hormone (MSH) was ineffective (Pang, 1973). Extending the investigation, killifish pituitary glands were split into three fractions, anterior (containing mainly prolactin and ACTH), median (containing mainly GH, gonadotrophin and thyroid stimulating hormone (TSH)), and posterior (containing mainly MSH). The fractions were homogenised and injected into pituitary-ablated hypocalcaemic male killifish. All three fractions elevated plasma calcium levels, but the median fraction was the least effective. The posterior fraction may have been effective because of inherent ACTH-like activity associated with this region (Pang, 1973).

2.2 Non-osmoregulatory effects of prolactin in teleosts

2.2.1 *Control of fat metabolism*

Most teleosts inhabiting the temperate zones are subjected to seasonal fluctuations in food availability (Nikolaisky, 1963). This results in a seasonal pattern of fat deposition which is usually inversely related to the gonadal cycle. Work by Meier and his co-workers (Lee and Meier, 1967; Meier, 1969) suggested that prolactin might be involved in fat metabolism in a number of teleost species (Table 2.5). Experiments on *Fundulus grandis* (Joseph and Meier, 1971) and *F. chrysotus* (Lee and Meier, 1967) revealed a circadian variation in the effects of prolactin on fat metabolism. In *F. grandis*, when prolactin was injected early in the photoperiod, lipid was mobilised from the fat stores, whereas injections given later in the photoperiod promoted the deposition of fat. These effects were shown to be independent

Table 2.5 Teleost species in which prolactin has been shown to effect lipid metabolism

Species	*References*
Fundulus chrysotus	Lee and Meier (1967)
Fundulus grandis	Joseph and Meier (1971)
Fundulus kansae	Merhle and Fleming (1970)
Fundulus similis	de Vlaming, Sage and Charlton (1973)
Notemigonus chrysoleucas	de Vlaming and Pardo (1974)
Oncorhynchus nerka	McKeown, Leatherland and John (1975)
Cyprinoden variegatus	de Vlaming and Sage (1972)

of the length of the light period, and this circadian rhythm in the response to prolactin has been further investigated. Injections of thyroxine (Meier, 1970) or cortisol (Meier *et al.*, 1971) may cause entrainment of the circadian rhythm. In general it appears for teleosts and other vertebrates that if prolactin is given 12 hours after an entraining injection of cortisol it promotes fat mobilisation; but if the same dose of prolactin is injected 24 hours after the cortisol injection then it promotes fat deposition (see Fig. 2.6). This change in the responsiveness to exogenous prolactin may be conditioned or phased by circadian rhythm in the secretion of adrenocorticosteroids (Meier *et al.*, 1971; Nicoll, 1974).

Circadian changes in prolactin secretion have been described in various teleosts (see Batten *et al.*, 1976). A number of papers have reported effects on lipid metabolism correlated with the circadian variations in circulating prolactin levels. Leatherland *et al.* (1973) demonstrated a rhythm in plasma free fatty acid (FFA) levels correlated with circadian rhythms in both plasma

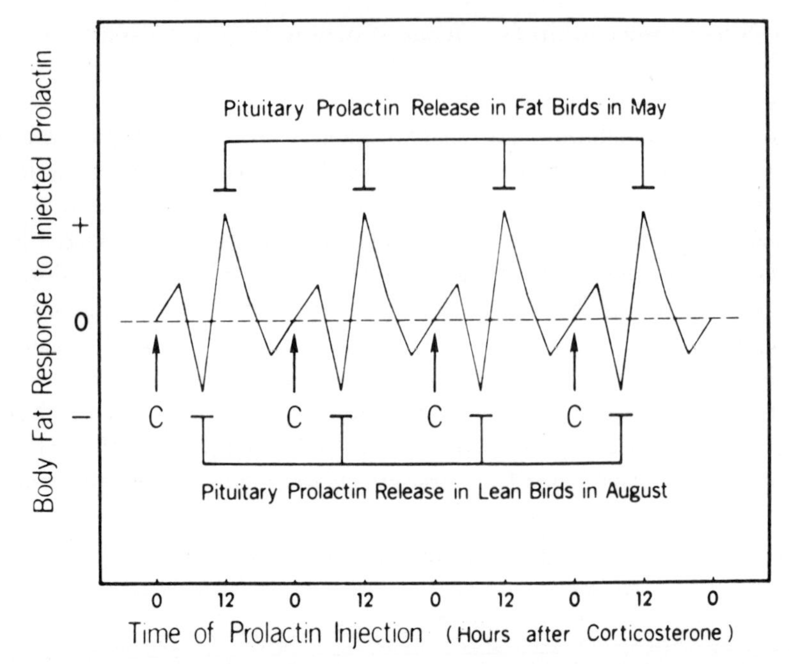

Fig. 2.6 Temporal synergism of corticosterone and prolactin controlling seasonal fat levels in the white-throated sparrow. This model relates the results of injection experiments with the analysis of daily rhythms of pituitary prolactin (Meier *et al.*, 1969) and plasma corticosterone (Dusseau and Meier, 1971) in May and August. C = time of rise of plasma corticosterone (about 0300 in May and about 2100 in August). (From Meier and Martin, 1971.)

prolactin and plasma growth hormone in *Oncorhynchus nerka*. Further work (McKeown *et al.*, 1975) has shown that injections of prolactin and growth hormone elevate muscle FFA and decrease muscle water content in the same species.

There is evidence that prolactin secretion varies seasonally in several teleosts, including *Gasterosteus* (Lam and Hoar, 1967; Lam and Leatherland, 1969). This seasonal pattern appears to be caused by changes in the photoperiod. This evidence has led de Vlaming and his co-workers to investigate in detail the relationship between photoperiodic variation and the fat deposition response to prolactin. They have shown (de Vlaming *et al.*, 1973; de Vlaming *et al.*, 1974) that melatonin, an indolamine produced by the pineal, has marked effects on fat deposition and on pituitary prolactin levels. The rate of melatonin synthesis, in teleosts as in mammals, is high during the dark phase and low during the light phase. In *Fundulus similis* and *Cyprinodon variegatus* melatonin injection reduced body lipids in fish exposed to long photoperiods. This effect is independent of the time of injection in *Fundulus* but not in *Cyprinodon*. However, in *Fundulus* adapted to a short photoperiod, the response to melatonin did vary seasonally, in that melatonin caused reduced lipid deposition in May but increased lipid deposition in July. The mechanism by which melatonin influences fat deposition is not clear, but de Vlaming *et al.* (1973) showed that melatonin reduces the level of prolactin in the pituitary of *F. similis*. Corroborative evidence indicates that in this species prolactin is stored in the pituitary during the dark phase and released, possibly as a result of melatonin stimulation, shortly after the onset of the light period (Sage and de Vlaming, 1973).

Further work by de Vlaming and Pardo (1974) suggests that the environmental influence may persist into *in vitro* experiments. Studying incubated liver fragments from *Notemigonous chrysoleucas*, these workers found that the hepatic response to prolactin is governed by the photoperiod and temperature to which donor fish were adapted.

2.2.2 *Prolactin and pigmentation*

There are two reports of prolactin affecting pigmentation in teleosts. Pickford (1956) reported that in hypophysectomised killifish, *Fundulus heteroclitus*, prolactin induces darkening of the dorsal surface and potentiates the development of new melanophores induced by MSH. Prolactin has little or no effect on the melanophores of the gobiid fish *Gillichthys mirabilis*, but in this species it causes dispersion of the yellow pigment within the xanthophores (Sage, 1970). This latter response, also shown by *Gobius minutus* (Ball and Ingleton, 1973), has been used as the basis of an assay system for fish prolactin (Sage and Bern, 1972), but more recent work indicates that the xanthophore-dispersing factor, while closely associated with prolactin in pituitary extracts, is an independent moiety and not a part of the fish prolactin molecule (see Farmer *et al.*, 1977).

2.2.3 *Prolactin and visual pigments*

The retinal rods of many freshwater and euryhaline teleosts contain both a rhodopsin and a porphyropsin, the relative proportions of which change according to a seasonal or migratory pattern (Bridges, 1972). In Pacific salmon, *Oncorhynchus* spp., there is a change from a rhodopsin- to a porphyropsin-dominated retina during the upstream migration from the sea to freshwater (Beatty, 1966). A study of another salmonid, the rainbow trout (*Salmo gairdneri*), showed that either prolactin or thyroxine injections promoted increased proportions of porphyropsin, the response to each hormone being dose-dependent (Cristy, 1974). The trout were intact, and as the author pointed out prolactin could have affected the retinal pigments *via* a TSH-releasing activity (see section 2.2.6), with consequent elevation of circulating thyroxine levels. However, in view of the dominant role of prolactin in the freshwater-adaptation of teleosts (see p. 16), it could well be that prolactin directly promotes the retinal changes during upstream migration.

2.2.4 *Prolactin effects on parental behaviour*

The part played by prolactin in controlling parental behaviour in teleosts is still unclear. The literature contains a number of reports of effects associated either with suppression of reproductive behaviour or with stimulation of true parental behaviour. Probably the best-documented of these prolactin effects is promotion of the skin sloughing which produces 'discus milk' in the cichlid *Symphysodon discus* (Blüm and Fiedler, 1965). The process involves division of the epidermal cells, which are then shed in association with copious mucus, forming a milky substance upon which the young fish feed.

Other reports of behavioural changes caused by prolactin suggest that a major role of the hormone may be to suppress aggressive and sexual behaviour during the parental period. Thus Blüm and Fiedler (1965) reported that prolactin inhibits aggressive behaviour in parental cichlids, a response which may be due directly to the suppression of androgen production or indirectly to blockade of gonadotrophin secretion. Prolactin has been shown to suppress follicle stimulating hormone (FSH) secretion in birds (Thapliyal and Saxena, 1964) and amphibians (Mazzi and Vellano, 1968). Prolactin has also been shown to cause a decline in the size of the nuchal hump in the cichlid *Cichlasoma citrivellus* (Bleick, 1975). The nuchal hump, a prominent protruberance from the forehead region, is developed by both males and females during the mating season. High concentrations of either ovine prolactin or of *Cichlasoma* rostral pars distalis extracts produced involution of the hump, but low doses were ineffective. Bleick (1975) concludes that the main role of prolactin may be to prevent the development of mating characteristics and behaviour during the parental period, rather than to cause the actual involution of the hump.

Another aspect of parental behaviour which may be affected by prolactin is the fanning of the nest by the pectoral fins. Prolactin has been shown to stimulate fanning behaviour in cichlids (Blüm and Fiedler, 1965; Blüm, 1974) and in a wrasse (Fiedler, 1962). However, Smith and Hoar (1967) found that prolactin did not stimulate fanning behaviour in *Gasterosteus*, and in fact may have been slightly inhibitory. Since in this teleost fanning behaviour is primarily stimulated by androgens, the slight inhibitory effect observed may have been due to an antigonadal action of prolactin.

The most complete evidence for the participation of fish prolactin in parental care comes from an extensive series of investigations on the sea horse, *Hippocampus*, by Boisseau (1964, 1965, 1967, 1969). The male *Hippocampus* incubates developing eggs in a ventral pouch (marsupium), the development of which is under testicular and gonadotrophic control. ACTH and corticosteroids appear to be responsible for maintaining the connective tissue structure of the developed marsupium during the incubation period. During incubation, the epithelial lining of the marsupium proliferates and secretes a protease which digests proteins present in the marsupial fluid, which are derived from the yolk of eggs arrested during early development; the amino acids so released are probably absorbed by the embryos. To quote Ball (1969*b*), 'the embryo sits in a nutrient broth which is predigested for its consumption by a paternal enzyme'. Hypophysectomy of the father arrests secretion of the protease, and induces histological regression of the marsupial epithelium. Prolactin treatment of both intact and hypophysectomised fish produces marked histological stimulation of the marsupial epithelium, and in the intact fish it accelerates secretion of the protease. The physiological role of fish prolactin implied by these observations was confirmed by Boisseau (1967), who showed by partial hypophysectomy that the endogenous hormone responsible for maintaining the marsupial epithelium originates from the rostral pars distalis, the source of fish prolactin (see Chapter 1, p. 2). Boisseau also demonstrated that the lactotrophs (erythrosinophils) of the rostral pars distalis displayed cyclic variations closely correlated with the development and secretory activity of the marsupial epithelium.

A factor that complicates interpretation of these results is that ovine prolactin injected after the tenth day of incubation in intact *Hippocampus* causes histological stimulation of the adrenocortical tissue, though producing adrenocortical involution if given earlier. In hypophysectomised males, exogenous prolactin only slightly stimulates the involuted adrenocortical tissue. Ovine prolactin has also been found to produce some degree of adrenocortical stimulation in various other teleosts (Chambolle, 1964; Ball and Ensor, 1969; Fleming and Ball, 1972; Ball and Hawkins, 1976). Whether this is due to minute contaminating traces of ACTH in the preparations or to molecular similarities between ovine prolactin and teleostean ACTH cannot be resolved (Ball and Hawkins, 1976), but there is no evidence at all that *fish* prolactin stimulates the teleost adrenocortical tissue (Ball and Ensor, 1969; Ball and Hawkins, 1976). In summarising his work on *Hippocampus*, Boisseau (1969) considered the possibility that adrenocortical steroids might have mediated the effects of exogenous prolactin in his experiments. However, there are temporal differences in the actions of ACTH and prolactin on the marsupium in addition to all the evidence that the two act on two separate elements, connective tissue and epithelium respectively. This work on *Hippocampus* is extremely interesting, and it constitutes the most satisfactory demonstration of a parental role for fish prolactin, perhaps significantly in a marine teleost in which the hormone presumably is not concerned in osmoregulation (see p. 16).

2.2.5 *Sexual effects of prolactin in teleosts*

In the gobiid fish *Gillichthys mirabilis*, castration or hypophysectomy causes regression of the seminal vesicles (de Vlaming and Sundararaj, 1972). Regression of these structures has also been observed in various teleosts following hypophysectomy (Vivien, 1938, 1941; Tavolga, 1955; Sundararaj and Goswami, 1965); Sundararaj and Nayyar (1969) showed that prolactin synergises with androgens to promote growth and secretion in the seminal vesicles of the catfish *Heteropneustes fossilis*. Prolactin alone had no effect on seminal vesicle development, whereas androgens were capable of initiating growth, an effect which was potentiated by prolactin.

In *G. mirabilis*, either testosterone or prolactin prevented the regression of the seminal vesicles after hypophysectomy (de Vlaming and Sundararaj, 1972). However, as these authors point out, the possibility that prolactin stimulates endogenous testosterone secretion cannot be dismissed. As in *Heteropneustes*, the stimulatory effect of testosterone on the seminal vesicles was potentiated by prolactin.

Little is known of direct effects of prolactin on teleost gonads, but Blüm and Weber (1968) have reported that prolactin elevates steroid 3β-ol-dehydrogenase activity and RNA synthesis in the ovary of *Aequidens pulcher*.

2.2.6 *Thyroid-stimulating effects of prolactin in teleosts*

In 1966 Olivereau showed that injections of ovine prolactin in eels induced an increase in the activity of thyrotrophs (TSH cells) in the pituitary, involving an increase in mitotic activity, in number of cells, and in nuclear and nucleolar size, together with degranulation. This increase in TSH cell activity was associated with an increase in thyroid activity, demonstrated both histologically (Olivereau, 1966) and by increased thyroidal uptake of radio-iodine (Olivereau, 1968). Prolactin produced no thyroidal stimulation in hypophysectomised eels (Olivereau, 1966), indicating that it acts in some way at the hypothalamo-hypophysial level. Olivereau concluded that exogenous prolactin could act in the intact eel as a kind of 'TSH-releasing factor' (TRF), stimulating the release of endogenous TSH. Similar findings have been reported for another teleost *Pterophyllum scalare* by

Osewold and Fiedler (1968), and in amphibians for *Triturus cristatus* (Vellano *et al.*, 1967; Mazzi and Vellano, 1967) (see page 61). However, Higgins and Ball (1972) demonstrated that ovine prolactin does not release TSH in another teleost, *Poecilia latipinna*, and it may be that this 'TRF' effect of ovine prolactin is limited to only certain species.

2.3 Regulation of prolactin secretion from the teleost pituitary

There appear to be three major mechanisms which control the secretion of prolactin by the teleost pituitary.

2.3.1 *Environmental control*

The rapid adjustments of prolactin secretion seen in euryhaline teleosts during transfers between seawater and freshwater appear mainly to be direct responses of the lactotrophs to changes in plasma osmolarity or sodium levels.

When *Poecilia latipinna* is transferred rapidly from one-third seawater to freshwater, there is initially a rapid fall in plasma sodium levels from 168 mEq/l to 140 mEq/l over 6 hours. This rapid drop is then followed by a more gradual decline to a level of about 130 mEq/l at 18 hours after transfer. Subsequently plasma levels rise, reaching 150 mEq/l after approximately 72 hours (Ensor and Ball, 1972). This pattern has been shown to parallel cytological changes in the erythrosinophils (lactotrophs) of the rostral pars distalis (Ball and Ensor, 1969), changes which are reflected by alterations in the level of prolactin in the pituitary, measured either by bioassay (Ensor and Ball, 1968) or by disc electrophoresis (Ball and Ingleton, 1973).

Studies on a wide range of euryhaline teleosts have agreed with the results for *P. latipinna*. In a range of species adapted to relatively high ambient salinities, *G. aculeatus* (Leatherland, 1970), *M. cephalus* (Abraham, 1971), *C. auratus* (Leatherland, 1972), *P. latipinna* (Hopkins, 1969), *G. mirabilis* (Nagahama *et al.*, 1975*a*), the lactotrophs have been shown to be comparatively inactive, in contrast to their high activity in hypo-osmotic media. Leatherland and his co-workers studied several *Tilapia* species adapted to a range of salinities in East African saline lakes, and showed that there is a close correlation between ambient salinity, plasma sodium levels and lactotroph activity (Leatherland *et al.*, 1974, 1975). Electron microscope studies have confirmed the numerous light microscopic observations (Ball and Baker, 1969; Holmes and Ball, 1974; Batten *et al.*, 1975), and there is now a considerable body of evidence showing that the prolactin cells in many teleosts are more active, by all criteria, in freshwater than in seawater. Nonetheless, some prolactin synthetic activity continues even in full-strength seawater, a kind of 'residual synthesis' possibly of a storage or precursor form of the hormone (Nicoll, 1972).

Though not conclusive, all this work suggests that the lactotrophs might respond directly to changes in the osmolarity of sodium content of the blood. Evidence that this is so comes from studies on the teleost pituitary incubated or cultured *in vitro*. By incubating the pituitary of *Poecilia latipinna* and *Anguilla anguilla* in media of different osmotic pressures, Ingleton *et al.* (1973) were able to demonstrate an inverse relationship between ambient

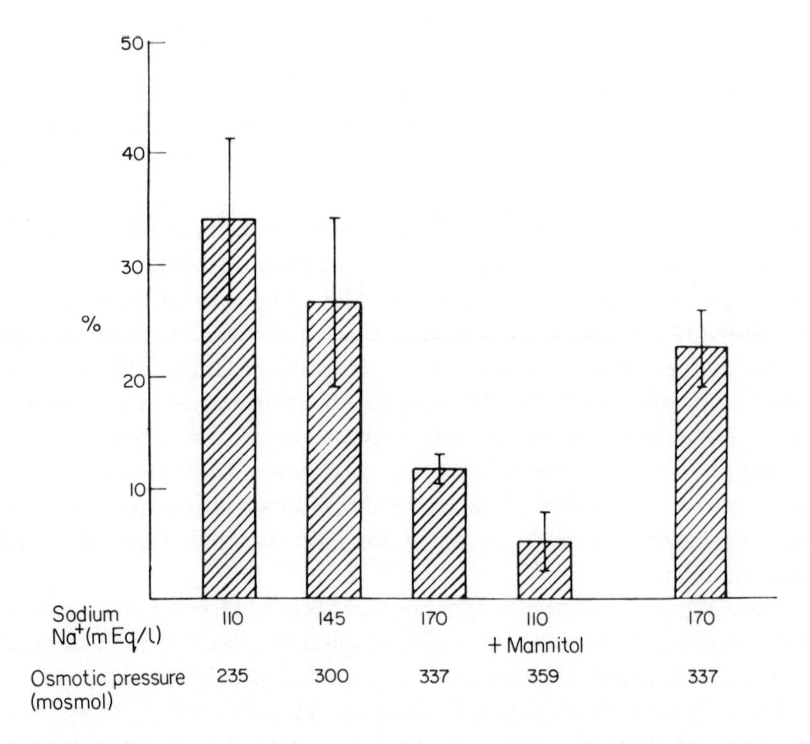

Fig. 2.7 Percentage of prolactin released into the medium. (From Ingleton, Baker and Ball, 1973.)

osmolarity and the rate of prolactin release; osmotic pressure, rather than sodium concentration, appears to be the factor to which the lactotrophs respond, since they were not stimulated by a medium of low sodium concentration but with a normal osmotic pressure maintained by mannitol (see Fig. 2.7). By incubating the pituitaries with ^{3}H-leucine it was possible to show that there was also enhanced *synthesis* of prolactin at low ambient osmolarities, and that this newly synthesised hormone was preferentially released from the pituitary. Longer-term culture experiments produced essentially similar results for *Anguilla anguilla* and *Salmo gairdneri* (Baker and Ingleton, 1975).

2.3.2 *Control by circulating prolactin levels*

Ball and Olivereau (1964) showed for the eel that prolonged injection of ovine prolactin causes degranulation and atrophy of the prolactin cells after about 20 days. Other evidence from intact fish bearing pituitary homotransplants, suggests that the prolactin released by the transplanted pituitary is sufficient to suppress release from the *in situ* gland (Olivereau *et al.*, 1971; Nagahama *et al.*, 1975). This suggests that high circulating levels of prolactin may have a direct inhibitory effect on the lactotrophs.

2.3.3 *Hypothalamic control*

An inhibitory hypothalamic control of prolactin was first indicated by work on *Poecilia formosa* involving pituitary homotransplants in hypophysectomised fish (Ball *et al.*, 1965; Olivereau and Ball, 1966). In these experiments it was shown that the lactotrophs in the transplants remained active though separated from the hypothalamus, and that the transplants permitted their hypophysectomised bearers to survive in freshwater, an attribute of prolactin (see Section 2.1). This evidence has now been extended to cover a wide range of teleost species (Olivereau, 1971; Leatherland, 1972; Peter, 1972; Zambrano, 1971; Abraham, 1971; Ball *et al.*, 1972; Chidambaran *et al.*, 1972; Leatherland and Ensor, 1973).

Recently Leatherland and Ensor (1974) have shown that osmotic regulation of prolactin release may operate in *Carassius* through the hypothalamus, rather than (or in addition to) acting directly on the pituitary as in *P. latipinna* and *A. anguilla*. The concentration of a prolactin release-inhibiting factor (PIF) was found to be higher in hypothalamic extracts from distilled water-maintained donors than in extracts from fish kept in one-third seawater (see Fig. 2.8). Other work on the goldfish indicates a dual control associated with the hypothalamic Nucleus Lateralis Tuberalis (NLT). Lesions in the lateral part of the NLT caused significant increases in serum prolactin levels without modifying the pituitary prolactin content. However, large lesions in the anterior-medial thalamus and anterior hypothalamus dorsal to the NLT resulted in decreased serum prolactin levels, again without affecting stored prolactin (Peter and McKeown, 1973). Evidence suggests that the NLT lacks Type A (peptidergic) neurosecretory neurons, but contains several bundles of aminergic neurons (Type B) which probably exert an inhibitory control over the pituitary (L'Hermite and Lefranc, 1972). It has now been shown for a number of species that prolactin release can indeed be modified by

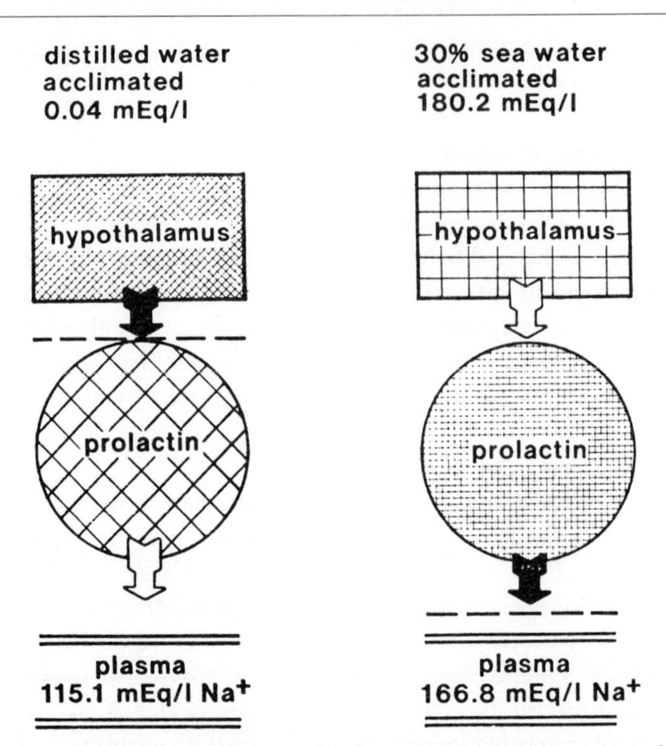

Fig. 2.8 Diagrammatic representation of the situation assuming the presence of prolactin-RIF in the hypothalamus. The closely cross-hatched hypothalamus and prolactin zones represent accumulations of secretory product. The black arrows represent a blocked release and the white arrows a release of material. (From Leatherland and Ensor, 1973.)

dopamine or by dopamine agonists such as ergocryptine and CB-154.

Zambrano *et al.* (1974), for *Tilapia mossambica*, have shown that the Type B fibres in the pars distalis can be destroyed by intra-cisternal injections of 6-hydroxydopamine (6-OHDA), confirming earlier work by Zambrano *et al.* (1972) on *Gillichthys mirabilis* and Follénius (1972) on *Gasterosteus aculeatus*, which had suggested that the Type B fibres were aminergic. Destruction of the Type B fibres in *Tilapia mossambica* (Zambrano *et al.*, 1974) increased the ability of the fish to withstand transfer to freshwater. Electron microscopic studies showed that in control fish the activation of the prolactin cells commenced at about 6 hours after transfer to freshwater, while in the 6-OHDA-treated fish degranulation and proliferation of the rough endoplasmic reticulum were already well advanced by this stage.

This view that dopamine might inhibit prolactin secretion in teleosts is supported by work on *Anguilla anguilla* by Olivereau (1975). Treatment of eels with L-dopa, which increases brain dopamine levels, or with ergocryptine, a known dopamine agonist, caused inhibition of prolactin release.

This agrees with the findings of McKeown (1972) on *Poecilia*, Kramer *et al.* (1973) on *Fundulus* (using CB-154, a synthetic bromo derivative of ergocryptine), and of Nagahama *et al.* (1975) who used L-dopa in *Gillichthys*.

Detailed studies on *Poecilia latipinna* (Wigham and Ball, 1974) have demonstrated more fully that the teleostean prolactin cell is directly inhibited by hypothalamic catecholamines, probably dopamine. Type B (aminergic) nerve fibres synaptically innervate the prolactin cells in this species, and are destroyed by 6-OHDA (Batten and Ball, 1976; Ingleton, Batten and Ball, 1977). Incubation of the isolated pituitary with catecholamine alpha- and beta-agonists and antagonists, and with dopamine, alone and in combination with DMPEA (dimethylphenylethylamine, a specific dopamine antagonist) showed that catecholamines can act directly on the pituitary to inhibit release of prolactin, and indicated that dopamine is probably the agent physiologically concerned (Wigham *et al.*, 1975). Further *in vivo* work on *P. latipinna* involved the injections of L-dopa, 6-OHDA, and the combination L-dopa with DMPEA, and the results confirmed and extended the *in vitro* work in showing that dopamine acts physiologically to reduce both synthesis and release of prolactin (Wigham and Ball, 1976).

For *P. latipinna*, there are indications that the dopamine and osmotic control mechanisms may act independently on prolactin secretion. Wigham and Ball (1976) found that destruction of the aminergic innervation of the prolactin cells by 6-OHDA (cf. Batten and Ball, 1976) resulted not only in increased prolactin secretion but also in a temporary rise in plasma osmotic pressure. If these results are set alongside those of Ingleton *et al.* (1973) for the same species (see p. 40), then it can be seen that the catecholaminergic control may act independently of osmotic influences. This might explain the existence of circadian rhythms in prolactin secretion (Leatherland *et al.* 1975; Batten *et al.* 1976), which are independent of environmental osmotic variation.

II Cyclostomes

The evidence for cyclostomes suggests that the pituitary does not produce prolactin (Aler *et al.*, 1971; Sage and Bern, 1972; Holmes and Ball, 1974). However, there is one report of an effect of prolactin in *Myxine* adapted to 60% seawater (Chester-Jones, 1963). Injection of prolactin under these conditions resulted in increased muscle water content, decreased blood potassium and decreased muscle electrolytes. Obviously more work is required to decide whether these results have anything more than pharmacological interest.

III Elasmobranchs

The osmoregulatory problems faced by elasmobranchs are different from those of teleosts. Retention of high levels of urea in the blood leads to the plasma being approximately isosmotic with seawater. Excess salt, mainly gained passively through the gills and the gut, is excreted largely through the kidney and rectal gland. Although the majority of elasmobranchs are marine their sodium turnover is considerably lower than that of teleosts, though water turnover may be relatively high (Payan and Maetz, 1970, 1971).

There is considerable evidence to suggest that the elasmobranch pituitary secretes a fish-type prolactin (Grant, 1961; Grant and Banks, 1968). Hypophysectomised *Raja erinacea* gain weight when transferred from full seawater to 50% seawater. This increase in weight is assumed to reflect increased water retention, and is opposed by injections of prolactin, suggesting that prolactin might act either by reducing water influx or by causing increased renal water clearance, or by both mechanisms (Grant and Banks, 1968).

More detailed work by Payan and Maetz (1970, 1971) has shown that in the dogfish, *Scyliorhinus canicula*, diffusional branchial water permeability is reduced by hypophysectomy and increased by both prolactin and ACTH. The osmotic water permeability of the gills was also reduced by hypophysectomy, but was not affected by either prolactin or ACTH. Following pituitary ablation there was no change in total sodium outflux, outflux across the skin and gills, or plasma sodium levels; however, urinary sodium loss was halved. Thus it would appear that hypophysectomy must increase sodium removal by the rectal gland. Replacement with prolactin was not tested in this experiment, but the same workers found that the reduced urine flow of hypophysectomised dogfish could not be restored by prolactin, though ACTH was effective (Payan and Maetz, 1971).

More recently, de Vlaming, Sage and Beitz (1974) reported that removal of the rostral lobe of the pituitary (the presumed source of prolactin) in the ray *Dasyatis sabina* resulted in increases in plasma calcium and sodium, which could be reversed by prolactin injections. This suggests that prolactin either promotes excretion of these plasma components or, by elevating water influx, promotes haemodilution. The idea of increased branchial water influx would agree with the findings of Payan and Maetz (1970). More work obviously needs to be done on the effects of prolactin on the elasmobranch kidney and rectal gland.

3 Prolactin in amphibians

The action of prolactin in amphibians appear to be similar in general to those in fishes, with pronounced effects on osmoregulation and the integument. In addition prolactin appears to be involved to a large extent in amphibian metamorphosis and growth processes, operating in a well-defined relationship with the thyroid, and it also seems to stimulate the activity of certain enzymes, in particular the enzymes of the ornithine cycle. These various 'new' effects are discussed in detail below. It remains true, however, as in fishes, that major targets of prolactin activity remain ion transport, the integument, and perhaps reproduction. A theme that will emerge in subsequent chapters is that in evolution there has been an increase in the significance of the reproductive actions of prolactin, coupled perhaps with a decline in the importance of its ion regulatory functions. For amphibians, information concerning prolactin comes from anurans and urodeles. The third living group, the worm-like apodans, have not been studied.

3.1 Prolactin and osmoregulation

The earliest reports of prolactin being involved in the osmoregulatory physiology of the amphibia stem from the work of Chadwick (1941) and Grant and Grant (1958) who showed that prolactin stimulated the breeding migration to water of the terrestrial juvenile stage of the eastern red-spotted newt (the red eft). Since the life history of most amphibians involves a period, either larval or adult, in which the animal is exposed to a hypotonic environment, it is reasonable to expect that prolactin might play an osmoregulatory role similar to that found in teleosts. This idea was formulated as early as 1968 by Bern and Nicoll, but it is only in recent years that much evidence has accumulated in its support.

Sampietro and Vercelli (1968) showed that hypophysectomy of adult *Triturus cristatus* resulted in a fall in plasma sodium levels when the animals were maintained in freshwater. Prolactin replacement therapy resulted in a rise in plasma sodium levels, and a similar effect was produced by aldosterone. This point will be of interest later, since in certain isolated tissue preparations the effects of prolactin appear to parallel those of aldosterone although the time course of action may be somewhat different (Howard and Ensor, 1977*a*). In hypophysectomised *T. cristatus*, prolactin and aldosterone were also shown to potentiate the effects of each other, the responses being additive (Sampietro and Vercelli, 1968).

Verification of these results has come from the work of Wittouck (1972*a*) on larval *Siredon* (*Ambystoma*) *mexicanum*, the axolotl. Hypophysectomy virtually doubled the sodium outflux into distilled water. Ovine prolactin, injected either 24 or 48 hours before flux measurements, corrected this defect and even reduced sodium outflux to below control values, a response qualitatively very similar to that shown by the gills of various teleosts (Ensor and Ball, 1972). Further work by Wittouck (1972*b*) has shown that prolactin can also stimulate sodium uptake by the gills of the axolotl, again opposing the effects of hypophysectomy. Similar results come from the neotenous urodele *Necturus maculosus* in which hypophysectomy causes significant reduction in plasma sodium levels (Pang and Sawyer, 1974). Treatment with prolactin which commenced 16 days after hypophysectomy was only partially successful in preventing this fall in plasma sodium. However, prolactin treatment starting immediately after hypophysectomy completely maintained normal plasma sodium levels. Flux measurements on hypophysectomised and sham operated *Necturus* showed that pituitary ablation resulted in a marked decrease in sodium influx and a large increase in outflux. Prolactin injections starting 20 days after the operation restored outflux to normal levels, but did not restore influx; but if injections were started just after hypophysectomy, both influx and outflux were returned to normal in 12–14 days. It may be that in addition to its immediate action on sodium fluxes, prolactin at low levels is necessary to maintain the condition of the transporting epithelia and that a prolonged withdrawal of prolactin impairs epithelial responsiveness to prolactin (or any other hormone) given subsequently.

In contrast, work by Crim (1972) on the adult anuran *Rana pipiens* and the adult urodele *Taricha torosa* demonstrated no effect of prolactin on plasma sodium levels. Hypophysectomy in both species decreased plasma sodium, and this change was opposed by neither prolactin nor aldosterone. Hormone replacement was started four or five days after hypophysectomy, and while it is unlikely that any irreversible changes in the condition of the transport sites had occurred in such a short time, the possibility cannot be ruled out. Certainly, the effects of hypophysectomy on amphibian osmoregulation are not consistent: for example, hypophysectomy of adult *Rana catesbiana* (Ridley, 1964) appeared to have no effect on plasma sodium levels, whereas hypophysectomy of *Bufo marinus* (Middler *et al.*, 1969) produced a severe hyponatraemia. One possible explanation for such discrepancies is that the degree to which a particular species is dependent on prolactin may depend on the proportion of its time spent in water; in other words, the more aquatic a species, the more dependent it may be on prolactin for maintenance of body electrolytes.

A rather complicated picture comes from the work of Brown and Brown (1973) who studied the effect of prolactin on plasma sodium levels in an aquatic urodele, the eastern red-spotted newt, *Notophthalmus viridescens.** Hypophysectomy resulted in a fall in plasma sodium levels, and although a slight rise followed replacement therapy with prolactin and/or corticosterone the levels remained significantly below those of sham-operated controls. Unexpectedly, when intact newts were treated with thyroxine their plasma sodium levels fell to 40–50% of control values. In this case (as in other amphibian examples), prolactin injections opposed the effect of thyroxine, (see also report by Grant and Cooper, 1965). Brown and Brown (1973) reported that after a week of prolactin treatment the animals appeared exceptionally slimy, indicating increased secretion of epidermal mucus. Further study showed that thyroxine appeared to increase urinary sodium concentration and enhance cutaneous water permeability, without altering the short circuit current across the skin. Unfortunately the responses of these various parameters to prolactin by itself were not investigated. However, prolactin administered together with thyroxine, while not affecting urinary sodium loss, did prevent the increase in cutaneous water permeability. Thyroxine and prolactin together reduced the cutaneous short-circuit current to about 50% of the level in control or thyroxine-treated animals. This contrasts with the results of Howard and Ensor (1977*a*) for the anuran *Rana temporaria* (see Fig. 3.1) in which prolactin increased the short circuit current across the isolated skin. It would appear from the results of Brown and Brown (1973) that prolactin may oppose the effects of high thyroxine levels in *Notophthalmus* but that prolactin is not itself primarily responsible for normal osmoregulatory processes. Perhaps in this urodele, normal sodium regulation is maintained by the balance of prolactin and thyroxine levels, although this would not satisfactorily explain the effects of hypophysectomy.

* Taxonomists have had a rare old time with this unfortunate beast. First *Diemyctylus*, it then became *Triturus*, then *Diemictylus*, and now *Notophthalmus*. Thus the same species (*viridescens*) appears on the scientific stage like a virtuoso actor in a bewildering range of disguises (including, incidentally, the mis-spelling *Notophthalamus*). Just to make matters worse, the terrestrial juvenile stage prior to assuming the adult aquatic habit has its own colloquial name, the red eft.

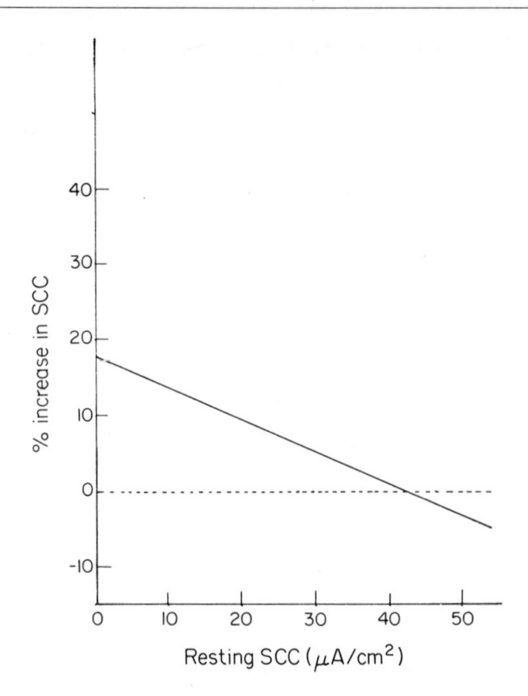

Fig. 3.1 The effect of prolactin on short circuit current (SCC) across the isolated frog skin. The results are plotted as a % increase over the initial short circuit current.

In contrast to the above results for a urodele, under certain conditions prolactin can increase the water-permeability of the skin of a larval anuran. Working with the African clawed 'toad' *Xenopus laevis*, Schultheiss *et al.* (1972) confirmed previous work of Leist (1970) in showing that cutaneous water permeability decreases during development and metamorphosis. In premetamorphic larvae (stages 48/50) hypophysectomy resulted in a decrease in tritiated water-turnover, during a two-week period after the operation. This reduction in water-permeability was corrected by injections of prolactin. Later in the life history, during metamorphosis and in young adults, removal of the pituitary had no effect on water-permeability.

In considering the various results, we see indications that responses to prolactin probably differ not only between species, and between urodeles and anurans, but also between different stages in the individual life history. Clearly it is unwise to make general statements about 'the effects of prolactin in amphibians' without bearing these differences in mind, and qualifying such statements accordingly.

Apart from *in vivo* experiments, there is now a growing body of information about prolactin effects gained from *in vitro* studies using the isolated frog or toad skin and the isolated toad bladder. The basic properties of these preparations have been known for some time (Ussing, 1951; Bentley, 1971; Erly, 1972; Cuthbert, 1973). The amphibian skin and, in some species, the urinary bladder, can transport salts against a concentration gradient by an active (energy-consuming) process. This process appears to involve primarily the movement of sodium ions and occasionally chloride ions against an electrochemical gradient. Potassium is not usually transported. The efficiency of the mechanism depends upon the prevailing osmotic conditions. Krogh (1939) showed that sodium can be accumulated from solutions as dilute as 10^{-5} M, and that if frogs are salt-depleted, by maintaining them in distilled water, the rate of sodium uptake on return to freshwater is greatly increased. The rate at which active transport is taking place can be assessed by measuring the resulting potential difference (PD) across the isolated skin, or by measuring the current required to short-circuit this PD, the short-circuit current (SSC). Occasionally direct measurements of sodium fluxes have been made using radiosodium as a tracer, and these results generally parallel the changes in PD or SCC. Figure 3.2 shows the apparatus originally produced by Koefod-Johnson and Ussing (1951) which, with a few modifications, has been used in most such studies on isolated amphibian tissues.

Early studies on the hormonal control of transport across these isolated tissues mainly concentrated on the actions of the neurohypophysial hormones and aldosterone. This field has been surveyed recently by Bentley (1971) and will not be dealt with here except insofar as these hormones have been shown to synergise with prolactin. The first indication that prolactin could stimulate active sodium transport *in vitro* came from the work of Dalton and Snart (1969) on *Bufo marinus*, showing that the SCC across the isolated toad bladder could be increased by as much as 60% when prolactin was added at dose levels between 10^{-6} and 10^{-8} M. They also showed that prolactin altered the temperature dependence of the SCC, shifting the temperature coefficient from 13·5 kcal/mol to 11 kcal/mol. Since the permeability of the mucosal surface is rate-determining for sodium transport under these conditions, the temperature coefficient may be taken as the permeability activation energy. Thus the shift in the temperature coefficient indicates that prolactin stimulates sodium transport by increasing the permeability of the mucosal surface. Later work by Snart and Dalton (1972) confirmed the earlier results and demonstrated that the increase in SCC displayed a dose-dependent relationship with the concentration of prolactin, with very high doses (7500 iu/l) of prolactin being required to induce the maximal increase in SCC. Prolactin also stimulated oxygen consumption by the toad bladder, subsequent to the rise in SCC. The increased oxygen uptake was presumably related to activation of a

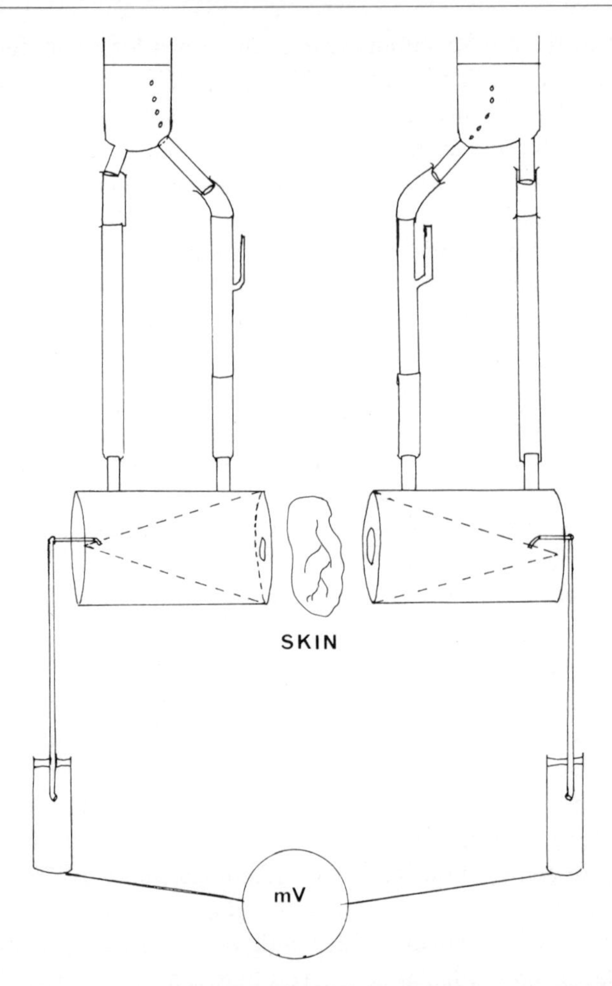

Fig. 3.2 This apparatus was originally used by Koefoed-Johnson and Ussing to determine potential differences across isolated amphibian skin. It consists of two independent hemispheres separated by the skin, electrical measurements being made *via* the connections shown.

sodium pump as a result of the larger tissue sodium pool which would result from the prolactin-induced enhancement of mucosal permeability.

In later work with the isolated bladder of *Bufo marinus*, Debnam and Snart (1975) found that prolactin can also increase the osmotic water flow across the bladder wall, in a dose-dependent fashion. Theophylline potentiated the response to prolactin, which suggests that prolactin acts on the adenyl cyclase system to elevate intracellular levels of cyclic AMP. Although vasopressin has been shown to act in a similar manner (Wright and Snart, 1971) the two responses are qualitatively different, which argues against the possibility that the activity of prolactin may result from contamination with vasopressin. Other arguments against this possibility is the finding that the response of the toad bladder to prolactin is insensitive to treatment with sulfhydryl reagent (*o*-hydroxy-mercuribenzoic acid) which causes an 83% inhibition of the vasopressin response (Debnam and Snart, 1975).

If we turn now to the transport activities of the isolated frog skin, it is interesting that Helbock *et al.* (1971) obtained results that illustrate the close relationship between prolactin and human growth hormone (hGH) discussed in Chapter 1. Despite much earlier dispute, human prolactin and

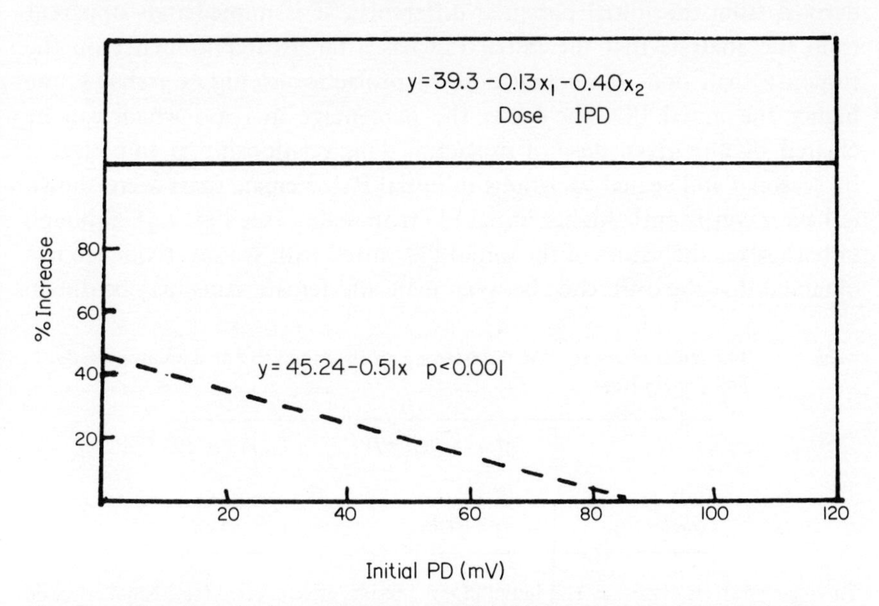

Fig. 3.3a The effect of ovine prolactin on potential differences across the isolated frog skin The results are expressed as a % of the initial potential difference.

hGH are now generally accepted to be separate entities (Nicoll, 1972), although with considerable overlap in their biological activities. Helbock *et al.* (1971) showed that clinical grade hGH could elevate the SCC across the isolated frog skin. Unfortunately, the preparation was contaminated with oxytocin and vasopressin, and no experiments were performed to separate the effects of these contaminating hormones from any response to hGH itself. Thus it is not clear how these results relate to those from experiments using ovine prolactin. Again, Rabbin *et al.* (1974) showed that an ovine growth hormone preparation stimulated sodium transport and increased the water permeability of isolated toad skin, but concluded that their results could be due to vasopressin contamination.

The effects of ovine prolactin on the frog skin have been studied in detail by Howard and Ensor (1976; 1977*a*,*b*,*c*). They showed that ovine prolactin increases PD, SCC and sodium influx in the isolated frog skin (see Figs 3.3*a*, 3.3*b* and 3.4). In contrast to the work on the toad bladder (Dalton and Snart, 1969) the PD response did not exhibit a simple dose-dependence, but displayed a complex relationship between initial PD, dose of prolactin, and the percentage increase in PD. When the results were subjected to partial correlation analysis an equation of the form $y = 39{\cdot}3 - 0{\cdot}13x_1 - 0{\cdot}40x_2$ was obtained (Howard and Ensor, 1977*a*), where x_1 represents the effect of the dose of prolactin and x_2 is a factor derived from the initial potential difference. It is immediately apparent from the analysis that the initial PD has a far greater influence on the response than does the actual dose of prolactin: as Fig. 3.3 shows, the higher the initial PD, the lower the percentage increase which can be elicited by any given dose of prolactin. This relationship is complicated by seasonal and sexual variations in initial PD. Female frogs were shown to have a consistently higher initial PD than males, (see Fig. 3.4), although in both sexes the values of the initial PD varied with season. Evidence was obtained that the difference between male and female skins may be due to

Table 3.1 The effect of 2×10^{-5} M testosterone on the initial PD of skins from male and female frogs

n	*Mean initial PD*	*% fall in PD*
Males [5]	58·3±4·22	3·6±0·90
Females [9]	83·8±8·61	13·1±3·48

Table 3.2 Effects of estradiol on initial PD; the results are corrected for ethanol effect on the control skin from male frogs

n	*Mean initial PD*	*Mean % increase in PD*
13	50·7±9·90	39·9±10·99

Table 3.3 The effect of varying dose of prolactin on % increase in transmembrane PD

Dose	*n*	*initial PD*	*% rise*
0·75 iu	12	70·3±10·12	12·7±9·65
1·50 iu	8	56·0±10·72	29·8±15·89
3·00 iu	10	57·3±8·20	11·1±4·72
6·00 iu	5	54·4±11·15	20·4±4·59
12·00 iu	4	52·0±11·89	14·8±7·39

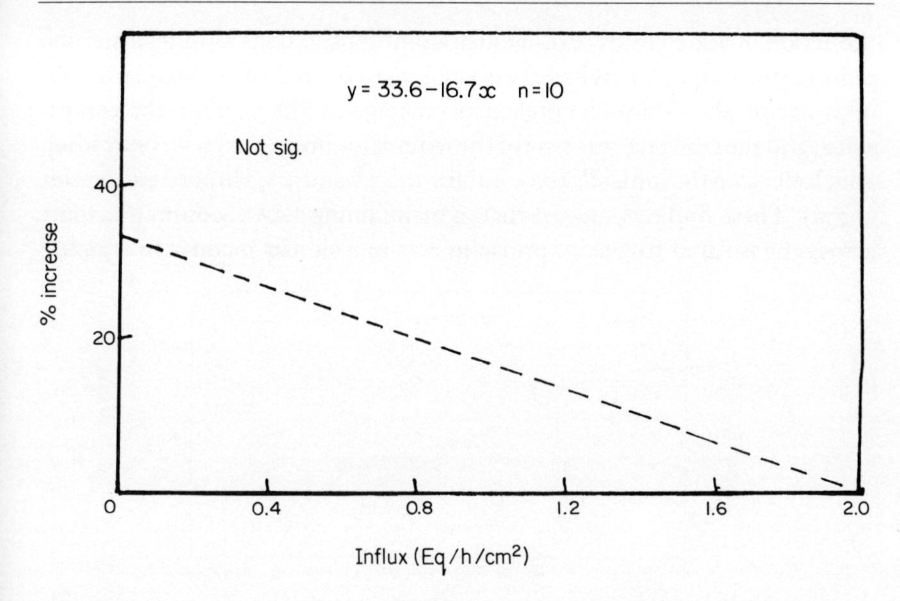

Fig. 3.3b The effect of prolactin on cumulative sodium influx across the skin of *Rana temporaria*.

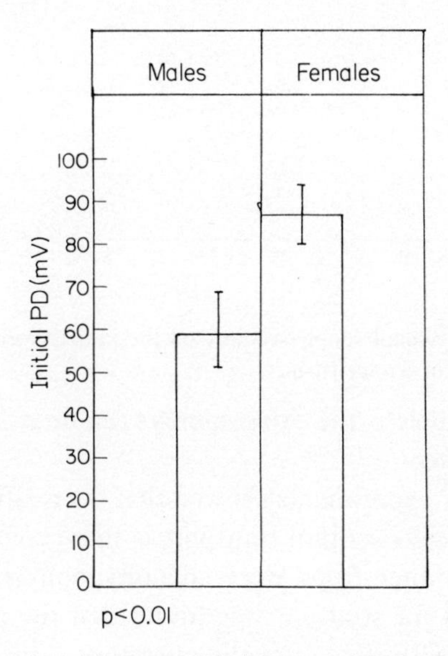

Fig. 3.4 The difference in initial potential difference (IPD) between male and female frogs.

the action of sex steroids. Pre-incubation of female skins with testosterone reduced the initial PD. When these skins were treated with a standard dose of prolactin, they showed a greater percentage in PD than did the control skins, and the converse was true of the male skins pretreated with oestradiol, which elevated the initial PD (see Tables 3.1, 3.2 and 3.3, Howard and Ensor, 1977*b*). These findings suggest that in maintaining active sodium transport across the isolated frog skin, prolactin acts in a similar manner to arginine

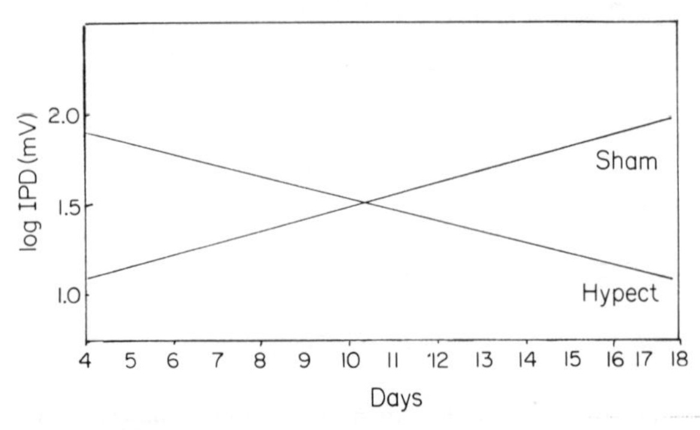

Fig. 3.5 The effect of adenohypophysectomy on the initial potential difference (IPD) across the skin of summer frogs.

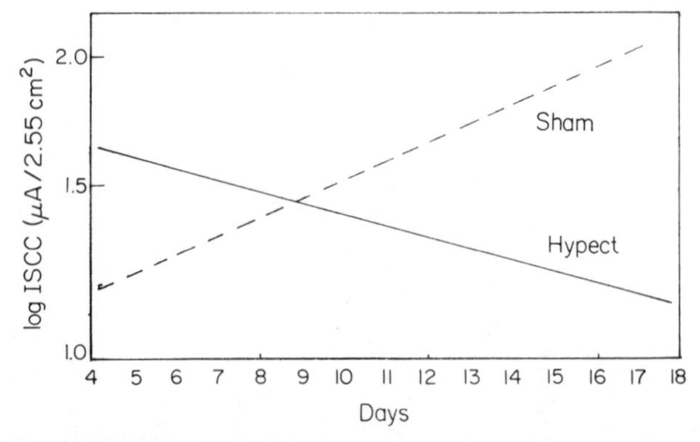

Fig. 3.6 The effect of adenohypophysectomy on the initial short-circuit current (ISCC) across the isolated skin of winter frogs.

vasopressin. Nevertheless, the two responses can be separated qualitatively, as will be shown later.

Hypophysectomy experiments showed that the relationship between the pituitary and cutaneous sodium transport is more complex than might be thought. When summer frogs were adenohypophysectomised and their skins later removed for study, it was found that the initial PD and SCC gradually declined with time after the operation. This effect was not seen in sham-operated animals (see Fig. 3.5). When this experiment was

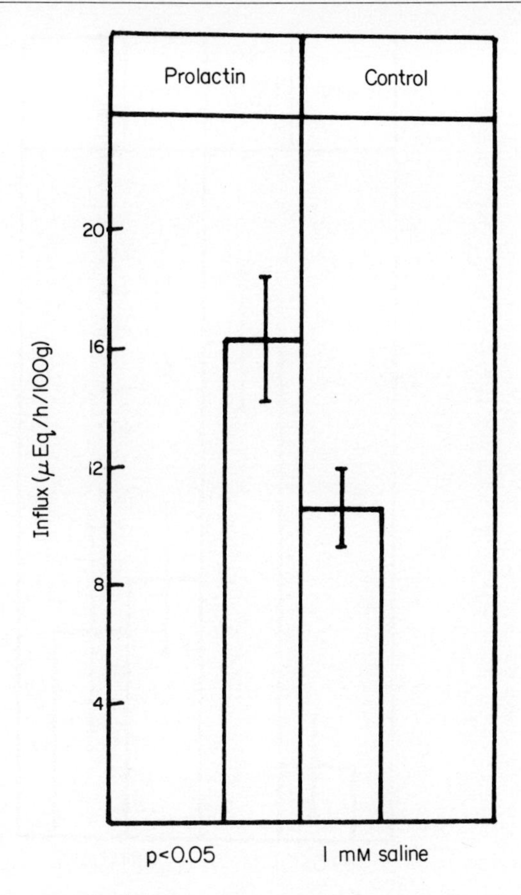

Fig. 3.7 The effect of prolactin on *in vivo* net uptake in intact frogs in 1 mM NaCl.

repeated with winter frogs there was no observable change in initial PD, but there was a decline in the SCC (see Fig. 3.6). Thus it would appear that the PD can be maintained independently of changes in SCC. The picture is further complicated by the finding that skins from hypophysectomised winter frogs were totally refractory to a standard dose of 6 iu ovine prolactin.

Further *in vivo* experiments have been aimed at clarifying these obscurities (Howard and Ensor, 1977*a*). Normal osmotic control by intact frogs placed in a dilute medium (250 μM saline) seems to be brought about by elevation of sodium influx and by limitation of sodium loss. Injection of ovine prolactin into intact frogs under these conditions caused a rise in the net flux which was subsequently shown to be due to an elevation of the sodium influx rather than a reduction in efflux. This response to

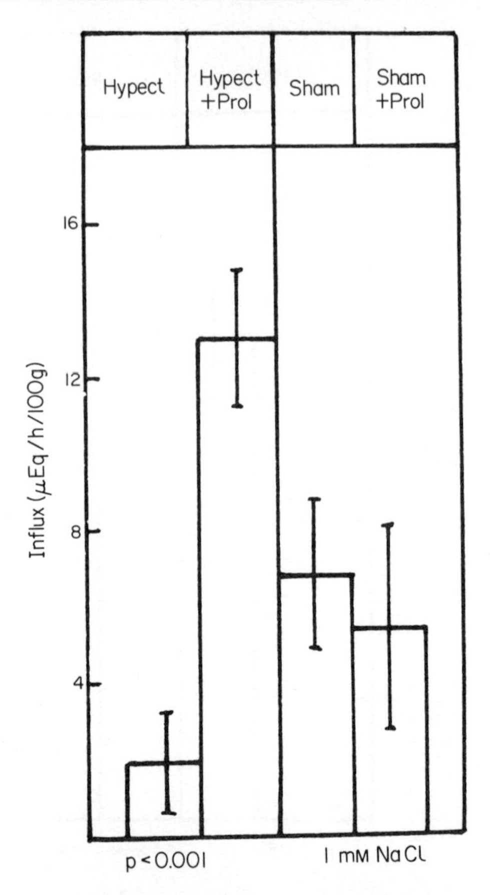

Fig. 3.8 The effect of adenohypophysectomy on the sodium influx of summer frogs. (Hypect = hypophysectomised; Prol = prolactin.)

prolactin is to some extent related to external sodium levels (see Figs 3.7, 3.8) in that the response declines as the external sodium concentration is lowered. Nevertheless, prolactin retains the ability to stimulate sodium influx down to an external concentration of 250 μM NaCl.

As in the *in vitro* studies, the effects of adenohypophysectomy were complex. Adenohypophysectomy in winter frogs slightly decreased sodium influx, but did not significantly alter the net flux. Summer frogs have a much higher rate of sodium turnover than winter frogs, and adenohypophysectomy of summer animals caused a marked decrease in influx coupled with a drop in efflux, the result being that the net flux remaining at the level of sham-operated controls (Fig. 3.9).

In summary, it seems that ovine prolactin can elevate sodium influx across the frog skin. The effects of adenohypophysectomy, though not

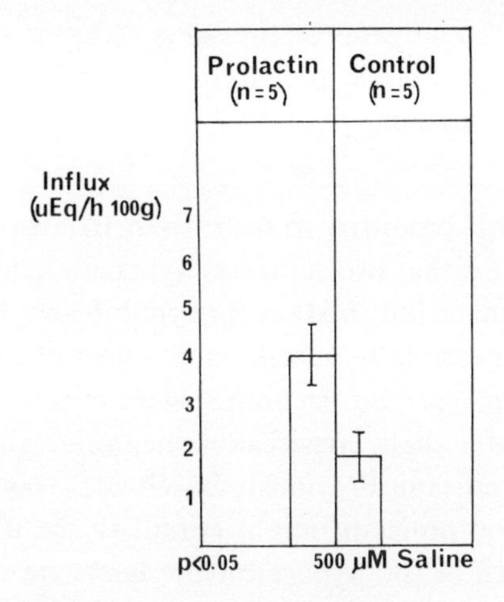

Fig. 3.9 The effect of prolactin on *in vivo* net uptake in intact frogs in 500 μM saline.

entirely clear, are compatible with the concept that this represents a physiological action of frog prolactin.

One remaining area of controversy centres around the possible contamination of prolactin preparations with neurohypophysial peptides. Dalton and Snart (1972) and Debnam and Snart (1974) demonstrated that although the two hormones affect the toad bladder in a similar manner, prolactin has its own intrinsic action, independent of any possible neurohypophysial contaminants. Howard and Ensor (1977*c*) have investigated this problem for the frog skin. There is no doubt that the response of the isolated frog skin to both arginine vasopressin and prolactin is similar in nature and has a similar time course. The relationship between initial PD and the percentage PD response, discussed on p. 54, has previously been shown to hold for vasopressin (Furham and Ussing, 1950; Kedem and Leaf, 1966). Additionally the seasonal variation in the response of the frog skin to prolactin is similar to that found with vasopressin (Herrera and Curran, 1963; Hong *et al.*, 1968). However, there are also certain differences in the actions of the two hormones. Pretreatment of the isolated frog skin preparation with prolactin or with vasopressin rendered the skin refractory to further treatment with prolactin but not to subsequent doses of vasopressin (Howard and Ensor, 1977*c*), and amiloride, which has previously been shown to be a potent inhibitor of vasopressin action on the frog skin (Nieber and Tomlinson, 1970; Cuthbert, 1973; Cuthbert and Shum, 1974) did not block the action of prolactin (Howard, personal communication). Thus, although there are great similarities between the actions of prolactin and vasopressin, the two hormones do not appear to act synergistically or to compete for the same receptor sites in the skin, and the effects of ovine prolactin preparations cannot be ascribed entirely to vasopressin contamination.

3.2 Prolactin and calcium metabolism in amphibia

We saw in Chapter 2 that prolactin has hypercalcaemic effects in teleosts. There is little cognate information for amphibians. Boschwitz and Bern (1971) reported a possible hypercalcaemic effect of prolactin in the toads *Bufo boreas* and *Bufo marinus*. In both species injections of ovine prolactin consistently caused a slight hypercalcaemia, although the responses were not statistically significant. Previously Boschwitz (1969) had suggested that large doses of ovine prolactin might stimulate the ultimobranchial body of toads, the source of the hypocalcaemic hormone *calcitonin*. Prolonged treatment with porcine calcitonin prolonged hypercalcaemia in these species, whereas a single injection caused a more typical hypocalcaemia (Boschwitz and Bern, 1971). Thus chronic injections of large doses of prolactin may perhaps cause hyperstimulation of the toad ultimobranchial body, resulting in the continuous hypersecretion of calcitonin. If the results of injecting porcine calcitonin are extrapolated, the consequence could be hypercalcaemia. The highly speculative nature of this interpretation will not escape the reader's attention.

3.3 Effects of prolactin on the amphibian integument

We have already seen that prolactin often appears to act in teleosts by modifying the activity of certain epithelial elements. Thus prolactin stimulates mucus production by the teleost skin as well as stimulating enzyme activity within the branchial epithelium. In the previous section the role of the amphibian integument as an osmoregulatory organ was discussed in some detail. Many of the osmoregulatory actions of prolactin in amphibia are undoubtedly brought about by alterations in the permeability of the skin and in the insulating effect of increased mucus production by the skin glands (Reinke and Chadwick, 1940; Vellano *et al.*, 1970). Nevertheless, the actions of prolactin on the amphibian integument are not limited to osmoregulatory functions. A number of effects on amphibian skin may be considered to be somatotrophic (see Section 3.4, p. 60) in that they involve increased cell division and proliferation (Berman *et al.*, 1964; Brown and Fry, 1968*a*,*b*; Schaumble and Mentwig, 1974). However, some actions of prolactin, namely moulting and the development of nuptial pads, are probably most reasonably classified as integumentary effects and will be considered here.

3.3.1 *Prolactin and moulting*

Dent (1975) considers that in most vertebrates moulting consists of three separate processes: the proliferative phase; cornification of keratinocytes; and ecdysis (sloughing) of the *stratum corneum.*

a. *Proliferation* It has been known for some time that prolactin has a specific mitogenic effect on the amphibian *stratum germinativum* (Chadwick and Jackson, 1948). More recently, Hoffman and Dent (1973) have shown that hypophysectomy reduces proliferation of the *stratum germinativum* in the red-spotted newt. They also showed that exogenous prolactin had a stimulatory effect in intact animals and can successfully restore proliferation in the epidermis of hypophysectomised newts.

b. *Cornification* There is considerable evidence that prolactin stimulates cornification in reptiles (see p. 86). Although it is known that prolactin can cause cornification of specific areas of the amphibian integument, particularly the nuptial pads (Dent, 1975), there is no reported evidence of a generalised cornifying effect in amphibians.

c. *Ecdysis* Epidermal proliferation and cornification in amphibia are independent of the actual act of ecdysis (Dent, 1975). A range of species have been studied, both anurans (Jørgensen and Larsen, 1964) and urodeles (Clark and Kaltenbach, 1961). In most of these species ecdysis is interrupted by hypophysectomy, and thyroidectomy also interrupts ecdysis in urodeles. In the hypophysectomised animal, new generations of keratinocytes continue to develop, although the old outer layer remains. The role of various hormones in ecdysis is still under discussion. For instance, Jørgensen and Larsen (1964) showed for the toad *Bufo bufo* that adrenal steroids are essential for sloughing, while thyroxine is ineffective. Thyroid hormones are usually considered to be essential for effective ecdysis in urodeles, but Vellano *et al.* (1970) found that the administration of exceptionally high doses of prolactin (120 iu) could bring about resumption of arrested ecdysis in hypophysectomised *Triturus cristatus*. Dent *et al.* (1973) have repeated this work, using a different urodele, *Notophthalmus viridescens*, and found that doses of TSH equivalent to the level of contamination in the prolactin injected by Vellano *et al.* (1970) could also cause resumption of moulting after hypophysectomy. However, Dent *et al.* (1973) were able to invoke moulting in hypophysectomised *Notophthalmus* with relatively low doses of prolactin after the animals had been maintained in subeffective doses of thyroxine. Dent (1975) concluded that although thyroid hormones are responsible for the initiation of ecdysis in most species, prolactin may play an important role by stimulating the production of mucus by skin glands, facilitating ecdysis by increasing cutaneous lubrication.

Interactions between thyroid hormones and prolactin in amphibian metamorphosis are discussed in Section 3.4.

3.3.2 *Prolactin and nuptial pad formation*

Prolactin appears to play a part in the keratinisation of special areas of the skin to form the so-called 'nuptial pads' in certain male urodeles. In autumn, the male red-spotted newt (*Notophthalmus viridescens*) undergoes a series of changes which involve the growth of a tail fin, and the appearance of cornified areas on the tips of the digits of the hind limb, and, later, on the ventral-caudal surface of the thigh (Dent, 1975). Forbes *et al.* (1975) found that throughout the summer the epithelium covering these areas is flattened and normal, but, as the pads start to develop, the *stratum germinativum* thickens and becomes organised into vertical columns stretching upwards from the basal layer to the pad surface. These pads are moulted following the cessation of reproductive activity in the spring. Structures similar to these nuptial pads have been reported in range of fish and amphibian species (Noble, 1931; Dodd, 1960; Wiley and Collete, 1970; Botte *et al.*, 1972). In those anurans which have been investigated in detail the pads disappear following castration. Replacement therapy with androgenic steroids induces redevelopment of pads in male anurans, and androgens also cause pad development in females (Dodd, 1960). In laboratory-conditioned male red-spotted newts however, high doses of testosterone failed to induce pad formation (Singhas and Dent, 1975), but testosterone given together with prolactin was effective although prolactin had no effect when administered alone. In hypophysectomised newts, thyroxine had to be added to prolactin and testosterone to produce the full response. Dent (1975) quotes unpublished observations of Dent, Zimmer and Reed which demonstrated that the combined dose of prolactin and testosterone was also capable of producing toe pads and rudimentary limb pads in female red-spotted newts.

3.4 Metamorphic, somatotrophic and metabolic actions of prolactin in the amphibia

Consideration of the integumentary growth-controlling actions of prolactin leads naturally to a wider discussion of the role of prolactin in larval growth and metamorphosis, and of the general growth-promoting action of this hormone in the amphibian. Most amphibians exhibit a *first metamorphosis* (primary metamorphosis) in which the aquatic larva (tadpole) changes into the semi-terrestrial young adult. In anurans, this is the only metamorphosis that occurs in the life-history. However, many urodeles undergo a *second metamorphosis*, in which the sexually-quiescent terrestrial adult changes into an aquatic form again for breeding purposes. At second metamorphosis the behaviour of the terrestrial animal alters, so that it moves towards water and develops a preference for being in water rather than on land. This behavioural change constitutes the so-called 'water-drive', and is an obvious prerequisite for courtship and breeding in the water (Chadwick, 1941; Chadwick and Grant, 1958). In addition, second metamorphosis involves a complex of many morphological and physiological changes.

3.4.1 *Prolactin and first metamorphosis*

a. *Development and regression of the larval tail* The tail is obviously a locomotory organ in the aquatic larva, and its efficiency is increased by the presence of a marginal fin. The tail fin initially develops and grows at a rate proportional to the increase in general body size. At first metamorphosis, either the tail fin only (urodeles) or the entire tail (anurans) is gradually resorbed.

Yoshizato and Yasumasu (1970; 1971; 1972*a*,*b*,*c*) have made a detailed study of the role of prolactin in promoting growth of the larval tail fin in anurans. It appears (Yoshizato and Yasumasu, 1970; 1971) that prolactin stimulates the incorporation of certain precursors into hyaluronic acid and collagen. Prolactin also inhibits the action of hyaluronidase, an enzyme which lowers the concentration of hyaluronic acid (Yoshizato and Yasumasu, 1972*b*). The effect on RNA levels within the larval tail fin is somewhat complex. Prolactin appears to inhibit RNA synthesis, but at the same time it also produces an inhibition of RNAase activity. According to Yashizato and Yasumasu (1972*b*) the net result of these processes is a rise in the RNA content of the tail fin which presumably results in increased protein synthesis. The resulting protein appears to be catabolised at a reduced rate, leading to an increase in the protein content of the fin. Prolactin also stimulates cell division, a change which can be measured by the increased incorporation of ^{3}H-uridine into new DNA. This synthesis of new DNA is paralleled by a fall in DNAase levels (Yoshizato and Yasumasu, 1972*c*).

Immediately prior to metamorphosis, in the late larva, there is a rise in the circulating levels of thyroxine. The overall pattern of changes is summarised in Figure 3.10. This pattern has been the subject of many reviews (e.g. Etkin, 1970; Dodd and Dodd, 1976) and is too well documented to require a detailed discussion. In brief, the metamorphic process consists of three stages: *premetamorphosis*, which is primarily a

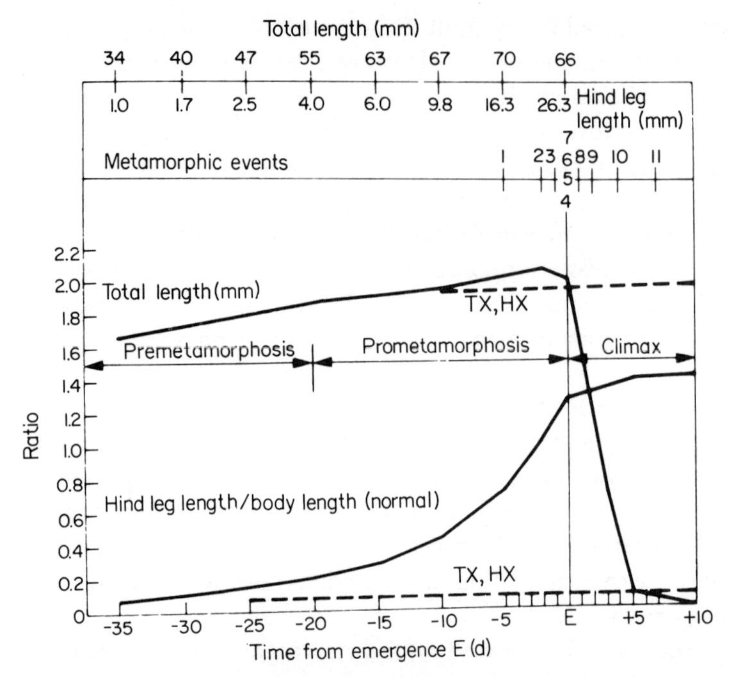

Fig. 3.10 Pattern of metamorphosis in *Rana pipiens*. Data from one batch of normal animals raised at 23°C shown by solid line. Comparable data for thyroidectomised animals (TX) and hypophysectomised animals (HX) are shown by broken lines. The metamorphic events are: 1, anal canal piece, first definite reduction; 2, anal canal piece reduction completed; 3, skin window for forelegs clearly apparent; 4, loss of beaks; 5, emergence of first foreleg; 6, emergence of second foreleg; 7, mouth widened to level of nostril; 8, mouth widened to level between nostril and eye; 9, mouth widened to level of anterior edge of eye; 10, mouth widened past level of middle of eye; 11, tympanum definitely recognisable. (From Etkin, 1968.)

growth phase and is hormonally-dominated by prolactin secretion; *prometamorphosis*, and *metamorphic climax* (Barrington, 1975). The last two stages are characterised by high circulating levels of thyroxine. The two hormones prolactin and thyroxine appear to act competitively during the later stages of metamorphosis, prolactin promoting the retention of larval characteristics, and thyroxine favouring the conditions leading to the metamorphic climax (Etkin, 1966). In anurans the climax is characterised by the reabsorption of the entire tail, but in urodeles (in which reorganisation at climax is generally less drastic), only the tail fin is resorbed. The two hormones may compete peripherally in blocking each other's action on the tissues, or prolactin may directly inhibit thyroid activity. In *Rana catesbiana*, Gona (1967, 1968) has shown that prolactin acts as a goitrogen, i.e. it inhibits synthesis of thyroid hormone. Additionally, there is evidence in the same species that prolactin can also exert its antithyroidal effect at the tissue level. This comes from a number of studies (Bern *et al.*, 1967; Etkin and Gona, 1967; Derby and Etkin, 1968; Jaffe and Geschwind, 1974*a*) demonstrating that the two hormones compete at the sites of tail resorption. Recent research by Giunta and his co-workers (Campantico *et al.*, 1972; Giunta *et al.*, 1972) has shown that in the tadpoles of *Xenopus laevis* and *Rana temporaria*, prolactin stabilises lysosomal membranes in the cells of the tail. This action presumably limits the release of autolytic enzymes from the lysosomes, so limiting the breakdown of the tail tissues.

Antagonism between thyroxine and prolactin also occurs at first metamorphosis in the tiger salamander *Ambystoma tigrinum* (Gona and Etkin, 1970), the newt, *Taricha torosa* (Cohen *et al.*, 1972; Licht *et al.*, 1972), and in frog tadpoles (Etkin and Gona, 1967, 1968; Bern *et al.*, 1967). In *Taricha torosa*, for example, Cohen *et al.* (1972) found that prolactin was capable of maintaining larval growth and totally blocking metamorphosis, a process previously demonstrated to be predominantly stimulated by thyroxine. A useful critical account of the endocrine control of amphibian metamorphosis is given by Dodd and Dodd (1976).

Table 3.4 Relative effects of prolactin and growth hormone in producing growth in larval amphibia

Anura species	Prolactin	Growth hormone
Rana pipiens	Stimulatory	Less stimulatory
R. catesbiana	Stimulatory	Less stimulatory
R. temporaria	Less stimulatory	Stimulatory
Bufo bufo	Non-stimulatory	Non-stimulatory
Urodela		
Triturus cristatus	Stimulatory	Less stimulatory
Taricha torosa	Less stimulatory	Stimulatory

Reports on the somatotrophic actions of prolactin in larval amphibians vary and are summarised in Table 3.4, which indicates the relative effectiveness of prolactin and growth hormone (GH). Basically this data shows that prolactin is relatively more potent than GH in stimulating larval growth, although there are exceptions to this rule (e.g. *Taricha torosa*, Licht *et al.*, 1972). In passing, it should be understood that the postmetamorphic adult amphibian, from all the evidence available, grows in response to a distinct GH and not to prolactin (Brown and Frye, 1969). Prolactin has also been shown to stimulate growth of specific areas of the larva, for instance the tail and tail fin (Dent, 1975) and the brain (Zipser *et al.*, 1969), while GH is more effective on growth of the trunk (Cohen *et al.*,

1972; Licht *et al.*, 1972). In opposing metamorphosis prolactin appears to increase the preference for an aquatic habitat, and promotes retention of the gills (Cohen *et al.*, 1972). The role of prolactin in preventing tail resorption has been discussed previously (see p. 61).

b. *Prolactin, thyroxine and enzymatic changes in organs other than the larval tail* We have seen that prolactin opposes lytic absorption of the larval tail and tail fin, and that thyroxine promotes these processes at metamorphosis. Recent reports indicate that these two hormones may act antagonistically on other enzymatic changes during first metamorphosis.

Campantico *et al.* (1972) and Giunta *et al.* (1972) found that prolactin elevates acid phosphatase levels in the intestine of *Bufo bufo* larvae. A similar enzyme response occurs in the prolactin-induced delay in the regression of the pronephros in these larvae (Vietti *et al.*, 1973). In addition, Giunta *et al.* (1973) have shown that prolactin inhibits the normal rise in cyclic AMP levels in the intestine of prometamorphic toad larvae, at the same time increasing the levels of non-cyclic AMP and ATP. Prolactin also causes a drop in adenyl cyclase levels in the intestine, and a concomitant rise in phosphodiesterase content. Both these changes would result in a reduction in cyclic AMP levels. Thus, prolactin opposes the intestinal enzyme changes characteristic of larvae undergoing thyroxine-dependent metamorphosis, in all probability another example of peripheral antagonism between the two hormones.

Further evidence of prolactin-thyroid antagonism was reported by Jaffe and Geschwind (1974*a*). In this study prolactin failed to block thyroxine-dependent increases in enzyme activities in the liver of *Rana catesbiana* tadpoles. However, although the hepatic levels of carbomyl phosphate synthetase (Blatt *et al.*, 1969), ornithine transcarbamylase (Jaffe and Geschwind, 1974*a*) and tyrosine transaminase (Jaffe and Geschwind, 1974*b*) – all hormones of the ornithine cycle – remained high, prolactin did reduce the rate of urea excretion in thyroxine-treated tadpoles, and it also caused a drop in ammonia excretion in non-thyroxinated tadpoles. Prolactin also increased liver LDH activity, which was reduced by thyroxine, but it did not prevent the thyroxine-induced elevation of glutamic dehydrogenase (Jaffe and Geschwind, 1974*a*). This evidence suggests that prolactin-thyroxine antagonism is widespread in the peripheral tissues of metamorphosing amphibians.

In ranid frogs, the aquatic larvae possess porphyropsin as the major visual pigment, but in the terrestrial adults rhodopsin predominates. Thyroxine induces rhodopsin synthesis in *Rana catesbiana* tadpoles. Although prolactin in these experiments antagonised the tail-resorption induced by thyroxine, it did not antagonise the thyroxine-induced increase in retinal rhodopsin (Crim, 1975*b*).

3.4.2 *Prolactin and second metamorphosis*

In those urodeles which undergo a second metamorphosis, not only does the behaviour alter (the water-drive), but the tail fin reappears and the skin changes from being relatively dry and cornified to become a mucus-secreting aquatic integument. Second metamorphosis may occur only once in the life history, as in the red-spotted newt *Notophthalmus viridescens*, which thereafter remains a water-dweller, or it may occur annually prior to the breeding season, as in *Triturus cristatus*, the terrestrial and aquatic phases alternating each year. Prolactin is the dominant hormone of second metamorphosis, stimulating appearance of the water-drive (Chadwick, 1940, 1941; Grant and Grant, 1958; Grant, 1961; Crim, 1975*a*) and nuptial pads (see pp. 58–60), and promoting growth of a new adult tail fin (Dent, 1975). The relationship between thyroid and prolactin during this second metamorphosis appears to be complex. Most work has been done on *Notophthalmus viridescens*, whose immature juvenile terrestrial phase is called the red eft. Gona *et al.* (1973) investigated prolactin-thyroxine interactions in inducing the water-drive in red efts. Prolactin treatment induced water drive in 30% of the treated intact efts. Given to hypophysectomised-thyroidectomised animals, prolactin caused water-drive in only 10%. However, prolactin given together with a low dose of thyroxine induced water-drive in 100% of the doubly-operated animals. Increasing the thyroxine : prolactin ratio in the double injections, progressively inhibited the water-drive activity of the prolactin (Gona *et al.*, 1970). Thus, while low doses of thyroxine synergise with prolactin, higher doses are antagonistic.

a *Development of the adult tail fin* The growth of the tail fin at second metamorphosis has been studied extensively by Dent and his co-workers (see Dent, 1975). The first evidence that prolactin may induce tail-fin growth in adult newts came from Tuchmann-Duplessis (1949), who demonstrated that exogenous prolactin would stimulate fin growth in intact and castrated *Triturus cristatus* and *T. alpestris*. These pioneer observations were later confirmed for *T. cristatus* by Vellano *et al.*, (1970). The adult tail fin in *T. cristatus* has an ambisexual character, i.e. it develops in both male and females at the time of the annual second metamorphosis (Galgano, 1942). The red-spotted newt *Notophthalmus viridescens*, as we have seen, remains in water all its adult life once second metamorphosis has occurred. Nevertheless, the tail fin of the male regresses at the end of the spring breeding season and grows again in the autumn (Adams, 1940), and the rise and fall of the male tail fin is accompanied by growth and shedding of the nuptial pads (see p. 60). However the tail fin of the female remains the same size throughout the year. Thus in contrast to *T. cristatus*, the tail fin in *N. viridescens* has a male sexual character, rather than an ambisexual one (Singhas and Dent, 1975).

Maintenance of adult *N. viridescens* in the laboratory with constant illumination at a temperature of 20–24°C induces retention of terrestrial characteristics, i.e. a rough skin secreting little mucus (Dent *et al.*, 1975) and regression of the caudal fin in the male (Singhas and Dent, 1975). Injection of those laboratory-conditioned animals with prolactin restored the skin to its typical aquatic state (Dent *et al.*, 1973) and increased the height of the male tail fin. Testosterone somewhat augmented the response of the male tail fin to prolactin but did not itself stimulate fin growth. Hypophysectomised male newts showed a fin-growth response to prolactin only about one-half of that in intact animals, but this response was returned to normal by simultaneously giving TSH or thyroxine (Singhas and Dent, 1975). Although the female tail fin undergoes no natural seasonal variations, surprisingly it responds strongly to prolactin under laboratory conditions (Dent, Zimmer and Reed, quoted by Dent, 1975). This perhaps suggests that in nature the basal secretion of prolactin in the female is low.

b. *Mucification of the skin* One of the features of second metamorphosis is an increase in cutaneous mucus secretion. We have seen that prolactin promotes the activity of the mucus glands in teleosts, and the same action may also be widespread in amphibians. Unfortunately, published work is confined to *Notophthalmus viridescens* and *Triturus cristatus*, in both of which prolactin stimulates skin mucification. However, in *N. viridescens* prolactin by itself does not stimulate mucus secretion in the hypophysectomised and thyroidectomised eft (Gona *et al.*, 1973), or in the hypophysectomised aquatic adult (Dent *et al.*, 1973); the stimulatory effect of prolactin is seen only when thyroid hormone (Gona *et al.*, 1973; Dent *et al.*, 1973) or TSH (Dent *et al.*, 1973) is given at the same time. In contrast, mucus secretion can be elicited in hypophysectomised *T. cristatus* by prolactin alone, with no accompanying thyroid hormone (Vellano *et al.*, 1973).

c. *Colour change at second metamorphosis* After the intact red eft enters the water precociously in response to injections of prolactin it assumes the adult olive coloration. However, if the eft is hypophysectomised prior to the prolactin injection, no colour change ensues (Grant and Grant, 1958).

Thyroid hormone given together with prolactin does not elicit the colour change in hypophysectomised efts, although we have seen that in the same experiments thyroxine synergised with prolactin to promote water-drive and mucus secretion (Gona *et al.*, 1973). Thus some pituitary hormone other than prolactin and TSH appears to promote the colour change.

d. *Visual pigments and second metamorphosis* In the red-spotted newt, retinas of the aquatic adult contain mostly porphyropsin, while rhodopsin predominates in the terrestrial juvenile eft. Crim (1975*a*) demonstrated that ovine prolactin not only initiated water-drive in the terrestrial eft, but also elicited the high levels of retinal porphyropsin characteristic of the adult retina. Prolactin also increased porphyropsin levels in the retina of the adult, whether kept in water or in a damp terrestrial environment; thyroxine did not oppose this retinal effect, although it did initiate moulting and antagonised the prolactin-induced water-drive behaviour.

We have seen that prolactin is capable of causing larval growth in amphibians. During first metamorphosis it acts as a thyroxine antagonist, either as a goitrogen or as a peripheral inhibitor, and in this way prolactin can be regarded as a kind of amphibian 'juvenile hormone'. In urodeles, prolactin promotes the changes at second metamorphosis, and thyroid hormone may either synergise with prolactin or oppose it, the difference possibly depending on the relative levels of the two hormones. A useful discussion of prolactin-thyroxine antagonism in amphibians will be found in Crim (1975*b*).

e. *Prolactin and reproduction in amphibians* When the male crested newt *Triturus cristatus* undergoes second metamorphosis and returns to water (Galgano, 1942), spermatogenesis is retarded and the formation of primary spermatocytes from spermatogonia is arrested, although previously formed spermatocytes complete their development normally. Mazzi *et al.* (1967) have shown that these changes can be brought about by exogenous prolactin, which also induced regressive changes in the B2 cells (basophils Type 2) of the pars distalis, the source of gonadotrophin (Mazzi *et al.* 1966). Thus, prolactin may affect the testis by inhibiting gonadotrophin (FSH) secretion, a mechanism reminiscent of effects of prolactin in reptiles (p. 88) and birds (p. 115).

More recently these findings were confirmed by Vellano *et al.* (1967), who also showed that prolactin stimulated the androgen-dependent sexual accessory glands of male *T. cristatus*. Vellano *et al.* suggested that prolactin might stimulate testosterone secretion by increasing LH output while at the same time (arguing from the simultaneous arrest of spermatogenesis) limiting the output of FSH. However, in view of the effects of prolactin on male sexual accessory structures in teleosts (p. 38) and mammals (p. 159), it is more probable that the injected prolactin in *T. cristatus* synergised peripherally with endogenous androgens. This interpretation finds confirmation in the studies by Kikuyama *et al.* (1976) on *Triturus pyrrhogaster*, which showed clearly that prolactin synergised with androgens to stimulate growth of the male cloacal glands.

In female amphibians prolactin has been shown to stimulate the secretion of oviducal jelly in *Bufo arenarum* (de Allende, 1939; Houssay, 1947). However, this response may not be widespread; Basu *et al.* (1965) were not able to demonstrate effects on oviducal secretion with either prolactin or growth hormone in *Rana pipiens* and *Bufo americanus*, although sex steroids were effective.

f. *Prolactin and limb regeneration in adult urodeles* Early work by Miwelenski (1958) had shown that both prolactin and growth hormone were essential for limb regeneration in *Triturus alpestris.* Pituitary hormones have also been shown to be important in limb regeneration and survival of *Notophthalamus viridescens* (Richardson, 1945; Dent, 1967; Connelly *et al.*, 1968), in which Schotté and Talon (1960) had shown that limb regeneration could also be induced by ectopic pituitary transplants. Connelly *et al.* (1968) found that prolactin given together with thyroxine could promote limb regeneration, and similar results were obtained with growth hormone by Richardson (1945) and Wilkerson (1963) both, however, using preparations which were highly contaminated with prolactin.

In 1969 Tassava showed that prolactin given together with thyroxine significantly increased growth of limb blastema in hypophysectomised adult newts. Growth hormone and ectopic pituitary homotransplants from newts or axolotls were also effective. Although growth hormone was able by itself to stimulate regeneration, prolactin was not effective unless combined with thyroxine, another example of synergism between these two hormones in an adult amphibian. Tassava also showed that nutritional status affected the regenerative response, in that newts which had been starved for two weeks prior to hypophysectomy regenerated more slowly in response to combined prolactin and thyroxine injection.

3·5 Prolactin cells in the amphibian pituitary

A tetrapod-type of prolactin (see Chapter 1) has been detected by bioassay in the pituitary glands of various anurans and urodeles, (Bern and Nicoll, 1968; Chadwick, 1968; Nicholl and Nichols, 1971; Sage and Bern, 1972) and has been shown to be secreted *in vitro* by the pituitaries of *Necturus* and *Bufo* (Bern and Nicoll, 1968; Nicoll, 1974).

The amphibian pituitary contains two acidophilic cell types, usually termed A1 and A2 cells. Physiological functions known to be promoted by prolactin have been used to distinguish between these two cell types. Prolactin appears to be the larval growth factor in *R. pipiens* (p. 60), even prior to the development of functional hypothalamo-adenohypophysial connections (Etkin and Gona, 1967; Brown and Frye, 1969), and Ortman and Etkin (1963) have shown that the larval pituitary in this frog contains only A1 acidophils, indicating that these are the lactotrophs. Ectopic pituitary transplants in larval *R. pipiens* are capable of inhibiting metamorphosis (Etkin, 1970), producing giant tadpoles (Etkin and Lehrer, 1960). Both these effects are attributable to the antithyroidal properties of prolactin, and in fact the A1 cells are the only active cells in these ectopic transplants.

The more general evidence (p. 60) that prolactin is the growth-promoting hormone in several anuran larvae (Etkin, 1970; Nicoll and Licht, 1971) correlates with the early appearance of the A1 cells in the larval pituitary, the A2 cells not appearing until metamorphosis or later (Holmes and Ball, 1974). This early differentiation of A1 cells has been demonstrated in *Xenopus laevis* (Kerr, 1966), *Bufo bufo* (van Oordt, 1966; Mira-Moser, 1969) and *Rana temporaria* (Doerr-Schott, 1968). In view of the anti-metamorphic properties of prolactin, it is significant that the A1 cells temporarily regress before metamorphic climax in *Bufo bufo* (van Oordt, 1966). In *Notophthalmus viridescens* the A1 cells greatly increase in size and number just before the prolactin-dependent second metamorphosis (Copeland, 1943; Dent and Gupta, 1967), and in the same newt the A1 cells are persistent and very active in ectopic pituitary transplants that secrete prolactin (Masur, 1969). Recent immunohistochemical studies, using fluorescent antibodies to ovine or bovine prolactin (Vellano *et al.*, 1970; Doerr-Schott, 1976), have further confirmed the lactotrophic nature of the A1 cells, while other evidence links the A2 cells with growth hormone secretion (see Holmes and Ball, 1974).

In staining properties the A1 cells vary from species to species, most notably as regards the relative affinities of their secretory granules for erythrosin (erythrosinophilia) and orange G (orangeophilia). Their distribution in the pars distalis also shows some variation. In summarising the literature Holmes and Ball (1974) drew attention to these variations, and also commented on the way in which some workers have introduced confusion by failing to adhere strictly to standard techniques and terminology; however, Holmes and Ball were able to collate the numerous accounts because of two reasonably constant properties which distinguish the A1 from the A2 acidophils: the A1 cells tend to be located rostrally or ventrally and have the larger granules, while the A2 cells, with smaller granules, tend to be concentrated in the posterior or dorso-caudal region of the pars distalis. In a classical account, Kerr (1965) described the A1 cells of *Xenopus laevis* as large and densely filled with typical acidophilic granules, staining with orange G, erythrosin and luxol fast blue, but entirely PAS-negative, i.e. without glycoprotein. They occur in most parts of the gland, but are most numerous in the rostro-central region. In *Rana temporaria* and *Bufo bufo*, the A1 cells are more generally scattered (van Oordt, 1968). Urodeles, too, have A1 cells scattered rather widely throughout the pars distalis, although they are rostro-ventrally located in *Necturus*, and are usually rare in the dorsal region (Doerr-Shott, 1966). Their principle staining affinity seems to vary. Described as strongly erythrosinophilic in *Rana temporaria* and *R. esculenta*, and erythrosinophilic and/or carminophilic in *Bufo bufo*, *Alytes obstetricans* and *Necturus maculosus* – the classical tinctorial features of lactotrophs (Chapter 1) – they have nevertheless been described as strongly orangeophilic in the bufonid *Nectophrynoides occidentalis* and in *Rana pipiens*, and in the urodeles *Triturus marmoratus*, *T. cristatus*, *Notophthalmus viridescens*, *Taricha torosa* and *Pleurodeles waltlii* (van Oordt, 1968, 1974; Doerr-Schott, 1968, 1976; Holmes and Ball, 1974). Thus the A1 cells have been described as either erythrosinophilic or orangeophilic in both anurans and urodeles. However, as workers on pituitary cytology are well aware, the erythrosinophilic-orangeophilic distinction is not always easy to achieve, being very sensitive to minor variations in fixation and staining techniques, (see Ball and Baker, 1969; van Oordt, 1974) and probably these reported differences in staining properties would disappear if the A1 cells in the whole range of amphibians were studied with standardised techniques in one laboratory.

3.6 Hypothalamic control of prolactin

Prolactin secretion by the amphibian pituitary appears to be under a predominantly inhibitory hypothalamic control. Ectopic pituitary transplants are able to induce precocious second metamorphosis in urodeles (Masur, 1969; Peyrot *et al.*, 1969), and in anurans such transplants sustain larval growth (Etkin, 1970), and inhibit metamorphosis, resulting in giant tadpoles in *Rana pipiens* (Etkin and Lehrer, 1960). All these are effects of prolactin, so that these observations point to a considerable degree of autonomy in prolactin secretion and raise the possibility of hypothalamic inhibition. Ectopically transplanted glands in *Triturus cristatus* were shown to secrete more prolactin than normal glands when excised and incubated *in vitro* (Peyrot *et al.*, 1969). The pars distalis of both *Necturus* and *Bufo* secrete prolactin autonomously *in vitro*, the latter at a rate comparable to many mammals (Bern and Nicoll, 1968; Nicoll, 1971). Using a heterologous radioimmunoassay technique, McKeown (1972) demonstrated that ectopic pituitary transplants in *Bufo bufo* maintain higher plasma levels of prolactin than the *in situ* gland. Thus in both anurans (Etkin, 1970) and urodeles (Mazzi, 1969) the hypothalamus probably secretes a prolactin inhibiting factor (PIF).

Etkin (1970) has postulated that hypothalamo-hypophysial interplay is virtually absent in the early larva, due to immaturity of the hypothalamic nuclei themselves and also to the undifferentiated state of the median eminence. On this view, prolactin would be continuously secreted at a high rate in the larva, free of any inhibition. When the hypothalamus and median eminence differentiate in the prometamorphic larva, prolactin secretion progressively becomes inhibited and TSH secretion stimulated. This concept, which is backed by considerable evidence, accommodates the importance of prolactin as a larval growth factor, and explains the control of metamorphosis as a balance between the antagonistic actions of TSH and prolactin. If this hypothesis is correct, then one would predict that pituitary prolactin content should be higher in the early larva than in the late (premetamorphic) larva of the adult, assuming a direct relationship between pituitary content and secretion rate (see Peyrot *et al.*, 1969). This prediction has to some extent been confirmed for *Ambystoma tigrinum* by Norris *et al.* (1973), who measured prolactin levels in the pituitaries of larvae, spontaneously-metamorphosed adults, and adults resulting from artificial thyroxine-induced metamorphosis. However, Norris *et al.* also reported exceptionally high levels of prolactin in the pituitary of sexually mature neotenic *A. tigrinum*, in which the hypothalamo-hypophysial axis must obviously be fully developed and functional, at least in respect of gonadotrophin secretion. One explanation of this anomaly may be that hypothalamic control of gonadotrophin develops independently of the control of prolactin, or, alternatively, that in this instance pituitary prolactin content is not directly related to secretion rate. There is clearly an urgent need for further studies to test Etkin's model of amphibian metamorphosis (see Dodd and Dodd, 1976).

Histological studies have yielded evidence of a seasonal pattern of prolactin secretion in adult *Rana temporaria*. Maximal secretion starts at the end of hibernation and continues during the reproductive period, a pattern which parallels the seasonal changes in gonads, thyroid, adrenal cortex and overall growth-rate (van Oordt *et al.*, 1968). A similar annual cycle exists in the magnocellular preoptico-neurohypophysial system in

R. temporaria (Dierickx and van den Abeele, 1959). Neurosecretory cells that control gonadotrophin secretion have been located by Dierickx (1974) in the basal hypothalamus, more precisely the *pars ventralis tuberis cinerei*. Cells in this region appear to synthesise peptides, and numerous monoaminergic nerve fibres also originate here (Dierickx, 1974). This area of the brain of *R. temporaria* has recently been tested for activities influencing the release of prolactin and TSH in rats (Kühn and Engelen, 1976). Fragments of tuber cinereum collected in spring were found to stimulate the release of rat prolactin and to inhibit release of TSH, while the opposite activities were displayed by fragments collected in November. Thus these findings to some extent support the suggestion put forward by Vellano *et al.* (1968) for urodeles, that the hypothalamus might produce both inhibitory and stimulatory factors for prolactin control. A similar system has also been postulated for mammals (e.g. Valverde *et al.*, 1972; see pp. 162–175). In urodeles, fluorescence techniques have demonstrated high catecholamine levels in the pre-optic nucleus of *Triturus vulgaris* during winter, and a much lower content during the breeding season (Weiss and Kabisch, 1973). It remains to be seen whether these observations relate to a catecholaminergic prolactin-inhibitory mechanism, such as exists in teleosts (p. 38) and mammals (p. 165), or whether they are unrelated to prolactin control.

As in certain teleosts (Chapter 2), ovine prolactin appears to stimulate thyroid activity in *Triturus cristatus* (Vellano *et al.*, 1967), the effect disappearing if the recipient is hypophysectomised, the hypothalamus lesioned, or the pituitary ectopically transplanted (Peyrot *et al.*, 1970). Thus, ovine prolactin appears able to stimulate TSH secretion by the newt pituitary, apparently by way of a hypothalamic mechanism. However, as in the case of teleosts there is no evidence that *endogenous* prolactin stimulates TSH secretion physiologically.

4 Prolactin in reptiles

Although the effects of prolactin in other vertebrates have been studied in considerable detail, our information for reptiles is restricted to a very few species, although ample evidence attests to the secretion of a prolactin by the reptilian pituitary.

The reptilian pars distalis can be divided into two regions, the cephalic and caudal lobes. Within the cephalic lobe a number of authors have described the presence of large cells, carminophilic and erythrosinophilic and adjoining blood capillaries. In *Sphenodon* and some lacertids these cells tend to be organised in pseudo-follicles, (Saint-Girons, 1963, 1965, 1967). These are A1 cells of Holmes and Ball (1974). The size and density of the A1 granules varies a good deal (Grignon, 1963, Eyeson, 1970) as does their degree of carminophilia (Wingstrand, 1966; Licht and Nicoll, 1969; Eyeson, 1970). Because of their localised distribution it has been possible to allocate prolactin secretion to the A1 cells, Bioassay of the cephalic and caudal lobes of various reptiles (*Pseudemys*, *Anolis*, *Dipsosaurus*, *Thamnophis* and *Pituophis*) has shown that prolactin activity is restricted to the cephalic lobe (Licht and Nicoll, 1969). With the subsequent allocation of other functions to the other cell types in this region, it was possible with certainty to identify the A1 cells as lactotrophs (Licht and Rosenberg, 1969). Anomalously, Eyeson (1970) reported a weak lactotrophic activity in the caudal lobe of *Agama* with none in the cephalic lobe, although the latter region contains all the A1 cells. This report clearly calls for further investigation.

Despite this detailed work on the pituitary of reptiles, our knowledge of prolactin physiology in these animals still remains very sparse. One critical factor which has hampered progress is that it has proved difficult to isolate reptilian prolactin using normal electrophoretic techniques, because the molecule is very readily degraded (Licht, 1974).

4.1 Prolactin and ion and water balance in reptiles

From the point of view of the comparative physiologist, reptiles pose some extremely interesting questions with regard to their ionoregulatory physiology. In evolutionary terms, they are the first vertebrates successfully to exploit a terrestrial environment. Reptiles inhabit most climatic zones, with the exception of the polar land masses and certain oceanic islands, and there exist quite closely related species which live in desert, marine and freshwater environments, for example within the order Chelonia (turtles and tortoises).

Some detailed studies of the ionoregulatory physiology of reptiles have been made, and endocrine aspects have been particularly well documented by Shoemaker (1969), La Brie (1972), Bentley (1971) and Bradshaw (1975). Dunson and Dantzler (1972) have examined in detail the active processes involved in maintaining ion balance in marine and desert species. It is now apparent that prolactin plays a part in the maintenance of electrolyte balance in most vertebrates (Chapter 1; Nicoll, 1974; Bern, 1975), but unfortunately our knowledge of this role of prolactin in reptiles, the pioneer terrestrial vertebrates, is very imperfect.

The first report of a prolactin effect on ion balance in reptiles came from Chan *et al.* (1970). These workers showed that hypophysectomy in *Dipsosaurus dorsalis* led to a rise in plasma water and sodium levels, and in muscle water and intracellular sodium. At the same time the intracellular potassium concentration decreased. Replacement treatment with corticosterone corrected the changes in plasma and muscle water and in intracellular potassium, while prolactin therapy restored all the parameters virtually to normal. Administration of both corticosterone and prolactin together achieved complete normality of the ionic balance.

Brewer (1976) has recently undertaken a detailed study of the ionoregulatory role of prolactin in the order Chelonia. Among turtles, tortoises and terrapins one finds species inhabiting a wide range of terrestrial and aquatic (marine and freshwater) environments. If the role of prolactin alters in relation to the environment, then it ought to be possible to demonstrate this phenomenon within the Chelonia. Brewer's work involved the comparative study of prolactin physiology in four aquatic terrapins (*Chrysemys picta*, *Pseudemys scripta*, *Emys orbicularis* and *Pelomedusa subrufa*) and in a terrestrial species, the Greek tortoise (*Testudo graeca*). The results for the aquatic and terrestrial chelonians are best described separately.

4.1.1 *Aquatic terrapins*

All four species of terrapin respond to a low ionic environment (freshwater)

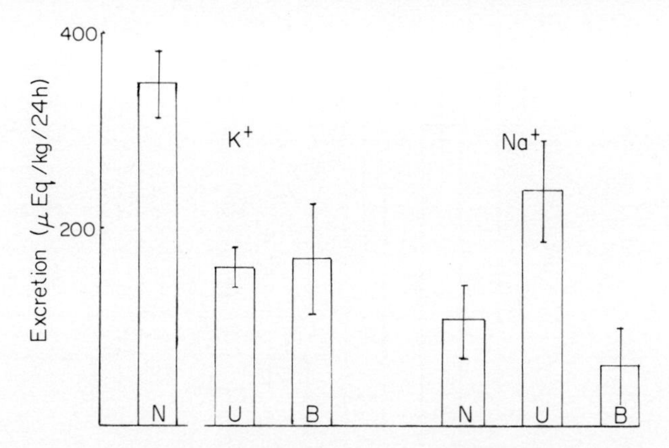

Fig. 4.1 Net (N), ureteral (U) and bladder (B) excretion of sodium and potassium in *C. picta*.

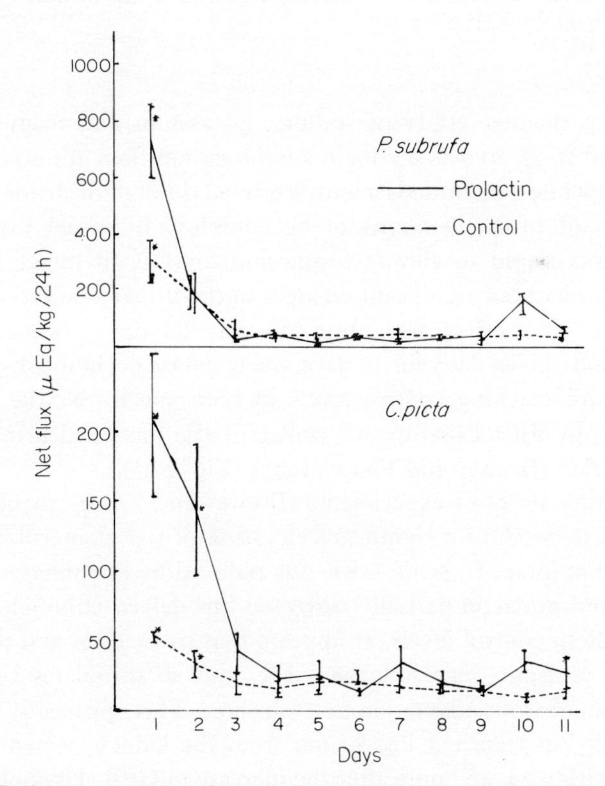

Fig. 4.2a Daily net magnesium fluxes in *P. subrufa* and *C. picta* receiving prolactin and control injections for 12 days. *C. picta*, n = 5; *P. subrufa*, n = 6. Results analysed by paired 't' tests and presented as mean ± S.E. (* p = 0·02.)

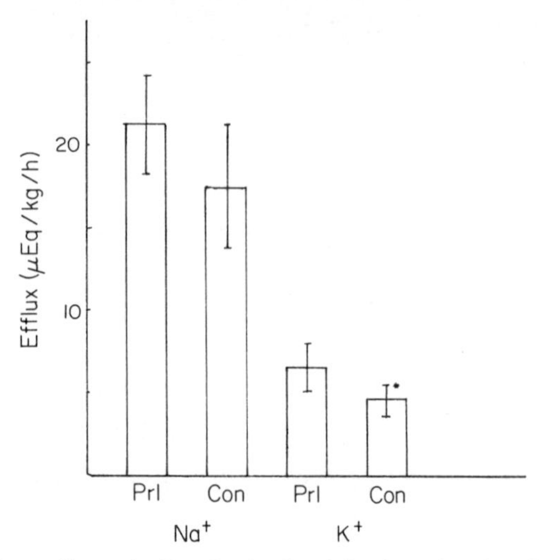

Fig. 4.2b Electrolyte effluxes in *P. subrufa* after injection of 20 iu of prolactin/kg body weight. n = 10; Pri = Prolactin; Con = Control. Results analysed by paired 't' tests and presented as mean ± S.E. (* p = 0·05.)

by reducing the net efflux of sodium, potassium and magnesium. In *C. picta* and in *P. scripta* the main site of sodium loss appears to be the integument while potassium is mainly excreted through the urine (Fig. 4.1). The action of prolactin seems to be complex. In initial experiments, prolactin was found to elevate magnesium outflux in intact *P. subrufa* and *C. picta*, without significant changes in the urinary excretion of other electrolytes. This effect was observed only on day 1 when prolactin injections were given daily for 12 days, the response declining over the next two days and reaching control levels in both species by the third day (Fig. 4.2*a*). In other experiments, prolactin also increased urinary potassium excretion (Brewer and Ensor, 1977) (Fig. 4.2*b*).

In a further series of experiments (Brewer and Ensor, 1976) prolactin was shown to produce a significant elevation of the glomerular filtration rate (GFR) in intact *C. picta*. GFR was reduced by hypophysectomy (see Fig. 4.3), and prolactin partially corrected this defect, although it did not return GFR to control levels. It appears that in *C. picta* and possibly in *P. subrufa* prolactin increases the GFR, and so stimulates renal water excretion, an obvious adaptation to freshwater. This apparently involves a concomitant but transient loss of ion from the kidney, which is rapidly corrected within 24–48 hours after the increase in GFR. Hypophysectomy of *C. picta* did, in fact, cause a decrease in fractional renal sodium reabsorption from 99·1% to 93·4%. This effect is very significant in terms of

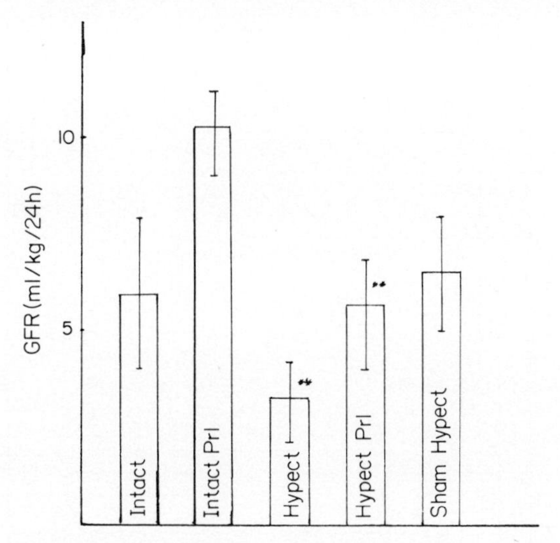

Fig. 4.3 The effect of 20 iu/kg body weight of prolactin on GFR in intact and hypophysectomised *C. picta*. Results analysed by 't' tests and presented as mean ± S.E. n = 8, intact; n = 8, hypophysectomised (Hypect); n = 6, sham-hypophysectomised (Sham-Hypect); Pri = Prolactin. (** p = 0·05.)

the animal's sodium balance. This defect was not corrected by prolactin, nor did prolactin affect sodium uptake in intact terrapins.

In contrast to the above results, in the closely related species *P. scripta* prolactin caused only a slight and not significant elevation of GFR, and had no marked effect on the renal handling of sodium and potassium, although it produced a significant increase in urinary magnesium excretion.

Aldosterone and possibly corticosterone appear to work synergistically with prolactin in maintaining a positive ionic balance in these terrapins (see Figs. 4.4 and 4.5). Aldosterone significantly increased potassium excretion and decreased the urinary sodium-to-potassium ratio in *P. subrufa*. It also caused a reduction in sodium, potassium and magnesium excretion and depressed the urinary sodium-to-potassium ratio in *C. picta* which had been previously loaded with 1·8% saline. Corticosterone also significantly depressed the sodium-to-potassium ratio in *P. subrufa* by increasing potassium excretion, a response which could be due to conversion of corticosterone to 11-deoxycorticosterone and thence to aldosterone, rather than to a direct effect of corticosterone on the kidney. Spironolactone (an anti-aldosterone agent) significantly depressed sodium excretion in *C. picta* without affecting potassium excretion. However, it has been shown (Kagawa, 1964) that spironolactone itself has sodium retentive powers when administered in the absence of mineralocorticoids.

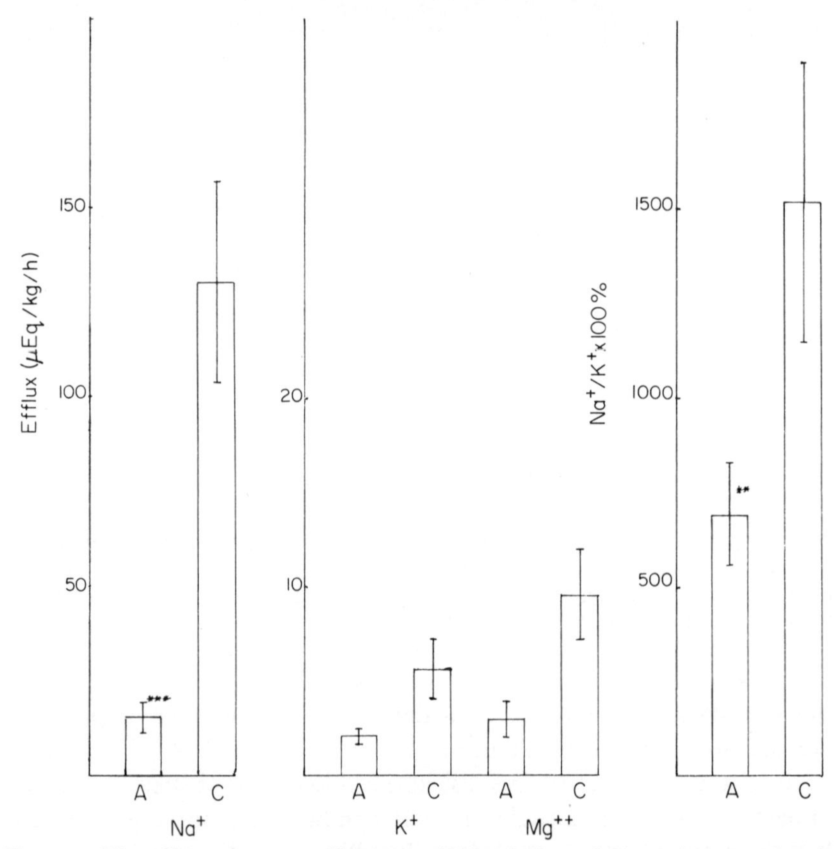

Fig. 4.4 The effect of 20 μg aldosterone (A)/kg body weight on ion excretion in *P. subrufa*, n = 7 (each group). Results analysed by paired 't' tests and presented as mean ± S.E. C = Control. (** p = <0·01>0·001; *** p = 0·001.)

From these results (see also Brewer and Ensor, 1977*a*, *b* and *c*), it would appear that prolactin acts primarily as a diuretic hormone in these fresh-water terrapins, inducing an increased electrolyte loss which is corrected by aldosterone and possibly by corticosterone. An interesting additional finding was that when prolactin was administered to dehydrated terrapins, its effect was to promote retention of ions, rather than their excretion (Brewer, 1976; Brewer and Ensor, 1977*a*).

4.1.2 *The terrestrial tortoise*

In the terrestrial chelonian *Testudo graeca*, somewhat different results were obtained. This species is found throughout North Africa, Southern Europe and Western Asia, and although fresh vegetation is available in its habitat, its natural environment could be described as xeric, if not frankly de-hydratory. The species has been observed to aestivate in certain areas of

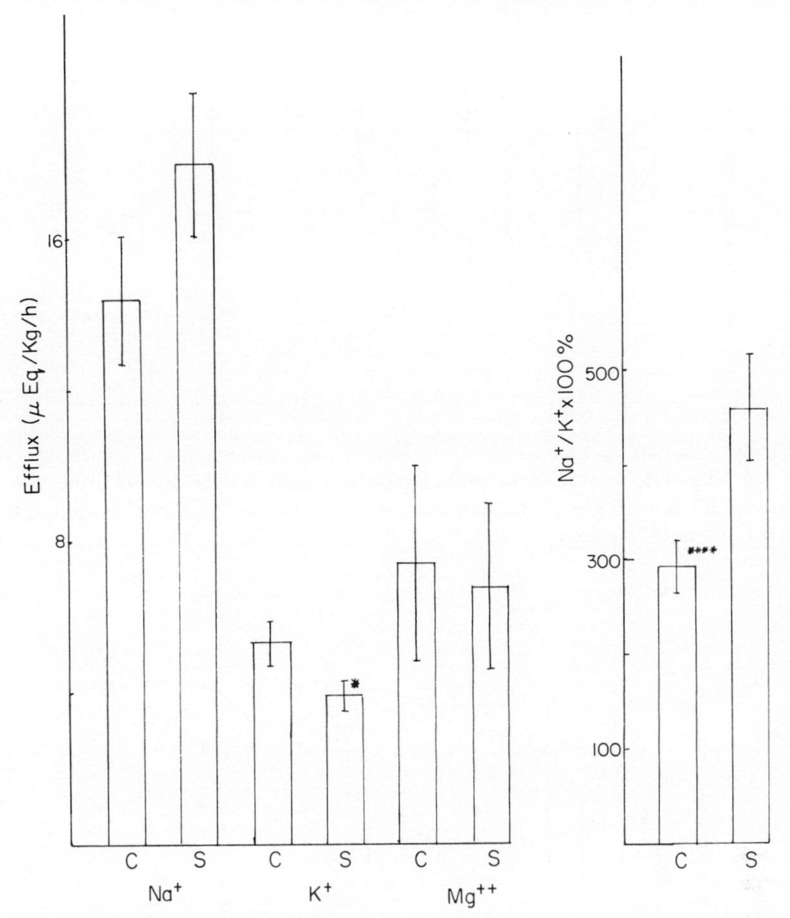

Fig. 4.5 The effect of 200 μg corticosterone (C)/kg body weight on ion excretion in *P. subrufa*. n = 11 (each group). Results analysed by paired 't' tests and presented as mean ± S.E. S = Saline. (* p = 0·05; **** p = 0·001.)

Southern Algeria, but presumably in general it faces severe problems of water loss with a consequent tendency towards increased plasma electrolyte levels. The bladder of *T. graeca* appears to be a major osmoregulatory organ (Bentley, 1971); it is permeable to ions and water, and sodium is reabsorbed and potassium excreted during water reabsorption across its walls. Injections of prolactin in intact *T. graeca* reduces the volume of urine excreted and increases the urinary levels of potassium and magnesium (see Figs. 4.6, 4.7 and 4.8). A consistent elevation of the sodium-to-potassium ratio is also observed, although it is not statistically significant. Hypophysectomy significantly reduces plasma sodium levels after eight days (Table 4.1), although other plasma electrolyte levels remain normal. Interestingly, prolactin significantly increases the urine volume in animals

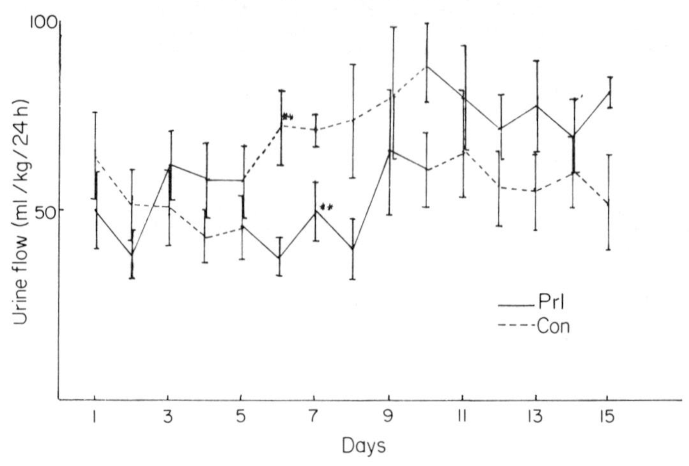

Fig. 4.6 The effect of 20 iu of prolactin (Prl)/kg body weight/day on urine flow in *T. graeca*. n = 6 (each group). Results analysed by 't' tests and presented as mean ± S.E. Con = Control. (** p = 0·05.)

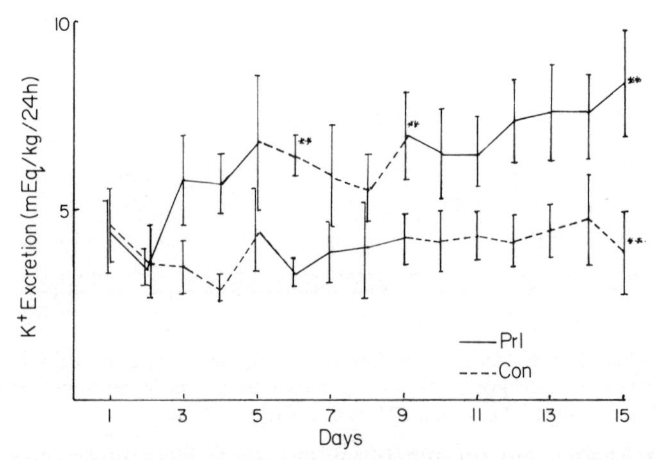

Fig. 4.7 The effect of 20 iu of prolactin (Prl)/kg body weight on bladder potassium excretion in *T. graeca*. n = 6 (each group). Results analysed by 't' tests and presented as mean ± S.E. Con = Control. (** p = 0·05.)

loaded with distilled water (Fig. 4.9), a reversal of effect which parallels the reversal seen in dehydrated terrapins (p. 77). Here we have a situation whereby prolactin in aquatic species acts as a diuretic hormone, but this action is reversed when the animal is dehydrated. Conversely, in the terrestrial species prolactin acts as an antidiuretic hormone but acts as a diuretic when the animal is given a water load. These results may have wide-reaching implications, for it may be that in looking for a unifying concept of prolactin action, comparative endocrinologists have under-

Table 4.1 Plasma and muscle electrolyte levels in *Testudo graeca*

Treatment		*Plasma*				*Muscle*							
	n	Na^+	K^+ (*mEq/l*)	Mg^{++}	% *water*	Na^+	K^+ (*mEq/kg muscle water*)	Mg^{++}	% *water*	Na^+	K^+ (*mEq/kg solids*)	Mg^{++}	Na^+/K^+ × *100%*
Sham-hypect. +saline	6	147·02 ±2·07 *	4·663 ±0·43	3·495 ±0·19	94·64 ±0·51	37·45 ±2·1	106·92 ±5·32	22·56 ±1·39	79·15 ±0·98	142·68 ±7·75	405·37 ±4·56	85·51 ±2·68	35·10 ±1·72
hypect. +saline	8	139·94 ±2·0 **	4·311 ±0·11	3·128 ±0·11	94·4 ±0·41	37·46 ±1·62	109·2 ±2·44	23·68 ±0·43	78·21 ±0·28	134·48 ±5·62	392·16 ±7·84	85·04 ±1·81	34·41 ±1·84
hypect. +cortico.	11	137·78 ±2·38 **	4·7 ±0·26	3·19 ±0·15	95·12 ±0·34	43·67 ±2·66	102·1 ±2·17	22·43 ±0·75	79·42 ±0·36	168·17 ±10·88	391·7 ±7·56 *	85·87 ±1·72	43·17 ±2·92
hypect. +PRL	8	138·75 ±2·33 ***	4·03 ±0·22	3·354 ±0·3	95·53 ±0·32	40·11 ±2·68	98·63 ±2·09	22·48 ±0·78	79·28 ±0·36	154·23 ±11·75	377·93 ±8·86	85·96 ±2·2	40·75 ±2·81
hypect. +PRL+ cortico.	6	125·02 ±3·5	4·54 ±0·27	3·19 ±0·17	95·71 ±0·19	35·74 ±2·42	101·93 ±3·13	22·910 ±0·85	80·15 ±0·37	143·88 ±8·01	411·17 ±6·89	92·78 ±4·49	34·93 ±1·85

Figures represent means ± S.E.
* – significantly different from sham-operated group at level of $P < 0·05$.
** – " " " " " " " $P < 0·02$.
*** – " " " " " " " $P < 0·001$.
Hypect. = hypophysectomised; cortico. = corticosterone; PRL = prolactin.

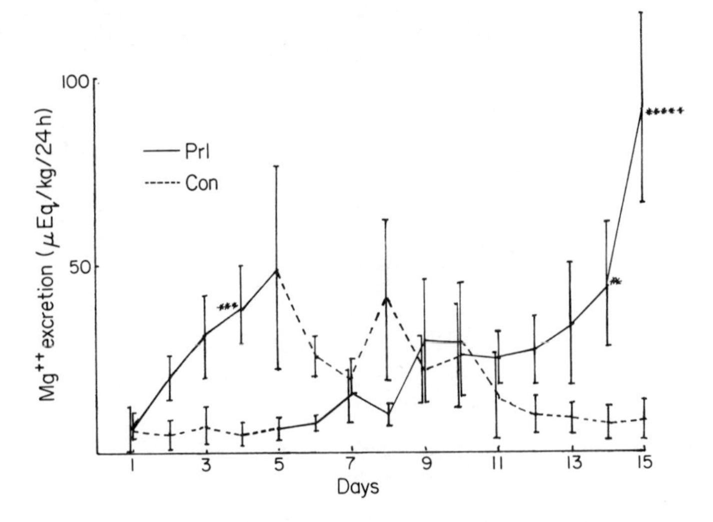

Fig. 4.8 The effect of 20 iu of prolactin (Prl)/kg body weight/day on bladder urine magnesium excretion in *T. graeca*. n = 6 (each group). Results analysed by 't' tests and presented as mean ± S.E. Con = Control. (** p = 0·05; *** p = <0·05>0·02; ***** p = <0·01>0·001.)

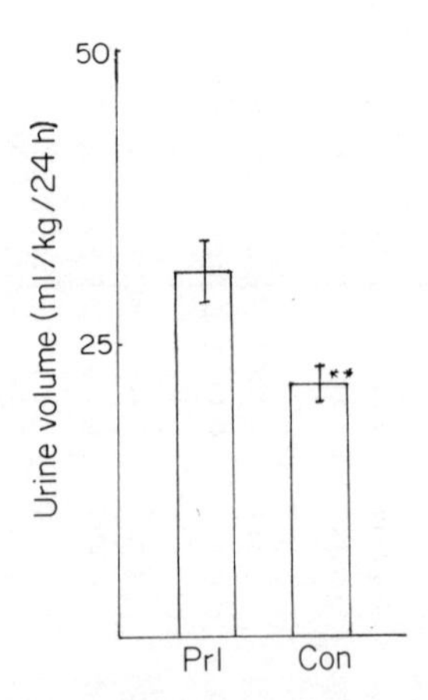

Fig. 4.9 The effect of 20 iu of prolactin (Prl)/kg body weight on urine volume in distilled water loaded *T. graeca* (cannulated). Values are the means of five day periods. n = 7 (each group). Results analysed by paired 't' tests and presented as mean ± S.E. Con = Control. (** p = 0·02.)

estimated the extent to which the effects of the hormone can be modified by changes in the physiological status of the recipient animal. Perhaps part at least of the great range of actions of prolactin, rather than being inherent in the molecule itself, reflects modifications of certain fundamental effects by collateral factors, such as changing conditions in the internal environment within which the prolactin molecule must act.

Table 4.2 Pituitary prolactin levels in the grass snake *Natrix natrix* following dehydration by exposure to 35% relative humidity at 21°C for 4 hours

	Prolactin concentration (iu/mg dry weight)
Control	0·170±0·025 iu/mg
Dehydrated	0·125±0·024 iu/mg

Other evidence for a role of prolactin in reptilian osmoregulation is summarised in Table 4.2. In this experiment grass snakes (*Natrix natrix*) were dehydrated by exposure to dry air, and this resulted in alteration of the pituitary prolactin level (Ensor, unpublished observation), suggesting that prolactin may act as an antidiuretic hormone in this species. However, Nicoll (personal communication) has warned that severe dehydratory stress might result in non-specific release of pituitary hormones. Thus these results must be viewed with caution, in the absence of correlated physiological studies. The advantage of studying a species such as *Natrix natrix* is considerable in relation to the previous discussion. *Natrix* is a semi-aquatic species, alternating between freshwater and terrestrial environments, and so might be expected to yield important information on the possible alteration of prolactin effects in response to environmental change.

4.2 Effects of prolactin on growth, metabolism and tail regeneration in reptiles

4.2.1 *Growth and metabolism*

As with so many aspects of reptilian physiology, there is little information on the role of prolactin in growth. Prolactin has been implicated in the growth of the green anole, *Anolis carolinensis* (Licht and Jones, 1967), and further studies by Licht (1967) have examined the interactions between prolactin and gonadotrophins in controlling various growth parameters in *Anolis*. Licht showed that prolactin greatly increased the appetite of green anoles on an *ad libitum* diet. The animals were maintained at 32°C on a light regime (6 hours light : 18 hours dark) which suppresses testicular recrudescence and prevents linear body growth. Three parameters were measured: the daily food intake of individuals; the numbers of animals feeding in each group; and the average daily food intake of these animals observed to be feeding. Prolactin caused a significant increase in all three parameters, with the average daily food intake increasing to between 190 and 350% of control levels. This increased feeding was reflected in an increase of about 21% in body weight, and by linear growth, expressed as an increase in snout-vent length. The simultaneous injection of gonadotrophin (pregnant mare serum, PMS) with prolactin prevented the increase in appetite and growth shown by the anoles receiving prolactin by itself. Certain visceral organs showed an isometric response to prolactin, growing relatively more than the body in general. This was most marked in the case of the fat bodies, and again it was prevented by PMS. Relative liver weight was not altered.

Daily variations in fat body development have been studied by Meier and his co-workers in *Anolis carolinensis* (Meier, 1969; Meier *et al.*, 1971), *Xantusia henshawi* (Trobec, 1974) and *Dipsosaurus dorsalis* (Trobec, 1974). In all these species the ability of prolactin to induce fattening varies with the time of injection, and Meier believes that the environmental cue for this effect is probably the onset of the daily photoperiod. However Meier *et al.* (1973) have also shown that the stress due to handling can act as an environmental cue in entraining the fattening response. Handling may well lead to release of adrenocortical steroids and prolactin (Nicoll *et al.*, 1960). Meier *et al.* (1973) have shown that repeated saline injections, with their concomitant stress, could induce changes in fat metabolism. They found that *Anolis carolinensis* was particularly prone to disturbance, and that handling shortly after the onset of a 16-hour light period caused an increase in fat stores, so that the weight of the fat bodies in *Anolis* disturbed four hours before the onset of the light period averaged only about 17% of those in animals disturbed four hours after the onset.

Unfortunately there is little information available about circadian secretion patterns of prolactin or corticosterone in reptiles. Birds usually exhibit a peak of prolactin secretion soon after the onset of the light period (Dusseau and Meier, 1971; Ensor and Phillips, 1971), and in the duck this peak can become entrained on such environmental variables as feeding or drinking. Whether a similar pattern exists in reptiles, with an early morning prolactin peak that can be shifted in time (entrained) in relation to handling and injection, is pure speculation at the moment. Meier (1975) has speculated more elaborately in suggesting for *Anolis*

'that the disturbance effect is not a result of stress-induced release of either corticosteroids or prolactin, but rather that the disturbance is interpreted as a recurrent environmental cue (such as the daily photoperiod) that may entrain some rhythms such as the plasma rhythm of corticosterone. Interactions between rhythms set by the daily photoperiod and those set by the disturbance could then account for changes in fat stores'.

If the pattern of fat deposition in reptiles resembles that in birds (Chapter 5), then it may be that fat is deposited as a consequence of the induced increase in appetite rather than as a consequence of a direct lipogenic action of prolactin. It is, however, notoriously unsafe to generalise from one vertebrate class to another. For instance splanchnomegaly of the kidney, pancreas, liver and intestine result from prolactin injection in birds and mammals (Riddle, 1963; Bates *et al.*, 1962, 1964), but these effects were not observed in *Anolis* (Licht, 1967), the increase in intestinal weight being probably simply due to increased food intake.

Riddle (1963) has concluded that in birds prolactin tends to increase the metabolic rate especially in the presence of thyroxine. However, in *Anolis*, prolactin has the opposite effect, injected prolactin reducing the metabolic rate in autumn and having no effect in spring (Licht, 1967; Licht and Jones, 1967). The effect is probably complex. Prolactin reduces the increased oxygen consumption brought about by gonadotrophins in sexually maturing *Anolis*, but is ineffective in sexually quiescent animals (Licht, 1967). Prolactin also stimulates lean growth of *Anolis*, an action that would appear to require some diversion of energy away from energy stores. Thus, since the fat bodies were greatly enlarged in prolactin-injected animals, the energy for lean body growth most probably came from the liver, which was reduced by prolactin.

In summary, prolactin increases appetite in *Anolis carolinensis*, increases fat deposition in *Anolis*, *Xantusia* and *Dipsosaurus*, and stimulates lean growth in *Anolis*. In *Anolis* prolactin appears to antagonise the diversion of energy stores to the testes which is normally brought about by gonadotrophins. Gonadotrophins also suppress the prolactin-induced increase in appetite in this lizard. Relationships with thyroid hormones are not clear. Prolactin has been shown to synergise with thyroxine in the induction of moult (Chiu and Phillips, 1971), and the reduction in metabolic rate in autumn *Anolis* might suggest an anti-thyroidal action. However Licht (1967) studied the thyroid of prolactin-injected *Anolis* histologically, and found no evidence of suppression. Chiu and Phillips (1971) also could detect no effect of prolactin on the thyroid of *Gecko gecko*.

Licht and Hoyer (1968) have compared the effects of prolactin and growth hormone (GH) in juvenile *Lacerta s. sicula*. The effects of prolactin were essentially similar to its effects on *Anolis* (Licht, 1967). Prolactin stimulated food intake by as much as 300%, and this was associated with weight gain and an increase in linear growth. The degree of fattening and growth was similar with both prolactin and GH, although GH was slightly more potent. However GH promoted marked splanchnomegaly and fat deposition in the liver, an effect not produced by prolactin. Given together the two hormones acted antagonistically on the liver. This work suggests that both prolactin and GH are potent somatotrophic hormones in lizards, but that their roles differ in certain essential respects.

The only report that prolactin can stimulate growth in non-lacertilian reptiles comes from recent work by Nichols (1973) on juvenile snapping turtles (*Chelydra serpentina*), which gave results similar to those of Licht and Heyer (1968) reported above. Both prolactin and GH accelerated the growth of the turtles, the linear growth response being similar to both hormones. However, prolactin stimulated spleen growth more strongly than GH. Whether this indicates a specific physiological action of prolactin on the reptilian haemopoietic system is uncertain; it recalls the erythropoietic effect of prolactin in the mouse reported by Jepson and Lowenstein (1964). As in the lizards GH was more potent than prolactin in stimulating liver growth, indicating that the contrasting hepatic effects of the two hormones are not confined to lacertilians. The responses to mammalian prolactin and GH in the lower vertebrates suggests considerable functional overlap (Nicoll and Nichols, 1971; Nicoll and Licht, 1971) and further research is needed in order to define the physiological roles of the endogenous hormones, particularly in reptiles and amphibians.

4.2.2 *Tail regeneration*

Licht (1967) showed that prolactin markedly increased the weight of the regenerated tail in *Anolis carolinensis*. This response was diminished by gonadotrophin, possibly because the gonadotrophin directed energy supplies to the developing testes rather than because of a direct peripheral antagonism between the two hormones (see the comparable action in growth promotion, p. 84). The regeneration of the reptilian tail appears similar to regeneration of the amputated fore limb in amphibia. The wound is sealed by a clot, this is followed by blastema development and finally by the growth of the tail. Prolactin appeared merely to increase the weight of the regenerating tail and not its length or water content.

4.3 Integumentary effects of prolactin in reptiles

The main changes undergone by squamate skin during a sloughing cycle

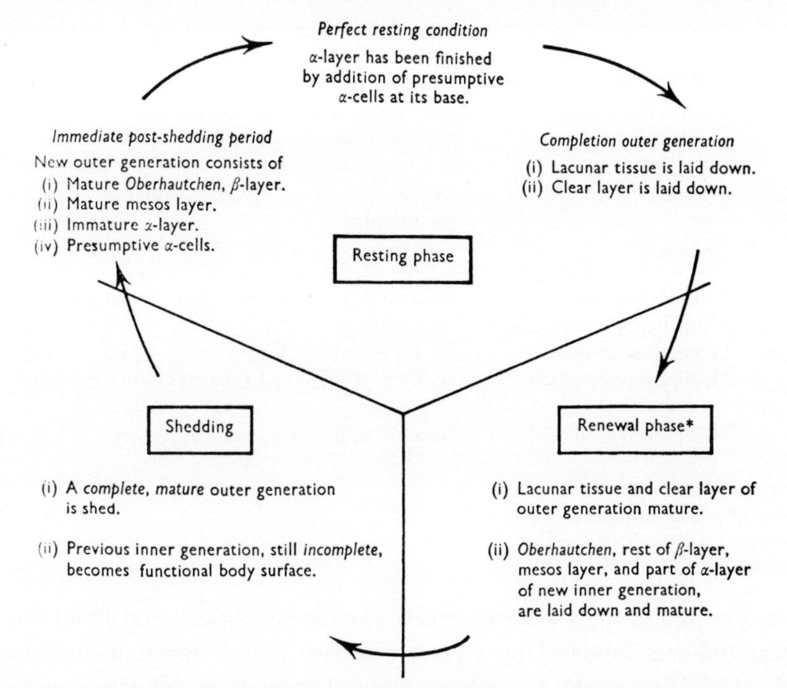

Fig. 4.10 Summary of the major histogenic events in the epidermal sloughing cycle of squamates. The cycle essentially consists of two parts, the resting phase and the renewal phase, the duration of which is measured in days. The act of shedding usually only occupies 1–12 h, depending on the species, but is indicated here to emphasise the changes which occur in general epidermal topography. The renewal phase (*) has been divided into five arbitrary stages covering various aspects of the differentiation sequence of the inner generation. (From Maderson, Chiu and Phillips, 1970.)

are outlined in Fig. 4.10. When a lizard or snake moults it loses a complete mature 'epidermal generation' (Maderson *et al.*, 1970). The epidermis then enters a resting phase in which only a single epidermal generation is present above the *stratum germinativum* (Maderson, 1967; Maderson and Licht, 1967). Eventually the resting phase ends as new cells are laid down which represent the final components of the outer generation (Maderson *et al.*, 1970). Completion of the outer layers is followed by a 'renewal phase' in which the inner layer develops. Once the new inner layer is complete the outer layer is shed.

In *Gecko gecko* Chiu and Phillips (1971) found that hypophysectomy reduced the capacity of the skin to slough. Prolactin restored the sloughing capacity in the first cycle to that of intact and sham-operated animals (see Table 4.3). However an increase in the prolactin dose during the second cycle significantly reduced the cycle length by nine days. The epidermal changes in the sloughing cycle in both control and prolactin-treated hypo-

Table 4.3 The effects of prolactin on the length of the sloughing cycle in hypophysectomised geckos (means ± S.E.M.)

		Length of sloughing-cycle days		
		Before treatment	*After treatment*	
Group no.	*Treatment*	*1st cycle*	*1st cycle*	*2nd cycle*
1	Untreated control	29·6 ± 1·8 (5)	36·0 ± 3·3 (5)	37·3 ± 1·5 (4)
2	Sham-hypophysectomised	33·0 ± 1·2 (5)	38·0 ± 3·0 (5)	38·8 ± 3·0 (5)
3	Hypophysectomised	33·4 ± 1·9 (7)	78·5 ± 7·1 (2)	78·0 (1)
4	Hypophysectomised + prolactin	37·6 ± 1·5 (7)*	35·3 ± 0·4 (7)	30·0 ± 1·2 (7)†
5	Hypophysectomised + vehicle	35·2 ± 2·3 (6)	—	—

Number of animals in parentheses.
* Group 4 *v.* group 1, $P < 0{\cdot}02$.
† Group 4 *v.* group 2, $P < 0{\cdot}05$.

physectomised animals were identical. The resting phase lasted about three weeks, and was followed by a renewal phase lasting approximately two weeks. Only two of the hypophysectomised animals in this study underwent a sloughing cycle, and both animals showed a markedly prolonged cycle in which the resting phase was prolonged to five–six weeks (Chiu and Phillips, 1971). Thyroidectomy of geckos also prolonged the cycle, and this was prevented by prolactin. However ^{131}I uptake was unaltered in the prolactin-injected animals, indicating that this reduction in the length of the cycle was due to a direct effect of prolactin, not to an indirect effect mediated *via* the thyroid (Chiu and Phillips, 1972).

The effects of hormones on the sloughing cycle of lizards is summarised in Table 4.4. It seems in *Gecko gecko* that synergism between prolactin and thyroxine is important in maintaining the length of the cycle, although it is likely that the two hormones exert separate effects on the dermal tissues. In the other species which has been studied extensively, *Anolis carolinensis*, a similar control mechanism appears to operate but the prolactin response is augmented by gonadotrophins (Maderson and Licht, 1967). In both species ACTH has an antagonistic effect on the process. Interestingly, in the work of Chiu and Phillips (1971) the prolactin-injected animals had significantly heavier adrenals. Only further work will elucidate the physiological meaning of this finding.

4.4 Prolactin and reproduction in reptiles

Early attempts by Licht and Jones (1967) to demonstrate a reproductive

Table 4.4 Summary of results of endocrinological investigations of the sloughing cycle in the tokay (*Gekko gecko*)

Histological examination of biopsy samples showed that in all cases where the cycle was increased or decreased in length, only the resting phase was affected: the renewal phase remained the same. In those treatments where subsequent sloughs were inhibited, no renewal phases were ever observed after the beginning of the treatment.

	Treatment		*Dosage**	*Conclusion†*
1	Intact plus T_4	i	0·08 μg daily	Enhanced
		ii	0·16 μg	Enhanced
2	Intact plus TSH		0·04 USP unit	Enhanced
3	Intact plus prolactin		0·4 iu	Enhanced
4	Intact plus prolactin plus T_4		0·4 iu	Enhanced‡
			0·16 μg	
5	Intact plus corticosterone		0·03 μg daily	No effect
6	Intact plus metopirone		0·06 μg daily	No effect
7	Thyroidectomised	i	—	Decreased
		ii	—	Decreased
8	Thyroidectomised plus T_4		0·16 μg daily	Enhanced
9	Thyroidectomised plus prolactin		0·4 iu	Enhanced
10	Hypophysectomised		—	Decreased§
11	Hypophysectomised plus TSH		0·02 USP daily	Enhanced
12	Hypophysectomised plus prolactin	i	0·26 iu	‖
		ii	0·40 iu	Enhanced
13	Hypophysectomised plus ACTH		0·02 iu daily	Inhibits
14	Hypophysectomised and thyroidectomised		—	Inhibits
15	Hypophysectomised and thyroidectomised plus prolactin		0·4 iu	Enhanced
16	Hypophysectomised and thyroidectomised plus TSH		0·04 USP unit	Inhibits
17	Intact plus testosterone (female)		2·00 μg daily	Enhanced
18	Hypophysectomised plus testosterone propionate (male)		2·00 μg daily	Inhibits
19	Hypophysectomised plus testosterone		2·00 μg daily	Inhibits

* Dosage 1 g body weight and on alternate days unless otherwise stated.

† Due to the fact that exact means for control animals vary slightly from experiment to experiment, and some of the treatments produce relatively small percentage increases or decreases in shedding frequency, only a qualitative conclusion is reported here; sloughing frequency with reference to the control group in each case was either unaffected (no effect), more frequent (enhanced), less frequent (decreased), or completely stopped (inhibits).

‡ This treatment produced a greater percentage increase in shedding frequency than either T_4 or prolactin given alone at similar dose levels, indicating synergistic action.

§ Some animals never shed after surgery.

‖ This dosage caused the hypophysectomised animals to shed with the same frequency as intact and sham-operated animals.

role of prolactin in reptiles were unsuccessful. Male *Anolis carolinensis* have a well-marked reproductive cycle (Dessaur, 1956; Fox, 1958), but injections of exogenous prolactin failed to increase testicular weight, the epithelial height of the epididymis or the development of the sexual segment of the kidney. Somewhat surprisingly, because of the known relationships of the two hormones in other situations, growth hormone did cause a significant increase in all three parameters (Licht and Jones, 1967).

Later work by Callard and Ziegler (1970) on the iguanid lizard *Dipsosaurus dorsalis* suggests that prolactin may exert an antigonadal effect in the female. Ovarian growth can be stimulated in the intact animal by PMS, which possesses inherent FSH like activity. The stimulated ovaries showed an increase in the number of developing follicles and an increase in follicular diameter. Prolactin blocked the increase in follicular diameter due to PMS, but did not reduce the number of developing follicles. In the hypophysectomised animal PMS caused a small increment in ovarian growth, but for the development of the full gonadotrophic response it was necessary to treat the animals simultaneously with GH. Prolactin did not block gonadal growth in hypophysectomised animals treated with PMS alone, but it reduced the response when PMS and GH were administered together. This suggests two interpretations which unfortunately have not been tested experimentally. Firstly, GH, but not prolactin, is essential for the development of the full gonadotrophic response, a hypothesis which is supported to some extent with the results for male *A. carolinensis* (Licht and Jones, 1967), and which implies that GH may have a more potent metabolic effect than prolactin under these circumstances. The second interpretation, which is particularly interesting, is that prolactin antagonises the synergistic relationship of GH and gonadotrophin. This antagonism between prolactin and GH would be a novel relationship, and does not seem to have been adumbrated by other work in lower vertebrates.

More recently Callard *et al.* (1975) have studied the efforts of prolactin on *in vitro* steroidogenesis by the ovary of *Chrysemys picta*. A series of papers by Licht and his co-workers (see Licht and Papkoff, 1974) have suggested that FSH is the primary gonadotrophin in reptiles, and that the different responses of the reptilian gonad to this single gonadotrophin could be explained to some extent by the effects of environmental temperature (Licht, 1972). However, Callard and Chan (1974) suggested that LH is important in reptiles as a steroidogenic hormone, particularly in stimulating ovarian steroidogenesis. Prolactin appeared to have no effect when administered alone, but it reduced the stimulatory effect of LH on progesterone synthesis. This observation indicates a possible antisteroidogenic action of prolactin in the intact animal, which would agree with the antigonadal effect of prolactin seen in PMS-treated *Dipsosaurus dorsalis* (see above, p. 90). This effect of prolactin in inhibiting progesterone synthesis is in obvious contrast to its luteotrophic activity in mammals (Chapter 6, pp. 130–133) and to its possible synergistic relationship with progesterone in birds (Lehrman, 1963; Jones, 1971).

In contrast to the work on *Chrysemys*, *Dipsosaurus* and *Anolis*, Burns and Richards (1974), working with the lizard *Phrynosoma cornutum*, showed that prolactin had a gonadotrophic effect in hypophysectomised females. Although not as stimulatory as FSH, prolactin was as effective as LH in increasing the diameter of the ovarian follicular cells, and more effective in increasing the nuclear diameter of these cells. Prolactin did not increase the size of the follicles, but nor did FSH or LH when administered alone. Unfortunately the possibility that prolactin might synergise with the gonadotrophin was not explored in this study. Prolactin was ineffective in increasing ovarian weight in hypophysectomised lizards, but it did stimulate growth of the oviduct.

These results on the reproductive effects of prolactin in reptiles are fragmentary, scattered taxonomically, and have yielded results, often contradictory, which cannot be welded satisfactorily into any general scheme. As yet no definite effect of prolactin on the male reproductive system has been reported. In the females, some evidence favours an antigonadal/antisteroidogenic role, while other data favour a weak gonadotrophic action. The differences between the species studied are marked: in *Dipsosaurus*, FSH and GH are required to induce a full gonadal response, whereas in *Phrynosoma* FSH and LH act synergistically, with no obvious GH involvement. It is obvious that generalisations at this stage can be of little value.

4.5 Hypothalamic control of prolactin secretion in reptiles

Evidence for regulation of the prolactin cells by the hypothalamus is remarkably scant in reptiles. Prolactin was secreted continuously by the pituitary of *Malaclemys terrapin* during a 10-day incubation (Bern and Nicoll, 1968) which indicates considerable autonomy. However more recent data on *Pseudemys* suggests the possibility of a stimulatory control (Nicoll and Fiorindo, 1969). In addition, work on fattening rhythms (Meier *et al.*, 1971; Trobec, 1974) points to a circadian pattern of prolactin release similar to that shown by other vertebrates, and this would in turn imply some degree of hypothalamic control. It may be, as in mammals, that the reptilian hypothalamus secretes both stimulatory and inhibitory factors controlling prolactin; only further work can decide (Nicoll and Fiorindo, 1969).

5 Prolactin in birds

In surveying the effects of prolactin in fishes, amphibians and reptiles a certain general pattern or theme has emerged, in that prolactin in these groups may be regarded as a general metabolic hormone, with major effects on salt and water regulation and on growth. Admittedly these are blanket terms: in the case of growth, for example, such diverse responses as the proliferation of integumentary elements, growth of larval amphibian brain tissue, fat deposition and tail-regeneration in reptiles are included. Even recognising the heterogeneous nature of the elements of the 'pattern', we can see that it cannot accommodate all the known effects of prolactin: for instance, how should we classify the effects on fish behaviour, and the antithyroid actions in the amphibians? At this point, as our study of prolactin physiology passes from the ectothermic vertebrates to the endothermic classes Aves and Mammalia, we encounter a sudden expansion in the range of actions of prolactin, and the hormone appears to acquire a new major role concerned with the control of reproduction and with parental behaviour and physiology. Obvious effects on growth processes, general or specialised (for example, the pigeon crop-sac), and on integumentary structures (for example, the mammary gland) remain, but the osmoregulatory actions, so important in aquatic vertebrates, now, as it were, take second place to new functions. Comparative endocrinologists interested in the evolution of this fascinating hormone may discuss interminably whether it arose primarily as an osmoregulatory hormone with secondary reproductive effects, or as a reproductive hormone with an extra capacity to protect the individual against osmotic stress. Given the nature of the fossil record the debate, like all good debates, will probably go on forever. There can be no doubt, however, of the primacy of the reproductive and parental roles of prolactin in the birds and mammals alive and available for investigation today.

5.1 The role of prolactin in ion and water balance in birds

Like other vertebrates, birds vary in their osmoregulatory abilities. In general the main site of salt exchange is the kidney. The avian kidney possesses two distinct types of nephron, the reptilian type with little concentrating power and the more convoluted mammalian type (Braun and Dantzler, 1972). On the whole the concentrating power of the avian kidney is not very high, although some arid-zone species such as the savannah sparrow can produce sodium urine-to-plasma ratios of around 3 : 1 (Bartholomew and Cade, 1964). In addition, Skadhauge, (1974) has shown that a number of avian species have a secondary concentrating mechanism located in the rectal-cloacal complex, involving the retrograde flow of urine and faeces into the rectum where secondary reabsorbtion of salt and water takes place. It is probable that this mechanism, acting as an adjunct to the limited concentrating capacity of the renal tubules, suffices to fit most birds for their normal environments. There remain, however, a number of avian species which apparently are faced with excessive salt loads either occasionally or as a regular feature of their lives. These include certain Falconiformes (Cade and Greenwald, 1966) and members of the Anseriformes, Charadiiformes and Pelecaniformes among other groups. These birds have evolved an extra-renal site for salt excretion in the form of modified facial glands, known as the nasal or supra-orbital glands.

A review of these structures is beyond the scope of the present text, and an excellent review of salt glands, both in birds and reptiles, has been published by Peaker and Linzell (1972). A schematic diagram of a typical avian nasal gland appears in Fig. 5.1. It can be seen that the gland comprises a series of straight tubules, with walls of a typical transporting epithelium, which empty into larger collecting ducts. The tubules appear to be able to concentrate sodium across a single cell layer and there is no evidence of a countercurrent mechanism such as operates in the mammalian kidney.

The first indication that prolactin affects salt and water balance in birds came in the work of Peaker *et al.* (1970), during investigations into the hormonal control of the duck nasal gland. Small doses of prolactin produced an immediate increase in salt excretion by the minimally stimulated nasal gland. Further work on the duck (*Anas platyrhynchos*) showed that birds maintained on 300 mM NaCl for five days developed a characteristic sequence of changes in pituitary prolactin levels (Fig. 5.2). There was a marked increase in pituitary prolactin on days two and three followed by a decrease to a very low titre by day five (Ensor and Phillips, 1970). By day five the birds showed obvious signs of distress, and it is likely that this degree of salt-loading constitutes such an abnormally intense stress that the

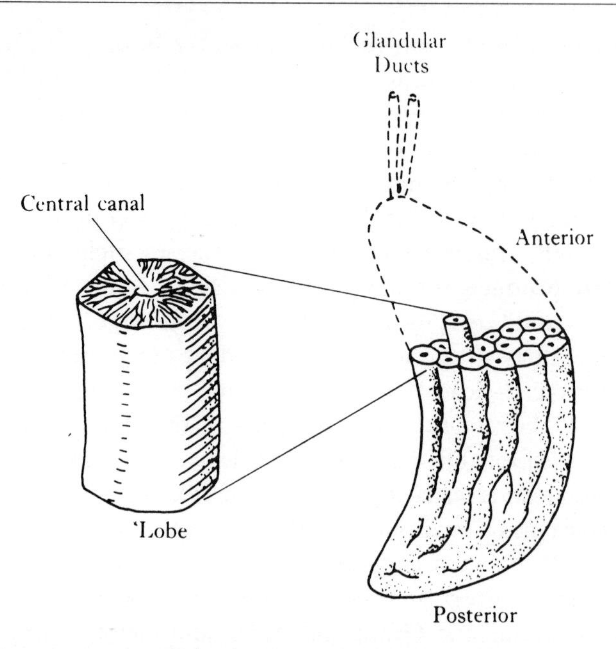

Fig. 5.1 The arrangement of lobes in the salt gland of the herring gull, *Larus argentatus*. (From Fänge, Schmidt-Nielsen and Osaki, 1958.)

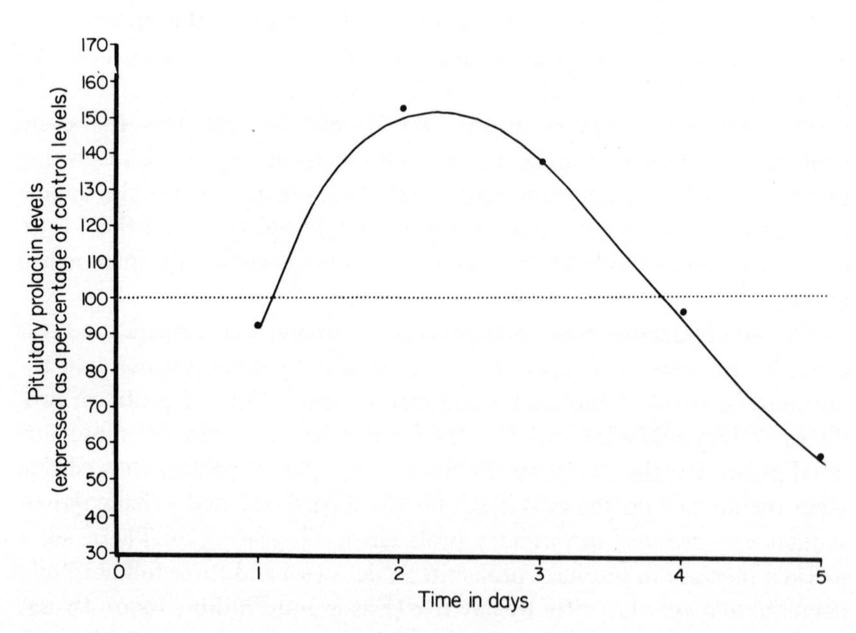

Fig. 5.2 The effect of saline maintenance on pituitary prolactin levels. Birds were given 300 mM NaCl to drink. Each point is the mean for six birds.

pituitary (and other) control mechanisms break down. The same changes in pituitary prolactin could also be demonstrated in juvenile gulls given 700 mM NaCl (Ensor and Phillips, 1972*a*). However, as Nicoll (1972) has shown, it is difficult to interpret changes in pituitary hormone levels, especially in response to dehydration, in terms of secretion rates.

The control of nasal gland function in birds is very complex. The actual initiation of the secretory response is brought about by direct cholinergic innervation (Ash *et al.*, 1969) probably triggered by osmoreceptors in the heart (Hanwell *et al.*, 1972). Modifying this is a complex of the endocrine factors which influence the pattern of salt excretion. These include corti-

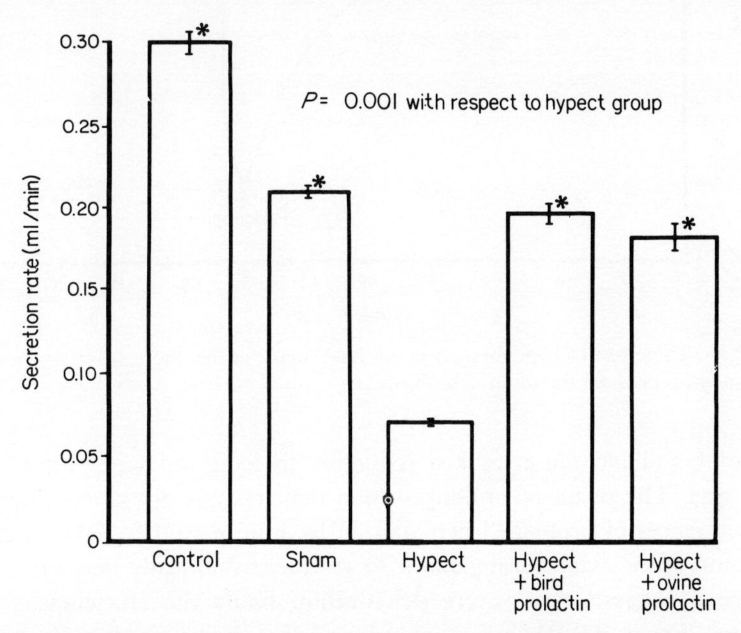

Fig. 5.3 The effect of hypophysectomy and replacement therapy on nasal gland secretion rate. Six birds per group: mean ± S.E. Hypect = hypophysectomised.

costerone (Holmes *et al.*, 1961; Peaker *et al.*, 1971) and thyroxine (Ensor *et al.*, 1971) as well as prolactin. Because of the implied roles of ACTH and TSH, interpretation of the effects of hypophysectomy in ducks is extremely difficult (Ensor *et al.*, 1972*b*). Hypophysectomy does reduce the nasal gland secretion rate (Fig. 5.3) and this reduction can be prevented equally well by either ovine prolactin or avian prolactin, which suggests that the ovine hormone acts in the same way as avian prolactin and not by imitating some other avian hormone such as ACTH. It should, however, be noted that prolactin failed to restore the secretion rate completely to normal. The many consequences of hypophysectomy in the duck are

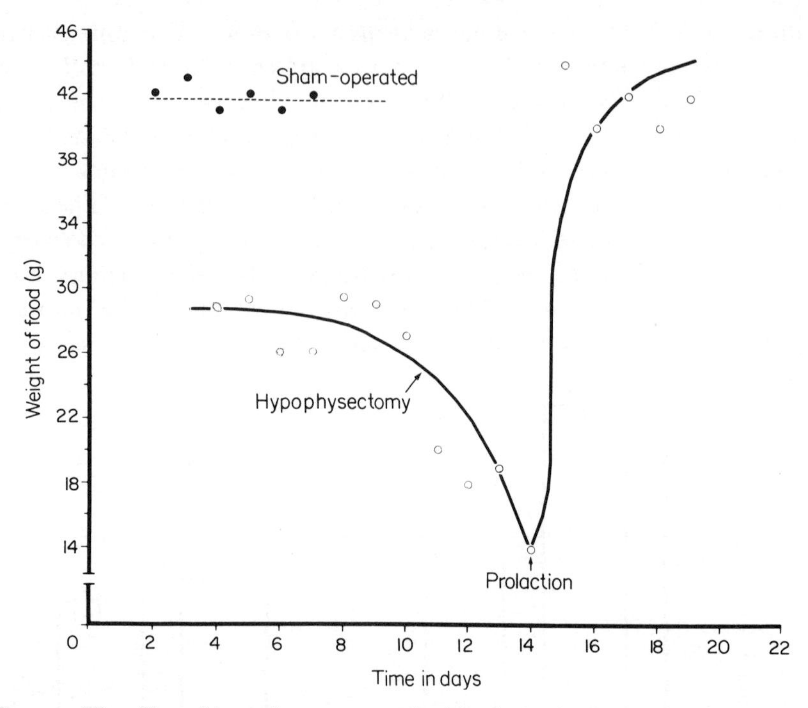

Fig. 5.4 The effect of hypophysectomy on food intake in the duck *Anas platyrhynchos*. Each point represents the mean of 12 birds.

complex and include a marked reduction in food and water intake (see Fig. 5.4). The result of prolonged food deficiency is not known, but the consequences of prolonged dehydration have been studied. The essential function of the avian nasal gland is to produce osmotically free water, and evidence suggests that severe dehydration limits the efficiency of this process. Ensor and Phillips (1972*b*) found that ducks which were given a test saline load, and then dehydrated for five days before being given a second test loading, showed a decreased response to the second test which was proportional to the degree of dehydration (see Fig. 5.5). This result suggests that the domestic duck may have only limited powers to maintain sodium balance in the face of extreme osmotic stress, and that a previous history of chronic dehydration renders the individual more vulnerable to an acute osmotic stress. Thus if we consider the water (W) balance of the domestic duck in terms of the simple equation:

$$W_{in} = W_{urine} + W_{nasal\ fluid}$$

and the difference between W_{in} water intake and W_{urine} is 54 ml/day, then we can see that for a 3 kg bird, an effective reduction in W_{in} of

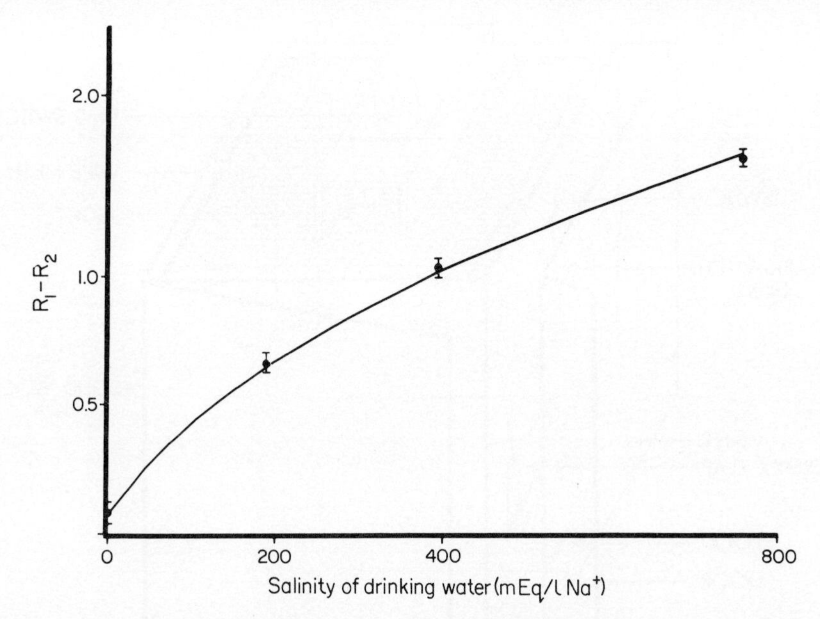

Fig. 5.5 The effects of increasing salinity of drinking water on the rate of production of nasal secretion. The secretion rates are expressed as the initial rate (R_1) minus the final rate (R_2). Five birds per group.

50 ml/day would severely restrict the maximum nasal gland output. Douglas and Neely (1969) have also suggested that dehydration might have a direct effect on the nasal gland by causing vasoconstriction of the arterioles supplying the gland, and have suggested that in some species the effect of dehydration could be opposed by alleviating vasoconstriction with adrenergic blocking drugs.

The possibility exists, therefore, that prolactin may not only act directly on the avian nasal gland, but may also indirectly promote the osmotic homeostasis of the bird by increasing food and water intake. There exists a considerable body of evidence to suggest that prolactin can indeed stimulate appetite in birds, particularly in the pre-migratory phase (see p. 121). For example, Bates *et al.* (1962) have shown that exogenous prolactin stimulates food intake in pigeons. There has, however, been little work aimed at relating endogenous prolactin levels with food intake in non-migratory birds. Ensor (1975) has attempted to measure these parameters in the domestic duck. With the use of the apparatus shown in Fig. 5.6, which basically separates the bird from its food and water supply by two perspex flaps fitted with microswitches which activate a pen recorder, it was possible to obtain a record of the feeding and drinking behaviour of individual birds. Both feeding and drinking showed a circadian rhythm,

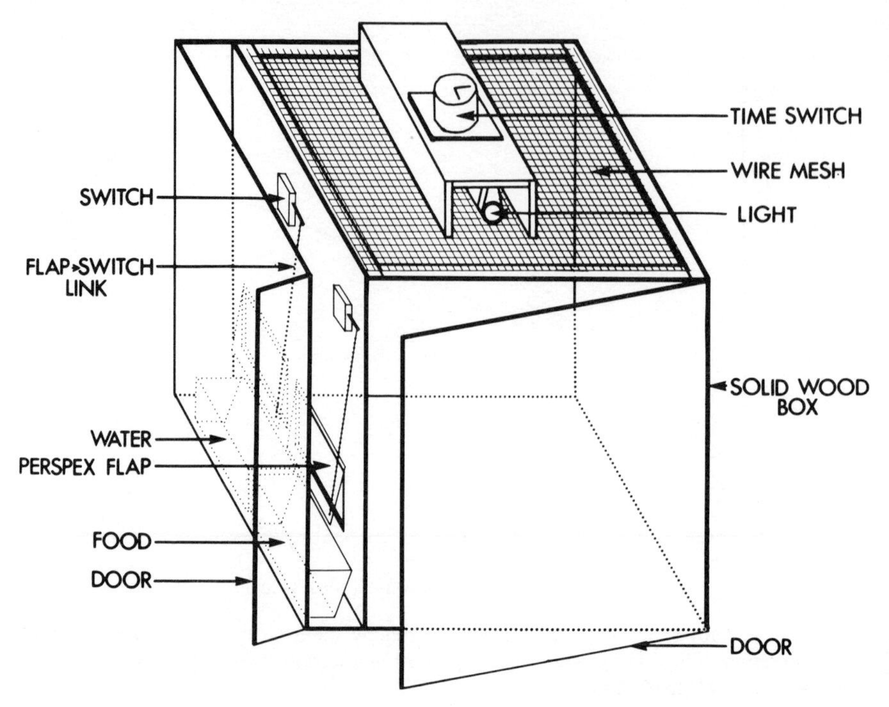

Fig. 5.6 Apparatus designed to study feeding and drinking behaviour in ducks.

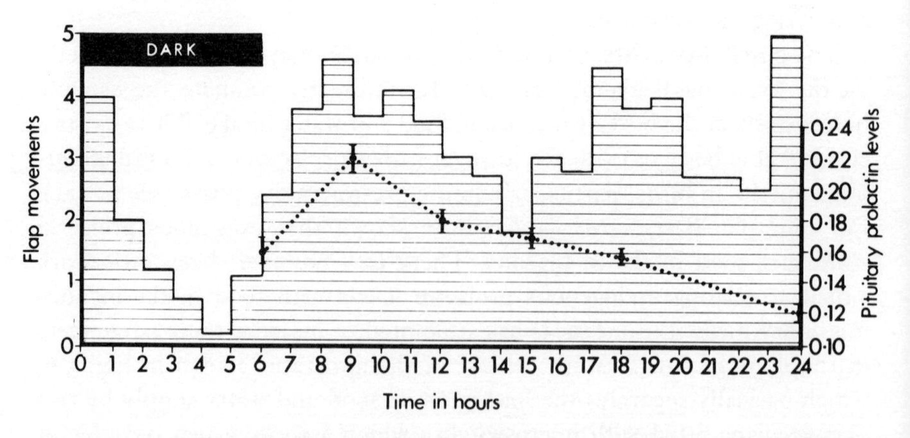

Fig. 5.7 Correlation between feeding behaviour and pituitary prolactin levels (mean ± S.E.; dotted line). The flap movements represent the number of times the bird fed and not the duration of the feeding periods. The histogram represents the mean of six birds.

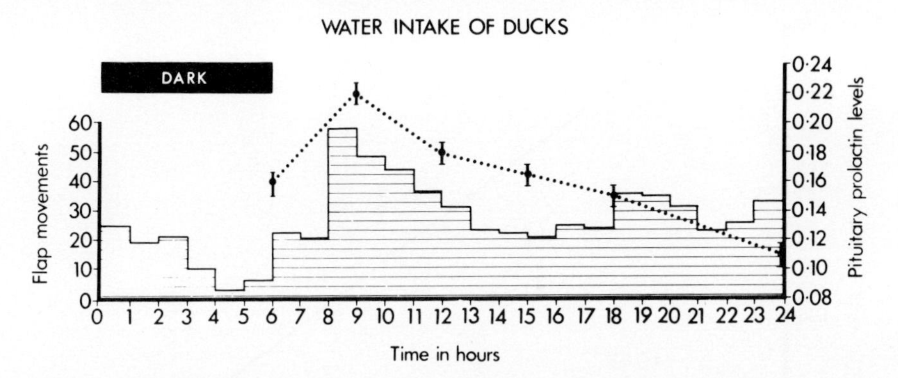

Fig. 5.8 Correlation between pituitary prolactin levels (mean ± S.E.; dotted line) and water intake. The flap movements represent the number of times the animal drank, not the duration of drinking. The histogram represents the mean for six birds.

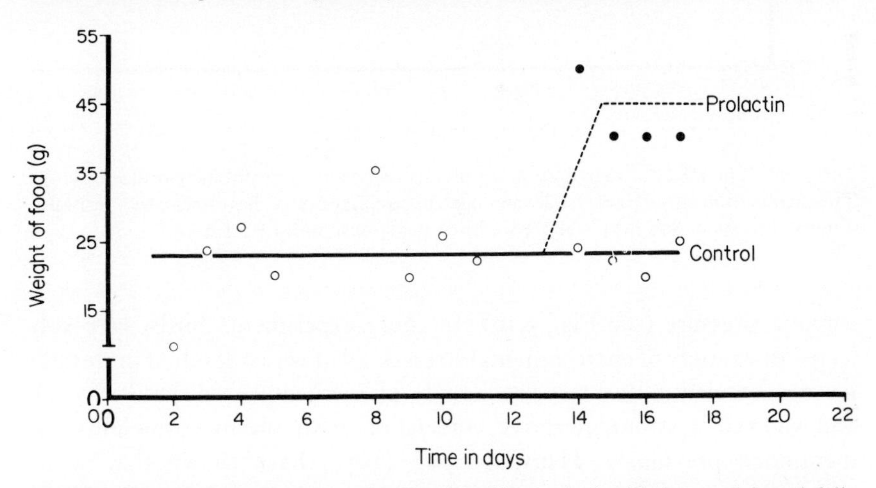

Fig. 5.9 The effect of prolactin on food intake of intact ducks. Birds were injected daily either with prolactin (ovine 0·4 iu/kg body weight) (closed circles) or 0·9% saline (open circles). Each circle is the mean for six birds.

with a peak shortly after the onset of the light period and a second peak later in the day (see Figs. 5.7 and 5.8). If the circadian pattern of pituitary prolactin levels is compared with these figures it can be seen that the early peak matches closely the increased levels of prolactin. Ensor (1975) further showed that serial injections of ovine prolactin in intact ducks significantly increased food intake to a high level at which it plateaued, despite subsequent injections (see Fig. 5.9). Possibly, therefore, one action of prolactin is to maintain the hydrated state of the bird by increasing the intake of water and food.

Interestingly, the release of prolactin from the pituitary appears to be proportional to the state of dehydration or, more precisely, to plasma

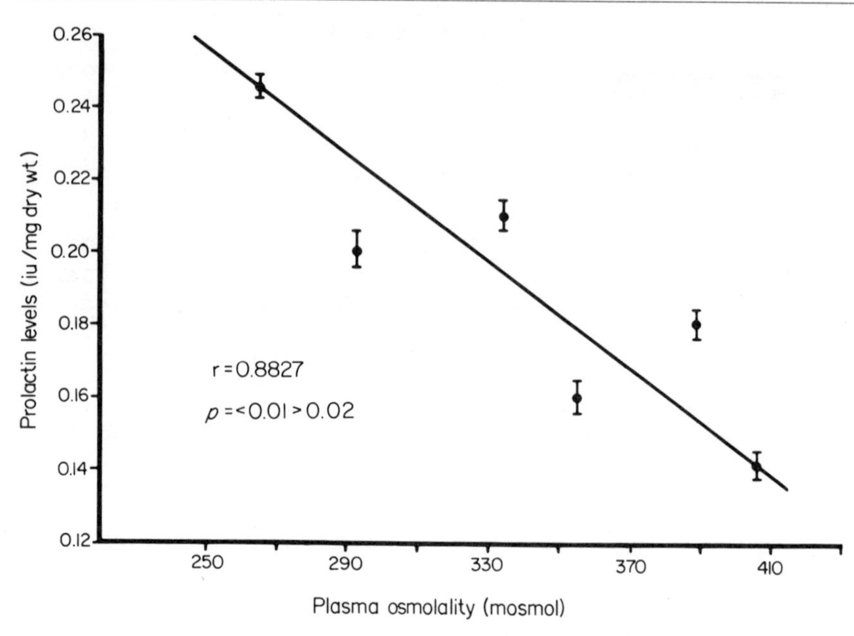

Fig. 5.10 The effect of dehydration on plasma osmolality and pituitary prolactin levels. Treatments from left to right: freshwater controls; 40% seawater; heat stress; 80% seawater; water deprivation; 800 mM NaCl. (Six birds per group; mean ± S.E.)

osmotic pressure (see Fig. 5.10). In these experiments birds were subjected to a variety of environmental stresses, all of which resulted in elevated plasma osmolality. Pituitary prolactin levels were measured simultaneously and showed a strong negative correlation with plasma osmolality. As mentioned previously, Hanwell *et al.* (1972) have shown that osmoreceptors in the heart are responsible for the initiation of secretion by the nasal gland. Ensor (1975) injected lignocaine into the pericardial cavity, thereby blocking impulses passed *via* the vagus, and demonstrated blockade of the prolactin response to increased plasma osmolality. This suggests that prolactin release is mediated *via* a neural reflex rather than by direct action of increased plasma osmolality on the pituitary, in contrast to the situation in teleosts (see pp. 39–42).

There are two ways in which prolactin may increase the state of hydration of a saline-loaded duck. Firstly, as has been previously discussed, prolactin stimulated water intake. Secondly, there is evidence that prolactin may oppose water loss by reducing urine flow. If juvenile herring gulls (*Larus argentatus*) are maintained on salt water, they show a marked and significant fall in urine flow rate (Ensor, unpublished). However, the same treatment of domestic ducks does not affect glomerular filtration rate or urine flow (Holmes *et al.*, 1968). In the duck, nevertheless, the more severe salt-loading produced by infusing 10% saline into the brachial vein did cause a slight fall in glomerular filtration rate (Brewer and Ensor, unpublished), suggesting that urine flow can be reduced under conditions of severe dehydration (Tables 5.1 and 5.2). Ensor (1975) demonstrated that prolactin injections in intact ducks produced a dose-related reduction in urine output at the cloaca (Fig. 5.11), and further showed that if a diuresis was induced by amiloride (which does not directly affect the nasal gland; Peaker, 1971), then pituitary prolactin levels fell, indicating increased release of prolactin (Fig. 5.12).

Table 5.1 Effect of severe salt-loading (infusion of 10% NaCl direct into the brachial vein) on the glomerular filtration rate (GFR) in *Anas platyrhynches*

	GFR (ml/kg/min)
Control	$1{\cdot}47 \pm 0{\cdot}12$
Saline-loaded	$0{\cdot}82 \pm 0{\cdot}10$

Table 5.2 Effect of injections of ovine prolactin (0·12 iu/kg) on glomerular filtration rate (GFR) in control and saline-loaded (0·7 ml/min 10% NaCl) ducks, *Anas platyrhynches*

	GFR (ml/kg/min)
Control	$1{\cdot}47 \pm 0{\cdot}12$
Control and prolactin	$1{\cdot}52 \pm 0{\cdot}16$
Saline	$0{\cdot}82 \pm 0{\cdot}10$
Saline and prolactin	$0{\cdot}96 \pm 0{\cdot}11$

Further work (Ensor, unpublished) has shown that injections of prolactin do not reduce glomerular filtration rate in either control or saline-loaded birds. This indicates an extrarenal site of action for prolactin, since it is unlikely that prolactin could produce the observed reduction in urine flow by means of a change in tubular reabsorption. Confirmation comes from recent work (Ensor, unpublished) showing that prolactin increases the uptake of tritiated water from the intact rectal-cloacal complex and from the isolated rectum of the duck.

It is difficult to interpret the physiological significance of these diverse observations. One tentative suggestion (Ensor, 1975) is that this particular group of prolactin actions may have evolved in order to protect brooding birds. As discussed later on pp. 105–109, prolactin is very much involved

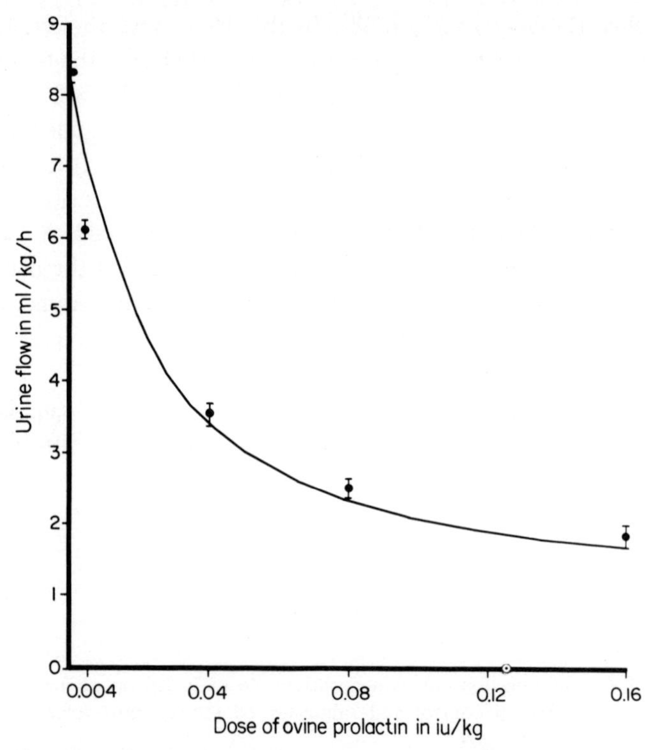

Fig. 5.11 The effect of prolactin on urine flow rate. Six birds per group.

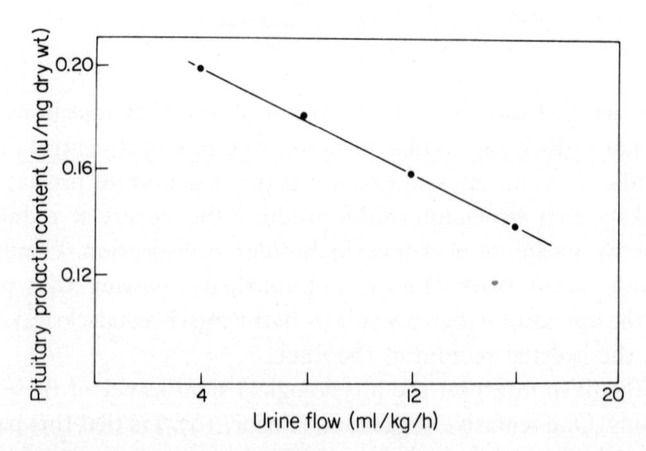

Fig. 5.12 The effect of urine flow on pituitary prolactin levels in the duck *Anas platyhynchos*. Six birds per group.

in parental care in many birds (Eisner, 1960). In marine birds the brooding of the eggs and nestlings might well entail certain osmotic difficulties, particularly in those oceanic birds which might brood the young for up to five days in a dehydrating environment, with no access to drinking water (Nelson, 1968). One can readily see that in these conditions a secondary action of prolactin on water retention by the kidney and cloacal region could have a strong selective advantage.

More recently Tai and Chadwick (1976) have shown that oral saline-loading in the chicken stimulates the pituitary prolactin cells (Tixier-Vidal and Assenmacher, 1966); the stimulation included an increase in the number of polymorphic granules and the appearance of small, recently synthesised granules in the Golgi apparatus. Exocytosis of granules and merocrine secretion of vesicles from the extended endoplasmic reticulum were also observed.

Further work by Scanes *et al.* (1976), using a newly-developed radio-immunoassay for avian prolactin, has confirmed these findings, showing that salt-loading in laying hens results in increased plasma prolactin levels. This has been followed up by Bolton *et al.*, (1976) who found that high circulating levels of prolactin, induced either by TRH injection or by saline-loading, were closely correlated with high circulating levels of potassium. This suggests that prolactin in the chicken induces potassium retention, possibly by the activation of a Na^{+}/K^{+}–ATPase exchange mechanism. If a major role of prolactin in birds is to enhance sodium excretion, then such changes would be expected.

5.2 The effect of cold stress on pituitary prolactin levels

In a programme of research on the effects of environmental stresses on ducks, Ensor *et al.* (1976) have recently demonstrated a fall in pituitary prolactin levels correlated with an increase in nasal gland function in the cold-stressed bird. The situation is intricate since cold stress was found to produce dehydration (see Table 5.3), with a rise in plasma osmolality and a fall in muscle water content, changes which elevated nasal secretion even in birds loaded orally with freshwater. At the same time there was a marked increase in the uptake of radioiodine by the thyroid gland, indicating presumably an increased rate of thyroxine secretion which would have a calorigenic effect.

The last observation raises the problem of whether prolactin is released in direct response to cold, or is released indirectly as a result of high TRH levels, TRH being known to stimulate prolactin release in birds (see pp. 126–128). It is also possible that the increased nasal secretion may be a result of more rapid gut absorption caused by high circulating thyroxine levels (Ensor *et al.*, 1971). However, cold-stressed birds which were thyroidectomised also showed an increased nasal secretion rate, although to a lesser degree than intact stressed birds, indicating that part at least of the prolactin response is independent of the thyroid, though not necessarily of TRH.

If prolactin is released in response to cold in birds, as it appears to be in rats (Mueller *et al.*, 1974), then the possibility exists of a wider metabolic role which remains largely unstudied, with the exception of the effects on food intake and fat deposition discussed elsewhere (pp. 121–125).

Table 5.3 Changes in blood and muscle parameters and in salt gland activity

	Control +fresh water	*Cold-stressed +fresh water*	*Control +hypertonic saline*	*Cold-stressed +hypertonic saline*
Plasma				
Osmolality (mosmol)	272·0 ±4·00	295·9 ±3·91	314·2 ±6·65	310·5 ±10·80
Na (mEq/l)	133·9 ±1·55	160·2 ±5·21	134·5 ±1·55	157·2 ±4·70
K (mEq/l)	3·2 ±0·21	3·7 ±0·30	3·1 ±0·24	3·3 ±0·33
H_2O (%)	96·5 ±1·72	95·2 ±0·41	96·2 ±0·42	95·2 ±0·69
Haematocrit (%)	42·6 ±1·72	47·3 ±2·13	43·6 ±1·20	48·2 ±1·00
Muscle				
H_2O (%)	76·2 ±0·69	75·1 ±0·81	76·5 ±0·66	76·0 ±0·81
Fat (%)	0·9 ±0·36	3·1 ±1·68	2·5 ±1·13	2·0 ±0·28
Na (mEq/l)	26·1 ±0·92	25·8 ±0·92	26·6 ±1·57	25·7 ±0·75
K (mEq/l)	101·2 ±4·60	102·4 ±2·30	101·6 ±9·78	103·5 ±8·02
ECFV	182·6 ±3·20	160·4 ±6·60	183·7 ±3·70	158·2 ±3·62
Salt gland secretion				
Secretion rate (ml/kg/hr)	nil	1·01 ±0·10	2·74 ±0·40	8·18 ±1·16
Na (mEq/l)	—	48·5 ±6·81	333·7 ±10·17	305·8 ±11·84
K (mEq/l)	—	8·57 ±0·46	15·7 ±0·66	13·7 ±0·68
Osmolality (mosmol)	—	220·2 ±7·60	770·5 ±25·10	678·5 ±30·10

5.3 Prolactin and parental behaviour in birds

Prolactin plays two major roles in the parental biology of birds. Firstly, the hormone is concerned with the maintenance of 'broodiness' in many species, although some controversy still exists regarding its importance relative to gonadal hormones. Secondly, and related to broodiness, in pigeons and doves (Columbiformes), prolactin stimulates the proliferation of the lining epithelium of the crop-sac. The fat-laden epithelial cells are shed and regurgitated by the parent birds as 'pigeon's milk', a cheesy mass which nourishes the nestlings during early development. This second parental adaptation, being specialised, will be discussed separately.

5.3.1 *Broodiness and incubation behaviour*

Early work by Lienhart (1927) showed that a humoral factor is involved in incubation of the egg-clutch by birds, since the injection of serum from incubating hens induced non-incubating hens to sit on eggs. The role of prolactin in promoting brooding behaviour was first shown by Riddle *et al.* (1935) who induced broodiness and incubation behaviour in laying hens with prolactin. Some strains of domestic hen, which show a low natural incidence of parental behaviour, proved however to be largely refractory to prolactin injections, although they showed some elements of broody behaviour such as clucking.

The main problem in assessing the role of prolactin in parental behaviour is that many of the behaviour patterns involved are induced equally effectively by progesterone. Riddle (1963) and his co-workers are firmly convinced that prolactin is the key hormone in parental behaviour. However, a more complex relationship is envisaged by Lehrman (1963), who suggests that in Columbiformes, progesterone initiates parental behaviour, and that prolactin merely maintains the condition. The major contention of Lehrman and others (see particularly Eisner, 1960) is that the main role of prolactin is to stimulate progesterone secretion.

Riddle's arguments, although not strictly limited to birds, are to a large extent based on stimulation of the crop-sac in pigeons and doves. Riddle (1963) emphasises that the earlier work (Riddle *et al.*, 1935) showed a specific action of prolactin, in that LH, FSH, progesterone and TSH all failed to stimulate broody behaviour in cocks and non-laying hens. However, if progesterone is involved in some crucial intiation phase of broodiness, it is possible that Riddle and his co-workers in their tests may have missed the crucial period. In another early paper prolactin was shown to stop egg production in one to four days, but progesterone was even more effective, arresting egg-laying in 0–3 days, (Bates *et al.*, 1935). This antigonadal role of prolactin will be discussed later (pp. 115–119). For the present discussion, it is important to note that some effects of prolactin on parental behaviour could in part be a result of its antigonadal action. Riddle contends that the reproductive cycle of birds can be divided into two phases, an ovulatory phase dominated by FSH and LH and an incubatory/parental phase dominated by prolactin, a pattern reminiscent of the reproductive cycle in some mammals (Schooley and Riddle, 1938). The critical questions to be answered are:

a. does progesterone and/or oestrogen induce incubation?
b. is there any evidence of increased prolactin secretion at the onset of the incubatory phase?

These two basic questions are too often confused with the action of prolactin on the crop-sac (p. 112) and on brood-patch development (p. 109). A survey of the literature reveals that the role of prolactin in *maintaining* broodiness is not in question; the point at issue is whether or not prolactin *initiates* broodiness.

Lehrman (1958) isolated mature sexually-experienced ring doves for three to five weeks in small cages. During the final week, birds were injected with progesterone or oestrogen, or acted as controls. All progesterone-injected birds incubated immediately. In the oestrogen-treated birds there was some incubatory behaviour after one to three days, but some birds did not commence incubation until 11 days after the final injection. Lehrman also states that there was no increase in crop-sac weight associated with the onset of incubation. However, earlier work (Riddle and Lahr, 1944) using progesterone implants, showed that under this experimental regime incubation did not commence for three to seven days following treatment, by which time prolactin secretion had increased and was maintaining a developed crop-sac. Riddle (1963) argues that the stimulation of the crop-sac in the short time-span of Lehrman's experiments would not be detectable as increased weight, but would be apparent only as an increase in the mitotic rate. Lehrman (1963) quotes the work of Sacki and Tanabe (1955) who showed for domestic chickens, that pituitary prolactin levels only remained high while the bird was sitting on the eggs. Even if the bird was sitting on sterile eggs for a considerable period of time, prolactin levels remained elevated. This suggests that stimulation of the brood-patch by contact with the egg may promote prolactin secretion. The implication for Lehrman's hypothesis is that prolactin production only increases markedly *after* incubation begins. Lehrman (1963) also argues that the increased mitotic rate in the crop-sac observed by Lahr and Riddle (1938) in incubating doves occurred later than would have been expected if prolactin were released at the onset of incubation, since prolactin causes increased mitotic division in the crop-sac as early as two hours after injection. More recently work by Komisaruk (1967) has tended to confirm Lehrman's view, by showing that progesterone implants in various sites in the brain can induce parental behaviour. Despite all the controversy about the initiation of broodiness, a *maintenance* role for prolactin would now be accepted by most workers in this field. Certainly, many studies have shown that prolactin levels increase during the earlier phases of incubation. As Fig. 5.13 shows, in the chicken, pigeon and pheasant, pituitary prolactin levels are highest in the egg-incubation period.

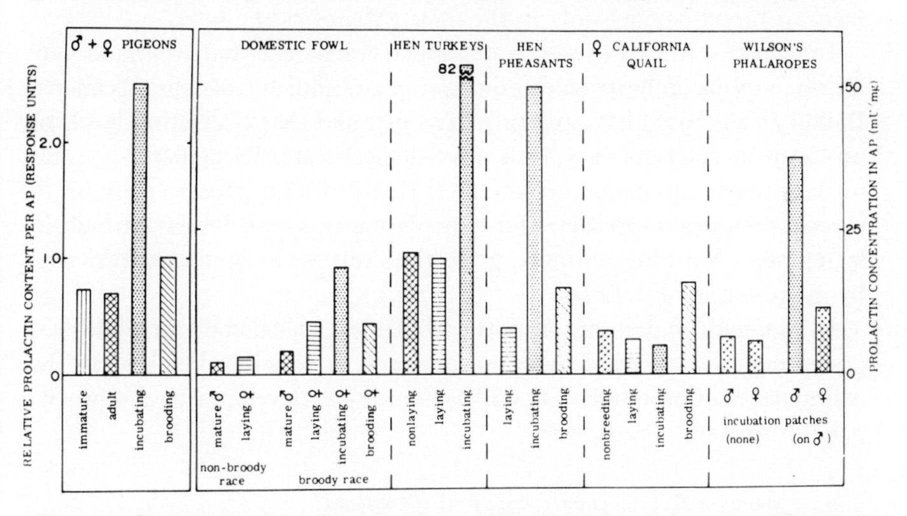

Fig. 5.13 Prolactin levels in adenohypophyses (AP) of birds in relation to incubation and brooding behaviour. The local crop-sac assay results in response units for domestic fowl and hen pheasants were converted to approximate mu/mg equivalents of prolactin, using the standard curve obtained with ovine prolactin. This was done to allow more meaningful comparison of the data among the species. In all cases except pigeons and phalaropes, results are expressed as milliunits of prolactin per milligram of adenohypophysial wet weight. The data on pigeon glands are in response units that could not be converted to milliunit equivalents, and results with phalarope glands are in milliunits per milligram of acetone-dried weight. In phalaropes prolactin levels are compared in animals of both sexes in breeding and nonbreeding seasons. In pigeons only males had incubation patches. (From Nicoll, 1974.)

Later, however, during the chick-brooding period, the pituitary content drops in all three species, so that even in the pigeon the pituitary levels are significantly lower when feeding the young than during the incubation period. This may seem somewhat surprising, since there is no question that prolactin stimulates the development of the crop-sac. However, the interpretation of pituitary hormone-content data is notoriously difficult without reliable figures for circulating levels and the reduced pituitary levels while feeding the young could perfectly well indicate an increased rate of release of prolactin in this phase. The problems in interpreting pituitary hormone levels are underlined by data from the Californian quail, in which pituitary levels were low during incubation and then increased rapidly after the young had hatched (Jones, 1971).

Wilson's phalarope, (*Steganopus tricolor*) presents an interesting case of reversal of the usual sexual roles. The female is larger and more brightly coloured than the male, and appears to play the more aggressive role in courtship behaviour. The male incubates the eggs by himself and plays the major role in brooding the hatchlings. This role is correlated with higher pituitary prolactin levels in the male, associated with the development of brood patches only in the male (Meites *et al.*, 1967).

Information from the recently developed homologous radioimmunoassay for avian prolactin bears on this discussion and adds further complications. Bolton *et al.* (1975) have recently demonstrated that circulating levels of prolactin in chickens vary with physiological state. Particularly relevant to the present discussion are the facts that prolactin appears to stand in inverse relationship to LH, and that plasma prolactin levels are high in laying hens. Most interestingly, prolactin levels were significantly elevated by progesterone injections.

In summary, at present the most reasonable conclusion is that while progesterone may initiate incubation, prolactin is certainly involved in the maintenance of broodiness, as well as in crop-sac development and brood-patch formation.

5.3.2 *Brood-patch (incubation-patch) formation*

For a detailed review of brood-patch formation in birds the reader is referred to the work of Jones (1971). Briefly, the brood-patch is a structure formed in certain species in relation to incubation of the eggs. Its formation involves the de-feathering of part of the ventral skin, followed by hyperplasia, increased vascularisation, and oedema of the area. The general pattern of changes involved are summarised in Fig. 5.14.

Although a number of earlier authors had suspected endocrine involvement in brood-patch formation, it was not until the work of Bailey (1952) that definite evidence became available. Bailey showed that oestrogen when given to intact birds produced changes resembling those in normal brooding birds. In hypophysectomised birds, however, oestrogen did no more than increase the vascularity of the area. Prolactin had to be given together with oestrogen in order to complete brood-patch development in hypophysectomised birds. Further work has shown that the hormonal control of patch formation to some extent varies with species, and three groups of birds can be distinguished:

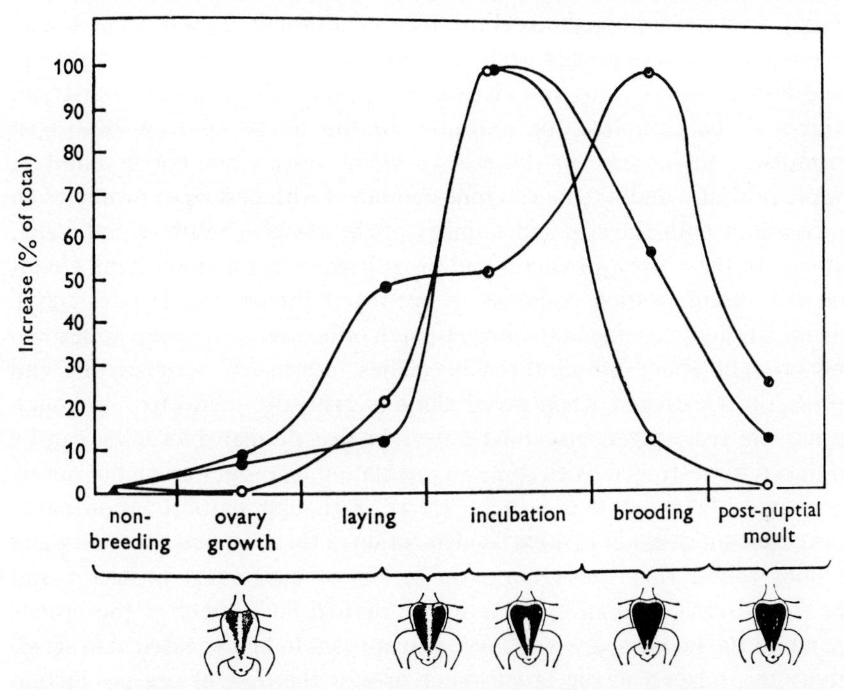

Fig. 5.14 Relative rates of processes during incubation-patch formation in the California quail. Increases over the nonbreeding condition are plotted as percentages of total increase. Degree of defeathering is represented by drawings of ventrum (clear regions medial and lateral to the stippling = down-like feathers of apteria; stippling = contour feathers; dark area = defeathered incubation patch; hatched area = refeathering patch). In typical patch formation in a passerine, the patch forms earlier in the cycle (i.e. during ovarian growth), and oedema occurs during incubation. ●, Thickness of epidermis; ○, number of blood vessels; ▲, diameter of blood vessels. (From Jones, 1971.)

a. *Only the female broods the eggs* In these species the skin of both sexes responds identically to hormone treatment, oestrogen causing a moderate de-feathering and prolactin augmenting the response (Bailey, 1952; Selander and Kinch, 1963; Steele and Hinde, 1963, 1964; Selander and Yang, 1966). Steele and Hinde (1964) showed that progesterone could also augment de-feathering in the domestic canary, but that the response to oestrogen by itself was the same in both intact and ovariectomised birds, which suggests that endogenous ovarian progesterone is not necessary for the full expression of the response to oestrogen. In other species, for example the red-winged blackbird, progesterone cannot augment the response (Selander and Yang, 1966). Testosterone has no effect in any of the species in this group. Epidermal hyperplasia in most species appears to be stimulated by oestrogen and the response is augmented by prolactin but not by progesterone (Selander and Kinch, 1963; Selander and Yang,

1966). In the red-winged blackbird, increased vascularity was promoted by oestrogen, and both prolactin and progesterone *opposed* this effect (Selander and Kinch, 1963). This vascularisation response shows species variation, and may be complex: for example, in the house sparrow oestrogen stimulates an increase in the size of blood vessels but not in number, while prolactin and/or progesterone combined with oestrogen promotes an increase in both the size and number of the vessels (Selander and Yang, 1966). In the canary, prolactin and progesterone had no modifying effects on the vascularisation response to oestrogen (Steele and Hinde, 1964). Stimulation of oedema in the brood-patch of birds in this group appears to be brought about by all three hormones, oestrogen, progesterone and prolactin. Oestrogen when given alone is generally ineffective, although again the red-winged blackbird differs in that dermal thickening can be induced by oestrogen, with some augmentation by progesterone but not by prolactin (Selander and Kinch, 1963). Although natural brood-patch development does not involve fat-deposition in the area, Selander and Yang (1966) found that the synergistic action of exogenous prolactin and oestrogen can cause an increase in the dermal fat content of the brood-patch in the house-sparrow. There also appears to be increased skin sensitivity to prolactin in the brood-patch area at the time of egg production (Hinde *et al.*, 1963), probably brought about by synergism of oestrogen (Hinde and Steele, 1964) with progesterone (Hutchinson *et al.*, 1967). All these findings, which indicate that ovarian steroids are responsible for the initial stages in brood-patch formation, could be taken as giving greater credence to Lehrman's views concerning the control of incubatory behaviour, since at the same time as they initiate development of the brood-patch, the ovarian hormones, according to Lehrman, would promote prolactin secretion by the pituitary (see pp. 107–108).

b. *Both sexes brood the eggs* Among birds in this category, detailed studies have been made on the European starling and the Californian quail, and the results for the two species are somewhat different. Oestrogen administered to the starling caused de-feathering, and the response was not altered by prolactin (Lloyd, 1965). However, in the quail Jones (1969*a*) found that oestrogen by itself caused only partial de-feathering, the response being augmented by prolactin. As Jones (1971) has emphasised, in the quail oestrogen does not cause the release of prolactin from the pituitary (Jones, 1969*a*), and pituitary prolactin levels are unusually low in this species (Jones, 1969*b*). Meier *et al.*, (1965) demonstrated in spring-migrating passerines that pituitary prolactin levels are high in prebreeding birds, and may remain high throughout the breeding season. This might explain the early development of the brood-patch in migratory passerines, such as the starling, oestrogens synergising with high endogenous prolactin levels during egg production. In the non-migratory quail, by contrast, prolactin levels would be high enough to synergise with oestrogens only later in the cycle (Jones, 1971). This concept would explain the difference between the two species in their responses to exogenous prolactin. Again, we may have here a situation in which high steroid levels in some birds *precede* any increase in prolactin secretion that may relate to brooding.

In both the starling and the quail, prolactin and oestrogen synergised to produce epidermal hyperplasia and increased vascularity, progesterone apparently playing no role in these processes; oestrogen appeared to play the dominant part, its effect being augmented to some degree by prolactin. The importance of prolactin is perhaps greater in the quail than in the starling (Lloyd, 1965; Jones, 1971).

c. *Only the male broods the eggs* Wilson's phalarope, *Steganopus tricolor*, is the only species in this category that has been studied. In this bird, testosterone, not oestrogen, synergises with prolactin in brood-patch development; oestrogen, when given together with prolactin and testosterone, appears to block the response (Jones and Pfeiffer, 1963). High prolactin levels have been found in the male phalarope during nesting (Höhn and Cheng, 1965; Nicoll *et al.*, 1967), and Jones (1971) has postulated that these high levels may be promoted by the presence of the female.

It would appear that fairly major changes in circulating hormone levels occur in birds at the onset of incubation. Prolactin/steroid relationsips at the onset of brooding behaviour have been discussed. Interestingly, Jones (1969*b*) has suggested that incubation and the physical contact of the ventral skin with the egg-clutch may stimulate prolactin secretion *via* a neural pathway. On this view, the system would resemble the mammalian mammary gland, in which the nipple, sensitised by oestrogen, responds to tactile stimuli which cause the release of prolactin (see Chapter 6, pp. 148–153). Support for Jones' suggestion comes from Holcombe (1970), who showed for the red-winged blackbird that individuals given model eggs will incubate bigger models for longer periods than smaller models. Furthermore, a number of studies have shown that the presence of even a single egg appears to suppress gonadotrophin release (Paluden, 1951; Weidman, 1956; Brockway, 1968). Thus the problem raised originally regarding the relationship between prolactin and progesterone in inducing broodiness (pp. 107–109) is now seen to be complicated by possible stimulatory effects of oestrogen on prolactin secretion, such as exist in mammals (Chapter 6), and perhaps by the existence of a direct neural release pathway for prolactin originating at the brood-patch and comparable to that in the mammary gland. Regarding the relationship between prolactin and progesterone, two possibilities exist: either prolactin can stimulate the production of progesterone by the postovulatory follicles, or progesterone can cause the secretion of prolactin. Either of these possibilities would explain the overlap in function between the two hormones. Lehrman (1963), for example, has postulated that progesterone might elicit prolactin release, but so could direct innervation from the oestrogen-sensitised brood-patch. On the other hand, Floquet *et al.*, (1963) have postulated a 'luteotrophic'-like effect of prolactin in birds, an idea opposed by Payne (1966). It would certainly appear that the relationship between prolactin and ovarian steroids in birds is as complex as in mammals (see Chapter 6).

Several authors have speculated about the possible evolutionary relationships of the avian brood-patch and the mammalian mammary gland, particularly having regard to the mammary/brood-patch complex in monotremes. There are certainly superficial similarities between the two structures (Long, 1969). Chadwick (1969) has suggested that the two reptile lines giving rise to the Aves and the Mammalia may have both possessed primitive brood-patches. This of course must remain as speculation, plausible, however, in view of the marked similarities in the developmental physiology of the two structures. These similarities may have evolutionary significance; they certainly illustrate the ability of prolactin to act as a wide spectrum integument-stimulating hormone.

5.3.3 *The crop-sac of Columbiformes*

The crop-sac of pigeons and doves is a remarkable structure, the source of 'pigeon's milk', the fatty mass of epithelial cells which both sexes regurgitate to feed the young. Riddle and Braucher (1931) first showed that maintenance of this process depends upon the pituitary gland. With the isolation of prolactin, Riddle *et al.*, (1933) were able to demonstrate that prolactin is the pituitary hormone involved. Beams and Meyer (1931) showed that changes in the crop-sac could first be detected, using histological techniques, on or about the eighth day of the incubation cycle. The results of Patel (1936) confirmed these observations. Presumably circulating levels of prolactin are high during this phase. In 1938, Lahr and Riddle applied the then new colchicine technique to study the proliferation of the pigeon crop-sac lining. They found that prolactin-induced mitoses are limited strictly to the basal layers of cells of the mucosa, and that the number of mitoses increases eight-fold in the 24 hours following laying of the first egg. Injections of exogenous prolactin were found to increase the mitotic rate about eight-fold within two hours. Females showed a higher mitotic rate than males.

Since this early work a number of reports have confirmed the role of prolactin in promoting proliferation of the crop-sac epithelium (Weber, 1962; Dumont, 1965; Meites and Nicoll, 1966). The response appears to involve increased phosphate uptake (Brown *et al.*, 1951; Damn *et al.*, 1961) and increased RNA and DNA content with a decrease in the RNA/DNA ratio (McShan *et al.*, 1950; Nicoll and Bern, 1968). Electron microscopic studies have demonstrated an increase in ribosomes and in fatty acid content of the epithelial cells 12 hours after prolactin injection (Dumont, 1965), and the increased fat content was also reported by Nicoll and Bern (1968). This response appears to be specific to prolactin, and has been shown to persist in the absence of gonadal, thyroidal or adrenal hormones (Riddle and Dykshorn, 1932; Schooley, 1937), although thyroxine, adrenocorticosteroids and GH, alone or in combination, augument the response in hypophysectomised pigeons (Bates *et al.*, 1962). More recently Frantz and Rillema (1968) have studied the uptake of ^{14}C-labelled amino acids and ^{3}H-labelled water by the pigeon crop mucosa. They found that a single prolactin injection stimulates water uptake by the mucosa in a dose-dependent manner, with two peaks: at four hours, probably related to increased capillary permeability, and at 15 hours, probably related to mucosal proliferation. Uptake of amino acids increases at 19 hours after the injection, and the amino acids are incorporated into peptides, indicating that protein synthesis is enhanced by this time.

Subsequent work by Garrison and Scow (1973) has shown that prolactin also stimulates lipoprotein lipase activity in crop-sac mucosa. The increase occurs on the third and fourth days of prolactin injection, a similar time-course of events to that in the mammary gland of the pseudopregnant rabbit (Falconer and Fidler, 1970). Lipoprotein lipase also occurs in the crop 'milk', but at a much lower titre than in mammalian milk. The epithelial cells shed in crop 'milk' contain numerous lipid droplets, mostly triglyceride (Dumont, 1965) and have a fat content (12%) about twice that in the actual crop-sac (Chadwick and Jordan, 1971). Dumont's work demonstrated a gradual increase in the fat content of the crop-sac after prolactin injection. At about 12 hours after injection small droplets of lipid appear in the nutritive layer. After two days these have increased in size and large droplets also appear in the proliferative layer. By 4 days all layers of the crop-sac are laden with fat droplets. These changes suggest that the mucosa abstracts triglyceride from the blood to produce the fat contained in the crop-sac 'milk' (Garrison and Scow, 1973). The enzyme changes presumably relate to the changes in RNA levels already mentioned. The response of the crop-sac to prolactin is blocked by actinomycin and puromycin, both agents which prevent RNA synthesis (Sherry and Nicoll, 1967). Thus, the first action of prolactin is probably to promote the production of new RNA, and indeed RNA extracted from a stimulated crop-sac will promote mucosal proliferation and fat deposition when injected into an unstimulated crop-sac.

Table 5.4 Effect of extracts of acetone-dried lungfish and tetrapod pituitary glands on crop-sac mucosa (from Nicoll and Bern, 1965)

Species	*No. of pigeons*	*Mg tissue per bird*	*Increase in mucosal parameter* Dry wt. (mg)	*Fat (mg)*	*RNA (μg)*	*DNA (μg)*
Lungfish (*Protopterus aethiopicus*)	7	7·3	12·1±1·8	2·5±0·42	940±126	137·9±18·8
Ambystoma tigrinum	6	2·0	9·8±1·1	1·2±0·05	441± 50	166·3±14·4
Necturus maculosus	6	3·5	9·1±0·8	1·9±0·17	731± 55	126·2±10·8
Rana pipiens	6	11·6	11·2±1·5	1·5±0·3	—	—
Bufo marinus	5	9·2	6·9±2·3	1·2±0·35	396± 38	76·1±20·5
Rana catesbeiana	6	6·2	9·2±1·7	1·0±0·3	437± 37	154·3±32·4
Turtle (*Pseudemys scripta elegans*)	6	3·3	9·0±1·1	1·8±0·27	682± 82	129·6±12·5
Pigeon (*Columba livia*)	5	2·2	7·7±0·7	1·9±0·03	538± 55	54·9± 5·3
♂ Rat	6	4·2	9·9±1·8	1·8±0·35	—	—
♂ Guinea pig	6	2·8	10·2±1·2	2·2±0·82	827± 53	122·2±10·2
Pigeon+Pollack	6	2·3	8·2±1·6	1·7±0·5	—	—

The response of the pigeon crop-sac is the favoured bioassay for prolactin, and is particularly used in work in lower vertebrates for which radioimmunoassay systems are generally not available. A wide range of tetrapod prolactins have been shown to stimulate the pigeon crop (Chadwick, 1966; Nicoll and Bern, 1968; see Table 5.4), but there is still considerable doubt that teleost prolactins can produce a true response (see Chapters 1 and 2).

In addition, Burns and Meier (1971) have shown that there are marked daily variations in the crop-sac response to prolactin. In pigeons kept on a 12-hour photoperiod, prolactin is most effective nine hours after the onset of the light phase, but prolactin given at dawn has virtually no effect. As with other avian responses to prolactin (see p. 118) the crop-sac response can be entrained by corticosterone injections, the greatest effect of prolactin occurring 18 hours after corticosterone administration (Meier *et al.*,

1971). The crop-sac rhythm appears to be truly circadian, in that it free-runs with a periodicity of 24 hours in continuous light (John *et al.*, 1972). As with other circadian rhythms it 'damped-out' after 10 days, but could be restored by injections of thyroxine. This suggests that both corticosteroids and thyroxine are involved in determining the diurnal variations in the response to prolactin, and it is obviously essential that these factors are taken into account in dealing with assay data.

5.4 Gonadal effects of prolactin in birds

As in other vertebrates, particularly mammals, the functioning of the avian pituitary-gonadal axis alters at the beginning of the parental phase. To what extent this is due to enhanced prolactin secretion in the parental

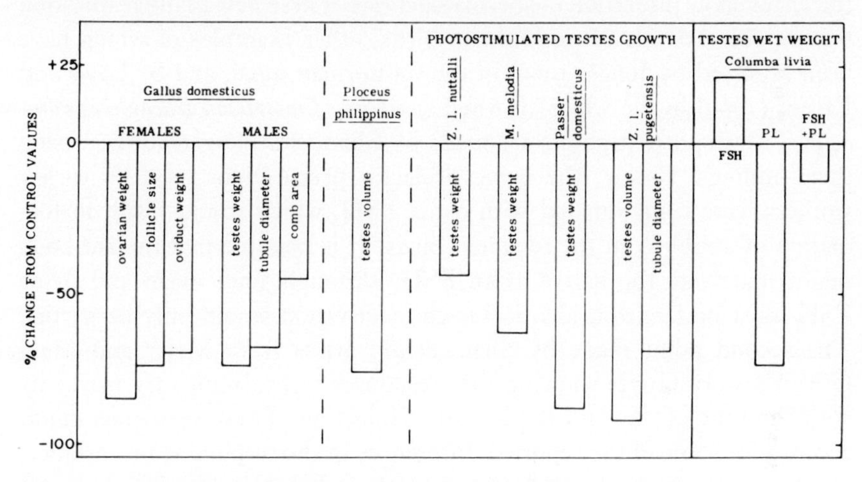

Fig. 5.15 Antigonadal effects of prolactin in birds. (From Nicoll, 1974.)

phase is not known. Jones (1971) has postulated that stimulation of the brood-patch by the clutch may directly inhibit gonadotrophin release. In addition, a number of workers have demonstrated both direct and indirect inhibitory effects of prolactin on the gonad. This anti-gonadal action was first postulated by Schooley and Riddle (1938), who found that the pituitary FSH content in the pigeon decreased during the early stages of incubation, at which time prolactin levels were increasing. Bates *et al.* (1935) had previously found that prolactin injections decreased the net weight of the testes and ovary, producing a reduction in the size of the ovarian follicles and a decrease in tubular diameter in the testes, responses indicative of a decrease in FSH secretion. Most of the antigonadal effects of prolactin, which are summarised in Fig. 5.15, can in fact be reversed by FSH injection. For example, Nalbandov (1945) reported that injection of prolactin into cocks (*Gallus domesticus*) causes a decrease in testis weight, tubular weight and androgen production, resulting in a decrease in comb size and cessation of crowing. A single injection of FSH opposed this testicular regression, suggesting that prolactin inhibits FSH secretion. However, Shani *et al.* (1973) recently re-examined these results using a more purified preparation of ovine prolactin, and they failed to confirm Nalbandov's findings. They used a variety of experimental procedures with injection periods ranging from 10 to 22 days and prolactin doses from 8 to 150 iu/day. None of the experimental regimes altered the body weight, testes weight, or comb development. Blood levels of testosterone were variable, but no evidence of decreased production was found in the prolactin-treated birds, and neither did histological examination of the testes show inhibition of spermatogenesis. These new findings question the antigonadal role of prolactin in birds, other examples of which have been reported by Jones (1969) in the Californian quail, and by Laws and Farner (1960) in the white-crowned sparrow, *Zonotrichia leucophrys gambelii*. Two important points are made by Shani *et al.* (1973) in discussing their findings. Firstly, the ovine prolactin preparations used by earlier workers were contaminated with ovine FSH, which would result in formation of antibody in the recipient birds. It is possible that the antibody might inactivate the native avian FSH, although since avian and ovine FSH must be far from identical such inactivation would only be partial. The second point made by Shani *et al.*, arises from Meier and MacGregor's work (1972) showing that responses to prolactin vary markedly with the time of day and the season of injection. These variations might account for some of the reported differences in the responses, for instance, of *Gallus domesticus* to prolactin, and the whole question of the antigonadal response, in this species particularly, is in need of further investigation.

Meier and Dusseau (1968) have reassessed the work of Law and Farner (1960) on the effect of prolactin on photoperiodically-induced gonadal growth in *Zonotrichia leucophrys gambelii*. They found that prolactin and ACTH (a possible contaminant of earlier preparations) had no inhibitory effect on gonadal growth. The response to prolactin was tested over a wide range of injection times and in combinations with ACTH, and all treatments proved ineffective. However these authors then went on to test the effect of prolactin on a range of species, classified with regard to their migratory behaviour as strong migrants, weak migrants, or non-migrants. The results showed that in three strongly migratory species, *Z. leucophrys gambelii*, *Z. albicolis* and *Junico hyemalis*, prolactin had no inhibitory effect on photoperiodically-induced testicular growth. However, in a weak migrant *Melospiza melodia merrelli*, and in two nonmigrants *Z. leucophrys ruttalli* and *Passer domesticus*, prolactin had an inhibitory effect on testicular development. Meier and Dusseau (1968) therefore concluded that prolactin is effective in blocking testicular growth only in non-migratory species. The relevant literature is surveyed in Table 5.5.

Table 5.5 The effect of prolactin on gonadal weight in birds

Species	*Photo-periodic gonadal growth*	*Mature gonad*	*Authority*
Strong migrants			
Zonotrichia leucophrys gambelii	None	N.T.	Laws and Farner (1960)
			Meier and Dusseau (1968)
Z. albicolis	None	?	Meier and Dusseau (1968)
Junco hyemalis	None	N.T.	Meier and Dusseau (1968)
Weak migrants			
Z. leucophrys pugetensis	Inhibitory	N.T.	Bailey (1950)
Melospiza melodia	Inhibitory	N.T.	Meier and Dusseau (1968)
Non-migrants			
Gallus domesticus	N.T.	Inhibitory	Nalbandov (1945)
	N.T.	None	Shani, Furr and Forsyth (1973)
Columbia livia	N.T.	Inhibitory	Bates, Riddle and Lahr (1937)
Fringilla coelebs	N.T.	Inhibitory	Lofts and Marshall (1956)
Passer domesticus	Inhibitory	Inhibitory	Lofts and Marshall (1956)
			Meier and Dusseau (1968)
Ploecus philipinus	N.T.	Inhibitory	Thapliyal and Saxena (1964)
Z. leucophrys ruttalli			

N.T. = not-tested.
Modified from Meier and Dusseau (1968).

Evidence discussed later implicates prolactin in the induction of the spring premigratory restlessness and fat deposition in migratory species of *Zonotrichia*. Since gonad development starts at this time, it seems obvious that prolactin should not be anti-gonadal in these migrants. However, Meier and Dusseau (1968) put forward an alternative hypothesis to explain the data in Table 5.5. They suggest that in the strong migrants gonadal growth may in fact be partially suppressed by *endogenous* prolactin, so that the effects of adding exogenous prolactin to the system would be undetectable. Prolactin has never been shown to produce *total* inhibition of photoperiodically-induced gonadal development, and they suggest that the natural situation may be different from the obvious interpretation of Table 5.5. Meier and Dusseau suggest that migratory birds, which have to delay full gonadal maturation until they reach the nesting sites (Wolfson, 1959), might rely on high levels of endogenous prolactin to retard gonadal development during the migratory period. In contrast, non-migrants, which breed as soon as environmental conditions are suitable, would have low prolactin levels, permitting rapid development of the gonads, and so allowing the anti-gonadal effect of exogenous prolactin to be detected. This ingenious hypothesis, which accommodates the concept of a general anti-gonadal role for prolactin, should be tested in order to clarify this confused field.

Another role which has been suggested for the anti-gonadal action of prolactin is in the induction of photo-refractoriness following the reproductive period (Farner, 1969). The photo-refractory period is essentially characterised by regression of the gonads, loss of body weight and a post-nuptial moult. In migratory species it is often found to precede the autumnal migration. Juhn and Harris (1956) showed that moult can be induced in capons by prolactin injections and they suggested that high levels of prolactin during the brooding period may result in moulting at the end of the nesting phase. Objections to this hypothesis have been raised, mainly on the grounds that such a mechanism could not operate in species which have repeated multiple nestings. More recent work by Meier and MacGregor (1972) and Meier (1973, 1975) has suggested that interactions between circadian rhythms of prolactin and corticosterone might play a major part in initiating photo-sensitivity and photo-refractoriness in some avian species. These authors envisage photo-sensitivity and photo-refractoriness in birds as ensuing from a mechanism similar to that previously postulated by Pittendrigh and Minris (1964), in which light acts both as an entraining agent for the circadian oscillation of photo-sensitivity and also as the photo-periodic inducer. In consequence, a rapid increase in light (dawn) may entrain a system to increased photo-sensitivity later in the day. Some hormones also appear to be capable of entraining light-sensitivity: for example, Meier and Dusseau (quoted by Meier and MacGregor, 1973) showed that corticosterone injected 18 hours before the onset of a six-hour light period caused gonadal development in *Zonotrichia albicolis*. Recent work by Murton and his co-workers (Murton *et al.*, 1969; Murton, Lofts and Westwood, 1970; Murton, Lofts and Orr, 1970) has suggested that two photo-inducible phases may occur in birds, one for LH secretion and one for FSH secretion.

Meier (1969) has demonstrated both anti-gonadal and pro-gonadal effects of prolactin in the white-throated sparrow, *Zonotrichia albicollis*. When prolactin was injected simultaneously with FSH and LH in photo-refractory birds, the gonadotrophin-induced increases in ovarian and oviducal weights were reduced or blocked. In contrast, if prolactin was injected 10 hours before the gonadotrophins, it augmented the ovarian and oviducal responses. More recently, Stetson *et al.* (1973) apparently failed to confirm Meier's work, although the disagreement may be more apparent than real. Working with white-crowned sparrows (*Zonotrichia leucophrys gambelii*), these workers found that the simultaneous injection of prolactin did not retard gonadotrophin-induced gonadal responses, nor did prolactin retard photo-stimulated gonadal growth. However, prolactin was injected at only one time in the 24-hour cycle; we are dealing here with a different species, and further extensions of this work may well produce agreement in essentials with the findings of Meier.

As a result of extensive experimental work, Meier and his school (Meier, Burns and Dusseau, 1969; Meier and Martin, 1971; Dusseau and Meier, 1971; Meier, Martin and MacGregor, 1971; Meier *et al.*, 1971) have produced a hypothetical scheme, based on temporal relationships between circadian rhythms in corticosterone and prolactin, which can reconcile many of the discordant findings in this field. Justification for using these experimental consequences of altering the time interval between injections of the two hormones to explain natural events comes from measurements of blood levels of corticosterone (Dusseau and Meier, 1971) and prolactin (Meier, Burns and Dusseau, 1969) in *Zonotrichia albicollis.* The results showed that in May during the spring migratory period, when the birds are photo-responsive, the interval between the daily rise in corticosterone levels and the release of prolactin is 12 hours. However in August, during the photo-refractory period, the interval between the two events is about six hours.

Most of Meier's work has been on the white-throated sparrow, *Zonotrichia albicolis* and generalisations must obviously be made with caution. However, many of his findings do justify re-examination of earlier controversial work on other species with particular reference to the temporal interactions of the various hormones involved.

Male birds exhibit a marked photo-stimulatory activation of the pituitary prolactin cells which is related to the testicular cycle. In the male duck pituitary prolactin content shows a pronounced annual cycle which parallels the cycle in testicular growth, the two parameters reaching simultaneous maxima, and furthermore exposure of ducks to continuous light in the autumn both increases the prolactin levels and induces testicular growth (Assenmacher *et al.*, 1962; Gourdji and Tixier-Vidal, 1966; Gourdji, 1970). Similar findings have been reported for two species of quail (Jones, 1969; Gourdji, 1970; Alexander and Wolfson, 1970). Castration tends to lower prolactin levels in ducks on long day-lengths, but has no effect on short day-length birds (Gourdji, 1970).

5.4.1 *Prolactin and accessory sexual structures in birds*

In most birds other than some raptors, the right oviduct and the right ovary remain vestigal. Like the left ovary, the left oviduct undergoes seasonal proliferation and growth prior to egg production. These changes have been studied in fowls (Richardson, 1935) and in great detail in the domestic canary by Hutchinson *et al.* (1967). Oviducal development occurs in parallel with the growth of the ovary and the weights of the two organs are linearly related (Hutchinson *et al.*, 1967). The major part of oviducal development appears to be stimulated by ovarian oestrogens (Witschi and Fuge, 1940; Brant and Nalbandov, 1956; van Tienhoven, 1961), and oestrogens have been shown to cause marked morphological changes in immature chick oviducts (Hertz *et al.*, 1947) and to stimulate the differentiation of mature oviducal cell types (Koller *et al.*, 1969). Other hormones may increase the response of the oviduct to oestrogens: thus, Lehrman and Brody (1957) have shown that prolactin and progesterone both augment the oviducal response to oestrogens in ring doves, and Hutchinson *et al.* (1967) made similar observations in the domestic canary. It must be noted, however, that Hutchinson *et al.* (1967) found little correlation between oviducal development and either brood-patch formation or nest-building activity, both of which may be regarded as indices of prolactin secretion by the pituitary (Jones, 1971). There was, however, a temporal correlation between albumin formation and brood-patch development, suggesting that prolactin may perhaps play a physiological role in augmenting the later stages of oviduct development and secretory function in response to steroids.

5.5 The role of prolactin in avian migration

We have seen in the previous section that rhythms in prolactin secretion entrained by corticosterone alter fat deposition and physical activity in white-throated sparrows. The possible relationships between gonad suppression, time of breeding and migration were also discussed briefly. The role of prolactin in avian migration must now be examined in some detail. Prolactin has been shown to be involved in movements of amphibians (Chadwick, 1941) and migration of sticklebacks to freshwater (Lam and Hoar, 1967). In both cases the process includes a preparative 'predisposition' to migration as well as certain specific behavioural changes.

The work of Meier indicates that prolactin may have a highly specific role in controlling migration in certain avian species. In the white-crowned sparrow, autumn migration is correlated with increased physical activity, and the rapid accumulation of lipid reserves (King and Farner, 1958). Similarly in the spring, migration follows as part of a well-ordered sequence of events involving pre-nuptial moult, gonadal development, pre-migratory fattening and increased activity, all of which factors appear to be stimulated by increasing day-length (King and Farner, 1960). Farner (1955, 1960) has shown that related non-migratory populations do not show the fattening response. Prolactin appears to be involved in at least two dissociable aspects of avian migration, the pre-migratory fattening and the pre-migratory physical behaviour.

5.5.1 *Control of pre-migratory fattening*

Dolnik (1961) found that photo-stimulated castrated *Carduclis flammea* had greater fat deposition than photo-stimulated intact birds. In contrast, Morton and Mewaldt (1962) showed that while fattening could be induced by photo-stimulation in castrated *Zonotrichia atricapilla*, the response was less than if the gonads were present. This earlier work suggested that although gonadotrophins could be directly involved in the fattening response, gonadal steroids are not essential, though in some species they may be permissively involved in fattening (Schildmacher and Steubling, 1952; Helmo and Drury, 1960).

It was not until 1964 that Meier and Farner were able to demonstrate that prolactin promotes the pre-migratory fattening in the white-crowned sparrow, *Zonotrichia leucophrys gambelii.* They showed that fattening could be induced by prolactin in both photo-sensitive and photo-refractory individuals, if the birds were maintained on a long day-length. Furthermore, fattening could be induced in photo-sensitive birds even on a short day-length (see Fig. 5.16). Prolactin was also shown to increase the diurnal gain in net weight (see Fig. 5.17). The fact that increased photo-

Table 5.6 Lipid indices of *Zonotrichia leucophrys gambelii* treated with hormones during the refractory period (from Meier and Farner, 1964)

Treatment	*No.*	*Mean*	*SEM*
Control	5	5·7	0·6
Prolactin[a]	5	12·2	3·2
Gonadotrophin[b]	4	4·6	0·4
Prolactin[a]+gonadotropin[b]	7	18·8	1·2
Prednisone[c]	5	8·8	0·9
Prolactin[a]+prednisone[c]	4	13·8	4·1
Gonadotropin[b]+prednisone[c]	4	9·2	3·8
Prolactin[a]+gonadotropin[b]+ prednisone[c]	3	16·1	4·0

Daily dosages (μg) of hormones (30 July–21 August, 1963): [a]300; [b]150 LH and 150 FSH, each; [c]200.

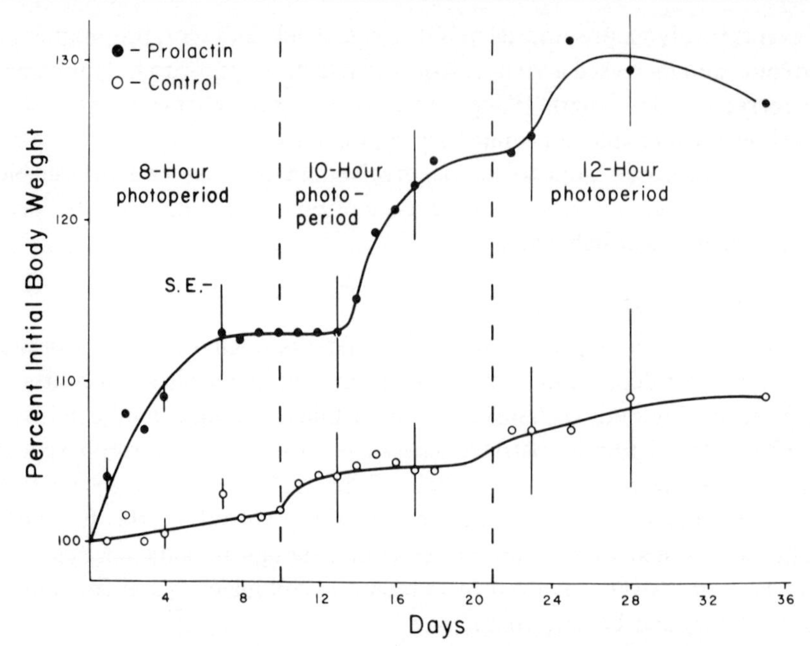

Fig. 5.16 The effect of photoperiod on weights of prolactin-treated photosensitive *Zonotrichia leucephrys gambelii*. Treatment consisted of daily intramuscular injections of 667 μg ovine prolactin beginning 25 February, 1963. ●, Prolactin; ○, control. (From Meier and Farner, 1964.)

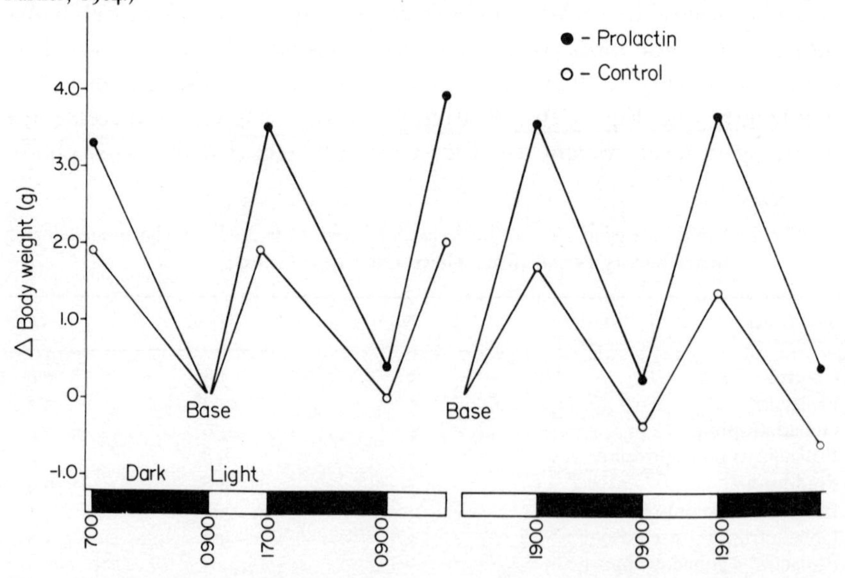

Fig. 5.17 The effect of prolactin on changes in body weight of *Zonotrichia leucophrys gambelii* between daily light and dark period. Treatment consisted of daily intramuscular injections of 667 μg ovine prolactin beginning 25 February, 1963. ●, Prolactin; ○, control. (From Meier and Farner, 1964.)

period augmented the response to prolactin led Meier and Farner to believe that gonadotrophins might also be involved, and they then showed that gonadotrophins, although not themselves the cause of fat deposition greatly increased the response to prolactin (see Table 5.6). This increase in fat deposition is associated with hyperphagia (King and Farner, 1956; Odum, 1960), and hyperphagia has also been induced in hypophysectomised pigeons by prolactin treatment (Schooley *et al.*, 1941). Thus, prolactin released from the pituitary under photo-stimulation may increase both food intake and fat deposition.

Not all species necessarily resemble *Zonotrichia* in the control of their fattening responses; for example, Dolnik (1961) showed that prolactin antagonised the fattening induced by gonadotrophins in *Carduclis flammea*.

5.5.2 *Fattening in non-migratory birds*

Although not directly related to bird migration, there is now considerable evidence that prolactin also alters lipid metabolism in non-migratory species. This has been more thoroughly studied than fattening in migratory species and a discussion of the findings might throw some light on the processes in migrants.

Schooley *et al.* (1937) at an early date had shown that prolactin could stimulate hyperphagia and weight increase in hypophysectomised pigeons. More recently Ensor *et al.* (1972*b*, see p. 98) have shown that a similar effect occurs in the duck in which species the response is also associated with hyperdipsia.

The mechanism of this response to prolactin is not clear. Results obtained by Goodridge (1964) suggested that prolactin cannot directly stimulate *in vitro* fat synthesis in the liver or fat-bodies of the white-crowned sparrow. In contrast, Goodridge and Ball (1967) found that prolactin can induce fatty acid synthesis in the liver of the pigeon. Taking livers from prolactin-treated and control birds and studying them *in vitro*, these workers showed that prolactin promoted a rapid incorporation of ^{14}C-glucose, pyruvate or acetate into fatty acids, and also produced an increase in liver size and elevated the levels of certain hepatic enzymes, especially malic dehydrogenase. These effects of prolactin appeared to be dependent on food intake, since livers from fasted birds were unresponsive; however, simultaneous fat changes in the crop-sac occurred independently of feeding. Thus, the liver responses may perhaps be produced not by the direct action of prolactin but by increased food intake induced by the hormone ('prolactin hyperphagia', p. 121). However Garrison and Scow (1975) have demonstrated a direct effect of prolactin on the enzyme lipoprotein lipase in pigeon adipose tissue. This enzyme regulates the uptake of triglycerides by extrahepatic tissues. The response appears to be tissue-specific in that although prolactin-stimulated triglyceride uptake by adipose tissue and by the crop-sac, it did not affect the levels of lipoprotein lipase in the oesophagus.

The observed responses to prolactin may involve initially an increase in food intake (hyperphagia), possibly as the result of the changed activity of specific higher brain centres (Soulairac, 1958). Consequent upon hyperphagia are non-specific changes in hepatic activity, although the possibility exists that prolactin could have direct hepatic effects, yet to be discovered. The hepatic changes presumably result in increased levels of triglycerides in the liver and in the blood, which are then removed from circulation into the adipose tissue as a result of prolactin stimulation.

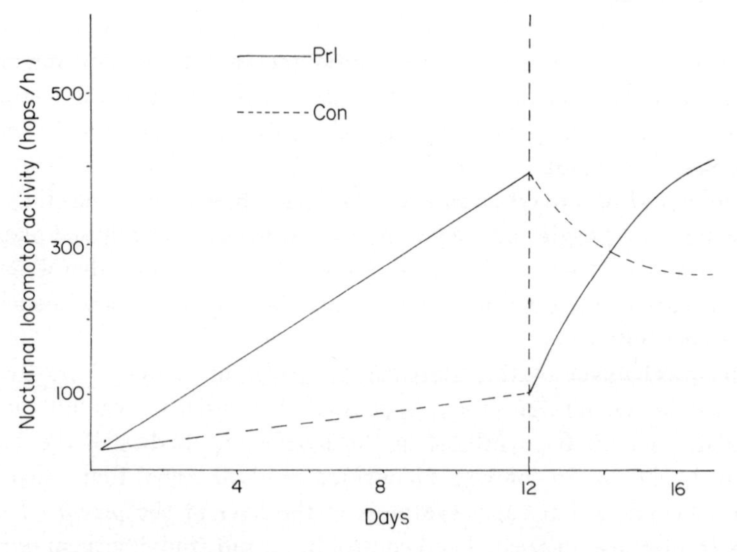

Fig. 5.18 The effects of prolactin on nocturnal locomotor activity in *Zonotrichia leucophrys gambelii*. (After Meier *et al.*, 1965.)

5.5.3 *The effect of prolactin on migratory behaviour*

Meier *et al.* (1965) found that prolactin stimulates the nocturnal physical activity of photo-sensitive white-crowned sparrows maintained outdoors in cages prior to the spring migration (see Fig. 5.18). Prolactin also stimulates nocturnal activity in photorefractory birds maintained indoors on a long day-length regime. These effects appear to be augmented by adrenocortical steroids which by themselves are ineffective. These effects may not be general to all birds, however: previous work by Wagner (1956, 1957) and

Wagner and Thomas (1957) had suggested that prolactin could *suppress* physical activity peaks in *Sylvia curruca* and *S. atricapilla*.

As discussed earlier, it appears that rhythmic prolactin secretion alters physiologically in relation to peak corticosterone levels (Meier, 1973). An interval of a few hours between corticosterone and prolactin injections produces a situation not unlike that in the autumn migration with a southerly orientation of the migratory activity and increased fat stores. With a 12-hour injection interval the fat stores are again increased, but the pre-migratory movements switch to a northerly direction, appropriate to the northward spring migration. This work suggests that prolactin not only controls activity in the white-throated sparrow but also has a directional effect depending on its temporal relationship with the corticosterone rhythm.

5.6 Cellular source of prolactin in birds

As in all vertebrates, prolactin lacks a specific endocrine target organ, which could be surgically removed to produce cytological changes in the pituitary to identify the prolactin cells. However, changes in certain pituitary cells can be correlated with variations in both pituitary prolactin content and in the development of various structures known to be stimulated by prolactin. As we have already seen, the prolactin content of the pituitary is elevated during care of the young and associated brood-patch development in the hen (Nakajo and Tanaka, 1956), the pigeon (Schooley, 1937), the Californian gull, (Bailey, 1952), the ring-necked pheasant (Breitenbach and Meyer, 1959) and the turkey (Cherms *et al.*, 1962). There are also alterations in pituitary prolactin levels induced by saline-loading in ducks (Ensor and Phillips, 1970) and cytological changes in the pituitary of salt-loaded hens (Tai and Chadwick, 1976). Najako and Tanaka (1956) showed that while in non-brooding hens prolactin potency was highest in the cephalic lobe of the adenohypophysis, the converse was true in brooding birds. Pituitary prolactin content varies with the reproductive cycle in relation to gonadal growth, and is elevated by photo-stimulation in a number of species (duck: Gourdji and Tixier-Vidal, 1966; white-crowned sparrow: Meier and Farner, 1964; Japanese quail: Gourdji, 1970). Pituitary prolactin content has also been shown to vary thoughout the day (Ensor and Phillips, 1971; see Fig. 5.19), in relation to the onset of the photo-period. These results are similar to those of Meier and Dusseau (1968).

In light microscopic studies, cytological changes relating to physiological events have been described in the pigeon (Schooley, 1937; Tixier-Vidal and Assenmacher, 1966), the quail (Tixier-Vidal *et al.*, 1968) and the hen

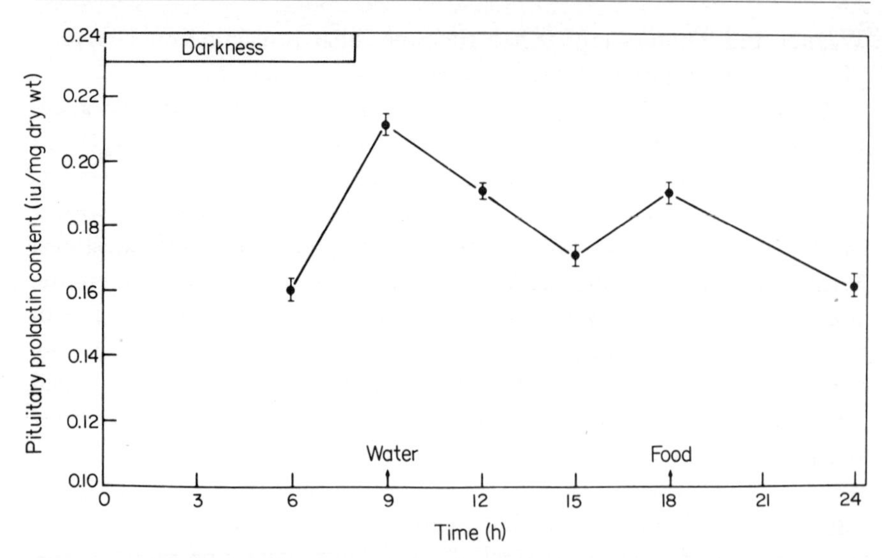

Fig. 5.19 The effect of daily photoperiod on the pituitary prolactin levels in the duck *Anas platyrhuchos*. Six birds per group.

(Payne 1955; Legait and Legait, 1955; Perek *et al.*, 1957; Inoguchi and Sate, 1962). Broodiness in the hen induces regression of the basophils and an increase in the acidophils of both the cephalic and caudal lobes. The cephalic acidophils appear to be most closely correlated with changes in prolactin release. These cells are active in pigeons undergoing crop-sac development and are erythrosinophilic (Tixier-Vidal and Assenmacher, 1966) like the prolactin cells in other groups (Chapter 1). Clumps of these cells have also been observed in the caudal lobe of the pituitary of some species.

Electron microscopic studies have shown in the duck (Tixier-Vidal and Gourdji, 1965) that the erythrosinophilic cells are ultrastructurally similar to rat prolactin cells, and that they release prolactin at a constant rate *in vitro* (Tixier-Vidal and Gourdji, 1972). Further studies showed that the prolactin cells could be photo-stimulated, and that they became degranulated following reserpine treatment (Tixier-Vidal, 1965). Mikami *et al.* (1969) have also demonstrated photo-stimulation of prolactin cells in *Zonotrichia leucophrys gambelii*.

5.7 Hypothalamic control of prolactin in birds

The main body of evidence suggests that in birds, in contrast to other vertebrates, prolactin release is primarily under the control of a hypothalmic stimulatory factor, prolactin releasing factor (PRF) (Meites and Nicoll, 1966; Pasteels, 1967; Nicoll and Fiorindo, 1969; Nicoll *et al.*, 1970).

This view mainly derives from *in vitro* studies: however, Baylé and Assenmacher (1965) have shown that autotransplantation of the pars distalis in pigeons results in the crop-sac remaining quiescent, indicating that there is no rapid release of prolactin from the transplanted gland such as occurs in mammals (see Chapter 6).

In vitro cultures of pituitaries from the pigeon (Meites *et al.*, 1963; Nicoll, 1965; Gala and Reece, 1965; Gourdji, 1970), the tricolored blackbird (Nicoll, 1965), and the quail (Gourdji, 1970) gave no evidence of autonomous prolactin secretion. In the case of the duck pituitary, (Tixier-Vidal and Gourdji, 1972). These observations correlate with the continued linear while in the pigeon the amount released into the medium was less than twice that found in the fresh gland (Meites *et al.*, 1963), contrasting with the usual rate for mammals of about 10 times the fresh pituitary content. Similar results were obtained for the quail and the tricolored blackbird (Meites *et al.*, 1963).

Ultrastructural studies on the cultured prolactin cells of the duck (Tixier-Vidal and Gourdji, 1965) indicated that granulation was still present, with exocytosis still occurring. The Golgi apparatus remained active, and the endoplasmic reticulum remained abundant. Autoradiography following pulse-labelling with radioactive leucine showed that the cells were still capable of synthesising protein (Tixier-Vidal and Picart, 1967; Tixier-Vidal and Gourdji, 1970). These observations correlate with the continued linear release of prolactin by such duck pituitary explants (Tixier-Vidal and Gourdji, 1972). In contrast, studies on pigeon and quail pituitaries *in vitro* suggested that in these species the prolactin cells remained quiescent (Tixier-Vidal, 1967, 1969, 1970), in agreement with their very low rate of prolactin release *in vitro* (see above).

Similar results were obtained with autografted pituitaries. In the pigeon (Tixier-Vidal *et al.*, 1968), after five months autotransplantation into the renal capsule the pituitary showed no evidence of prolactin cells. However, in the duck a similar experiment showed that the pituitary possessed numerous active prolactin cells (Tixier-Vidal *et al.*, 1968), although they did not hypertrophy to any marked degree. It would seem that in the pigeon, quail and tricolored blackbird the prolactin cells are definitely under a stimulatory control, whereas this is less clearly so in the duck, in which prolactin secretion seems more autonomous.

Hypothalamic extracts have been shown to stimulate prolactin secretion by the *in vitro* pituitaries of pigeon (Kragt and Meites, 1965), tricolored blackbird, (Nicoll, 1965), the pekin-duck (Gourdji and Tixier-Vidal, 1966), the quail and the chick (Kragt, 1966), and the turkey (Chen, 1967). Knight and Chadwick (1976) have also demonstrated an *in vivo* stimulation of the pigeon prolactin cells by hypothalamic extracts, assessed by increased development of the crop-sac. The evidence for the duck demonstrates that a degree of autonomy in prolactin secretion is not incompatible with hypothalamic stimulation. Tixier-Vidal and Gourdji (1972) have also demonstrated K^+-stimulated prolactin release from the duck pituitary in the presence of Ca^{++}. This is similar to K^+-induced release in the rat pituitary (Parsons, 1970; Macleod and Fantham, 1970) and suggests that the duck prolactin cell may have properties similar to those of its mammalian homologue. It also suggests that K^+ ions may depolarise the membrane as in mammals, where the main hypothalamic control is inhibitory. It may be that, like mammals, birds secrete a PIF as well as a PRF.

Hypothalamic extracts from photo-stimulated donors were found to stimulate strongly prolactin secretion by duck and quail pituitaries (Tixier-Vidal and Gourdji, 1972). It has also been shown that high circulating levels of testosterone inhibit prolactin release in the duck and the quail, without significantly altering the pituitary content, indicating an intracellular feed-back control (Tixier-Vidal and Gourdji, 1972).

More recently Hall *et al.* (1975), have shown that TRH can cause the release of prolactin from the pituitaries of hens and pigeons *in vitro*. These authors have also shown that TRH can cause prolactin release *in vivo* in the pigeon, as demonstrated by an increased proliferation of the crop-sac. Bolton *et al.* (1976) used a homologous RIA (radioimmunoassay) system to measure prolactin levels *in vivo* and have suggested that a dual control system (PIF and PRF) may exist for avian prolactin. Rat hypothalamic extracts known to contain high levels of PIF (dopamine) (MacLeod, 1975) were shown to be inhibitory in the presence of chicken hypothalamic extracts *in vivo* (Bolton *et al.*, 1976).

The suggestion has been made that prolactin in the duck may be released in direct response to plasma osmotic changes. This seems unlikely (Ensor, 1975) and it is more probable that prolactin release following saline-loading is mediated *via* a neuroendocrine reflex arc originating in receptors in the ventricles of the heart (p. 100).

6 Prolactin in mammals

In mammals prolactin is considered to be primarily a hormone concerned with reproduction, including the growth of the mammary gland and the secretion of milk. Other functions of prolactin which evolved in the lower vertebrates still persist, such as its influence in osmoregulation and parental behaviour. These, however, are of secondary importance compared to the major role now assumed by the hormone in reproduction.

The secretion of prolactin has been shown to vary during the life span of some mammals, notably the rat, and such variations may be important to the understanding of its role in the reproductive cycle. Circulating levels of prolactin in males are normally low (Amenomori *et al.*, 1970) and do not show cyclic changes, although there may be seasonal variations (Hart, 1974). The function of prolactin in males is considered on pp. 157–161. In the prepubertal female rat, serum levels are also low up to about day 37 (Voogt *et al.*, 1970) and then rise sharply with the onset of puberty and the establishment of the oestrus cycle. Prolactin secretion in the cycling female rat is characterised by a surge of prolactin at pro-oestrus (Sar and Meites, 1968), which may or may not persist into oestrus (Sar and Meites, 1968; Niell, 1970; Kwa and Verhofstad, 1967; Niswender *et al.*, 1969). A similar pattern has been observed in the goat (Bryant and Greenwood, 1968) and in cattle (Rand *et al.*, 1970). The pro-oestrus surge in the rat is believed to be triggered by increased secretion of oestrogen on day two of dioestrus. Prolactin secretion is also high during pseudo-pregnancy (Kwa and Verhofstad, 1967) and in the later stages of pregnancy prior to postpartum lactation. As will be seen later, prolactin levels also remain high throughout postpartum lactation in most species (Bryant and Greenwood, 1968; Karg and Schams, 1970; Frantz and Kleinberg, 1970).

In older female rats beyond reproductive age, normal oestrous cycles rarely occur (Clemens and Meites, 1971). Vaginal smears show a mixed pattern of constant oestrus interrupted by pseudo-pregnancies of irregular length. Anterior pituitaries cultured from old female rats released more prolactin *in vitro* than those from young rats (Meites *et al.*, 1961) and the pituitaries were heavier and contained more prolactin (Clemens and Meites, 1971).

This summary indicates the general pattern of prolactin secretion throughout the postnatal life of mammals, and in particular the rat. The rest of this chapter will deal in detail with the multifarious actions of prolactin in mammalian reproduction, with its other functions, and with its regulation.

I Prolactin in the female

6.1 The ovary

6.1.1 *Luteotrophic effects of prolactin*

The role of prolactin in the maintenance of the structure and function of the mammalian corpus luteum still remains controversial. The increase in prolactin secretion during the afternoon of pro-oestrus has long been considered to be a trigger for the development and/or activation of the new corpora lutea. On the other hand, recent evidence suggests that this pro-oestrus surge of prolactin may have as its main role a luteolytic action on the corpora lutea remaining from the previous cycle.

Luteinisation of the mammalian ovary is associated with an increase in progesterone production by the corpora lutea and a rise in various enzymes involved in progesterone synthesis, such as glucose-6PO_4-dehydrogenase and 3β-OH-dehydrogenase (Weist and Kidwall, 1969). Hypophysectomy in the rat reduces progesterone synthesis and impedes the incorporation of acetate into cholesterol (Armstrong *et al.*, 1970), as well as increasing the production of 20α-OH-pregn-4-en-3-one (20αOHP). Prolactin treatment of hypophysectomised rats increases progesterone synthesis and at the same time suppresses the rise in 20αOHP production, while reducing the levels of free and acetylated cholesterol. Prolactin also increases the incorporation of acetate into various sterols and triglycerides. Similar results have been obtained in studies of the cholesterol stores of the ovarian interstitial tissue of hypophysectomised rabbits (Hilliard *et al.*, 1969) and of the perfused bovine ovary (Bartosik *et al.*, 1967; Bartosik and Romanoff, 1969).

Prolactin has been shown to produce acute stimulation of progesterone secretion in rat (Grota and Eik-Nes, 1967), sheep (Demanski *et al.*, 1967; Hixon and Clegg, 1969) and cattle (Bartosik *et al.*, 1967, 1969). However it is necessary to hysterectomise sheep, pigs and cattle before a luteotrophic response to prolactin can be demonstrated (Denamur *et al.*, 1973).

Since the first observations of Astwood (1941) and Evans *et al.* (1941), several workers have claimed that prolactin is able to maintain the rat corpus luteum in a fully functional condition (MacDonald and Greep, 1968; MacDonald *et al.*, 1970). On the other hand, evidence discussed below suggests that this may not be a simple role of prolactin by itself, but rather of a luteotrophic complex involving prolactin and LH. Nevertheless, MacDonald *et al.* (1973) have shown that prolactin secreted by ectopic pituitary transplants can maintain progesterone synthesis in the presence of LH antiserum. This is contrary to the findings of Moudgal (1969) and Madhwa Raj *et al.* (1970), which showed that LH antisera administered to rats in early pregnancy could cause abortion due to progesterone insufficiency. Again, McNathy *et al.* (1976) have shown that low doses of prolactin can stimulate progesterone secretion by mouse ovaries *in vitro*, whether or not the ovaries contain corpora lutea. High doses of prolactin inhibited progesterone synthesis in ovaries without corpora lutea, but still stimulated progesterone production by the corpora lutea themselves. The observations confirm earlier work by McNathy *et al.* (1975, 1976), showing that prolactin can stimulate progesterone production by human granulosa cells.

The action of prolactin on the rat corpus luteum, apart from stimulating the rate of incorporation of acetate (see above), appears to be to maintain the enzymes essential for steroidogenesis (Behrman *et al.*, 1970). At the same time prolactin inhibits enzymes which catabolise progesterone, in particular the 3β- and 20α-hydroxysteroid dehydrogenases and 5α-reductase (Farmer, 1970; Zmigrod *et al.*, 1972). Hypophysectomy rapidly increases the levels of these catabolic enzymes. Recently, Guraya (1975) has shown that prolactin causes depletion of cholesterol and lipid stores in rat luteal cells, in both early and late pregnancy. This response could to some extent be blocked by the simultaneous injection of oestrogen. Fajer (1971) found that prolactin produces similar changes in the hamster corpus luteum, and also, as in the rat, blocks progesterone catabolism.

Despite the above findings, a large body of evidence from rats (Armstrong, 1968), mice (Choudrhay and Greenwald, 1969), hamsters (Greenwald, 1969) and domestic ungulates (Demanski *et al.*, 1967; Hixon and Clegg, 1969) indicates that prolactin requires the presence of either LH or FSH to maintain full luteal function. An involvement of LH has been demonstrated, for example, by Herlitz *et al.* (1974) who showed that LH caused a dose-dependent increase in cyclic AMP (cAMP) levels in rat luteal cells *in vitro*. Prostaglandin E_2 was also stimulatory, but the response was much shorter. Neither prolactin nor FSH were able, acting alone, to stimulate cAMP levels. cAMP levels were high in the young corpora lutea and the sensitivity of the cells to LH decreased with the age of the structure. Recently, Holt *et al.* (1976) have shown that administration of ergocryptine (an inhibitor of prolactin secretion to female rats caused a reduction in the LH-binding capacity of the rat corpus luteum. Prolactin injection into the ergocryptine-blocked rats restored the LH-binding capacity. Thus one action of prolactin on the luteal cells may be to increase the number of LH-binding sites, so indirectly stimulating progesterone synthesis. In the studies of Holt *et al.* (1976), progesterone synthesis altered in parallel with the LH-binding capacity. Gruinnich *et al.* (1976) have shown also that prolactin and prostaglandin $F_2\alpha$ may combine to control levels of the LH receptor in the rat corpus luteum.

Rolland *et al.* (1976) recently showed that porcine luteal cells possess specific sites which bind ovine prolactin with a relatively high affinity. In addition, Saito and Saxena (1975) found that the rat ovary has a highly specific receptor-effector complex for human prolactin, which is not modified by either LH or FSH.

The available evidence suggests that prolactin probably acts in the rat as part of a luteotrophic complex, stimulating the synthesis of progesterone by increasing the levels of a LH receptor protein, while at the same time acting more directly by preventing the catabolism of the progesterone newly synthesised under the stimulatory influence of the higher levels of LH which now have access to the luteal cells.

Evidence for the cow, pig, rabbit and guinea pig suggests that prolactin is without luteotrophic action in these species (see Rothchild, 1965), but the picture is not at all clear, and luteotrophic effects of prolactin have been described for a variety of mammals (Donovan, 1963; Greenwald and Rothchild, 1968; Spies *et al.*, 1968). The true position may be that prolactin plays a small but essential part in developing and maintaining the corpus luteum, and that the importance of this role differs in different species. For example, a recent report for the pig claims that prolactin has luteotrophic action (du Mensil, du Boisser and du Boisseau, 1973) and Rolland *et al.* (1976) has demonstrated specific binding of prolactin by corpus luteum cells in this species; while Kann and Denamur (1974) have demonstrated a luteotrophic effect of prolactin in nonpregnant sheep. On the other hand, there now seems to be convincing evidence that prolactin is not luteotrophic in the cow (Hoffman *et al.*, 1974; Karg and Schams, 1974; Cumming, 1975).

In conclusion, it seems safe to accept that prolactin is certainly luteotrophic, probably acting as part of a luteotrophic complex, in rat, mouse, hamster, pig and sheep; but evidence for other species is either negative, or at best controversial. Nicoll (1974) has pointed out that prolactin appears to have rather widespread effects among mammals in maintaining steroid-hormone precursor-pools, not only in the ovary but also in the testis and perhaps the adrenal cortex. It may be that, in part, its effect on the corpus luteum is a manifestation of this more general action, serving to increase and sustain the responsiveness of the corpus luteum to other components of the luteotrophic complex, such as LH. In addition, we have seen that prolactin blocks the induction of enzymes that catabolise progesterone, at least in the rat and the hamster.

6.1.2 *Prolactin in pseudo-pregnancy*

In the rat oestrous cycle, the corpora lutea normally persist for only two or three days (Feder *et al.*, 1968; Hashimoto *et al.*, 1968) whereas most spontaneously ovulating mammals, in contrast, have a prolonged luteal phase even without any external stimulus. However, in the rat cervical stimulation can produce a prolonged luteal phase termed *pseudo-pregnancy* (Long and Evans, 1922). In nature, copulation with a sterile male would cause pseudo-pregnancy but in the laboratory this response can be brought about by mechanical or electrical stimulation of the cervix during pro-oestrus or oestrus, and the luteal phase can be further prolonged by traumatisation of the endometrium, which produces a decidual reaction akin to the formation of the maternal placenta.

The role of prolactin in pseudo-pregnancy became clear with the observation that lactogenesis in rats is always associated with ovarian luteinisation, and that pseudo-pregnancy could be induced in normally cycling rats by the introduction of foster pups to the female (Rothschild, 1960), the suckling stimulus supplied by the pups then inducing and maintaining prolactin secretion (see pp. 148–153). Furthermore, in pseudo-pregnancy marked development of the mammary gland is observed, which is far in excess of that seen in the normal oestrus cycle but is similar to the development in the early stages of pregnancy (Freyer and Evans, 1923; Schultze and Turner, 1933; Weichert, Boyd and Cohen, 1934).

In addition to this direct pseudo-pregnancy, a delayed pseudo-pregnancy can be induced in rats if the cervix is stimulated on day two of dioestrus (Greep and Hisaw, 1938), an observation since confirmed by various authors (Everett, 1952, 1964, 1967; Quinn and Everett, 1967; Zeilmaker, 1965). Delayed pseudo-pregnancy apparently ensues if cervical stimulation takes place in the absence of new corpora lutea capable of progesterone synthesis.

Recently Smith *et al.* (1975) have compared blood progesterone levels in the normal oestrus cycle and during early pseudo-pregnancy. The levels diverge on the second day of dioestrus, when either maintenance or regression of the corpora lutea occurs. Previously Freeman *et al.* (1974) had shown that cervical stimulation produces on each successive day two daily surges of prolactin, a diurnal one on the afternoon and a nocturnal one in the early hours of the morning. The two peaks appear to be independent of one another, the diurnal peak being eliminated by 'stress' of various types and also disappearing fairly rapidly following ovariectomy early in pseudo-pregnancy. On the other hand the nocturnal surge persists when the animal is stressed and does not disappear until the tenth day of pseudo-pregnancy in the ovariectomised animals. Hypothalamic deafferentation obliterates both these surges and Freeman *et al.* (1974) concluded that the daily prolactin surges are induced *via* a neural pathway from the cervix to the hypothalamus, the two peaks presumably being mediated *via* separate pathways. The minimal effective cervical stimulus has been investigated by a number of workers and appears to be equivalent to ten intromissions by the male (Adler, 1969; Adler *et al.*, 1970; Chester and Zucker, 1970). The immediate consequence of cervical stimulation is a rapid rise in serum prolactin levels (Spies and Niswender, 1971) associated with depletion of pituitary stores (Herlyn *et al.*, 1965; Kuroshima *et al.*, 1966). The increase in progesterone secretion during pseudo-pregnancy appears to be induced solely by prolactin, the levels of oestrogen, LH and FSH remaining low throughout the period (Smith *et al.*, 1975). In fact in suckling-induced prolactin release (see pp. 148–153), Lu *et al.* (1976) have shown that LH release is probably directly suppressed by the neural input from the nipple. An analagous suppression of LH might follow cervical stimulation, though this has not been demonstrated.

Smith and Neill (1975) have shown that the induction of the daily dual peaks of prolactin secretion by cervical manipulation can operate best within a certain critical period (see p. 135). The peaks appear at the same time of day regardless of the time of cervical stimulation. Nor, according to Smith and Neill (1975), is the response related to ovarian or adrenal steroids, since it can be produced in the absence of either ovary or adrenal. However, other workers have shown that the presence of oestrogens certainly increases the sensitivity of the rat to cervical stimulation (Everett, 1964; 1966), and pseudo-pregnancy is induced more readily during oestrus than during dioestrus (Defeo, 1963; Greep and Hisaw, 1938). Injection of ergocryptine on the day of cervical stimulation (Smith *et al.*, 1975*a*) confirmed that prolactin is the major stimulatory influence on progesterone secretion in the pseudo-pregnant rat. Ergocryptine treatment resulted in suppression of prolactin release in the stimulated animals and this was correlated with a suppression of progesterone levels.

The maintenance of prolactin secretion during pseudo-pregnancy may fall into two phases. The first phase, lasting for approximately half of the pseudo-pregnancy, seems to be a direct consequence of cervical stimulation.

The second phase is apparently maintained by the action of progesterone in stimulating prolactin secretion. Prolactin injected for four days starting at oestrus causes pseudo-pregnancy (Alloiteau and Vignall, 1958*a*), and the injection of progesterone for several days also induces the onset of pseudo-pregnancy (Alloiteau and Vignall, 1958*b*; Everett, 1963; Rothchild and Schubert, 1963). These results have led a number of authors to postulate that a positive feedback relationship between progesterone and prolactin exists in pseudo-pregnant rats (Alloiteau and Vignall, 1958*a*,*b*; Everett 1963, 1967). Oestrogens will also stimulate pseudo-pregnancy, given either as high doses on the day of oestrus (Merckel and Nelson, 1940) or as small daily doses (Selye *et al.*, 1935; Wolfe, 1935). It is now known that oestrogen injections, particularly in dioestrus, can stimulate prolactin release, and it seems likely that this is what occurred in these experiments, leading then to the initiation of a positive feedback relationship between prolactin and progesterone.

During lactation in the rat, the corpora lutea are maintained in a pseudo-pregnant condition. Ford and Yoshinga (1975) have shown that ergocryptine treatment of lactating rats causes a drop in progesterone secretion, as it does in pseudo-pregnant animals. However the progesterone response occurred more slowly in lactating rats than in pseudo-pregnant females. This presumably is due to the higher initial circulating levels of prolactin in lactating animals, which a) would be less readily suppressed by ergocryptine, and b) would stimulate the uptake of steroid precursors into the corpus luteum. Armstrong *et al.* (1970) have shown that prolactin increased esterified cholesterol levels and stimulated long-term fatty acid synthesis in corpora lutea of immature psuedo-pregnant rats.

Prolactin secretion in delayed pseudo-pregnancy seems to follow the same pattern as in direct pseudo-pregnancy. Beach *et al.* (1975) have shown that in both conditions the daily peak of prolactin output is in the afternoon, in both cases the onset of prolactin secretion occurring about 5·5 to 7·5 hours after cervical stimulation (Niswender *et al.*, 1968; Wuttke and Meites, 1972; Butcher *et al.*, 1972). The difference in the two conditions lies in the fact that delayed pseudo-pregnancy follows on the completion of a new oestrus cycle and the production of a new crop of corpora lutea. This has been confirmed by Zeilmaker (1968) who showed that following ovariectomy of female rats bearing infantile ovarian grafts, pseudo-pregnancy could not be initiated until the infantile grafts had undergone luteinisation.

Pseudo-pregnancy can also be induced in other rodents, and in rabbits. In the rabbit however, the role of prolactin is not as clear as in the rat. Pseudo-pregnancy, for instance, is not associated with such marked mammary gland development in the rabbit as in the rat, though lactogenesis can be induced in pseudo-pregnant rabbits by injecting prolactin. This induced lactogenesis is blocked by the subsequent injection of low doses of progesterone (Assairi *et al.*, 1974). Thus the positive relationship between prolactin and progesterone during pseudo-pregnancy in the rat may not operate in other species.

6.1.3 *Luteolytic effects of prolactin*

In most mammals it is necessary that the existing corpora lutea should be destroyed before a new oestrous cycle can be initiated. This luteolytic process has been shown to depend upon a number of factors including FSH and LH (Rothchild, 1965) and uterine factors (Schemberg, 1969). However, prolactin also appears to play a part in the destruction of the old corpora lutea (Malven, 1969*a*,*b*; Malven *et al.*, 1969). Prolactin can promote luteolysis of non-functional corpora lutea in hypophysectomised rats (Malven and Sawyer, 1966; Piacsek and Meites, 1967), although the response is not simple, the effect of prolactin being dependent upon the time after hypophysectomy that it is administered. Prolactin injected within 56 hours after hypophysectomy still acts as a luteotrophic hormone, and stimulates the synthesis of progesterone. If the injection of prolactin is delayed for more than 80 hours then it has a luteolytic effect (Malven, 1969*a*). This, together with the observations of Lam and Rothchild (1973) that the corpora lutea of pregnant rats were exempt from the luteolytic action of prolactin, suggests that the manifestation of luteolysis may depend on the state of the corpora lutea rather than on the hormonal status of the animal. This idea is given support by the observation that destruction of the follicular elements of the rat ovary by x-rays do not alter the luteolytic effect of prolactin (Hixon *et al.*, 1973).

The luteolytic action of prolactin in the rat is not merely pharmacological. MacDonald, Yoshinga and Greep (1973) showed that rat prolactin injected into female rats could produce luteolysis. Increased oestrogen levels on day two of dioestrus in the rat stimulate the release of prolactin from the pituitary in the afternoon of proestrus, causing a surge in serum prolactin (Gay *et al.*, 1970; Wuttke and Meites, 1970*a*). It has been suggested that one function of this prolactin surge is to destroy the old corpora lutea prior to the production of a new 'crop'. Blockade of the pro-oestrous prolactin surge does not interfere with the oestrous cycle (Wuttke and Meites, 1970; Billeter and Fluckiger, 1971; Richardson, 1973), but it does lead to an increase in ovarian size due to the persistence of the old corpora lutea (Heuson *et al.*, 1970; Nagasawa and Meites, 1970). Injection of prolactin into the ergocryptine-blocked cycling female rat produced luteolysis of the old corpora lutea. A similar response to prolactin has been observed in the mouse (Grandison and Meites, 1972). Thus it seems that one role of prolactin in the rat, mouse and possibly the hamster may be to remove the old corpora lutea from the ovary and, together with the gonadotrophic hormones, to initiate hypertrophy and progesterone synthesis by the new corpora lutea. Its role in the reproductive cycle in other female mammals still remains ill-defined.

6.2 Direct effects of prolactin on the female reproductive tract

In addition to its action on the corpus luteum (pp. 133–136), prolactin has been found to produce a number of more direct effects on the female reproductive tract. Traumatisation of the uterine wall causes a local reaction leading to the formation of deciduomata, an essential part of the implantation process. This response has been shown to depend on synergism between low doses of oestrogen and progesterone (Yochum and Defeo, 1963). Indications that prolactin may influence the response came from the finding that lactation, when stimulated by large litters, was associated with a decrease in the size of the deciduomata formed during the lactational pseudo-pregnancy (Lyons and Allen, 1938), and from the observation that suckling delays implantation in the rat (Krehbiel, 1941; Wishart, 1942). These effects were originally thought to be due to inhibition of ovarian oestrogen secretion (Lyons, 1939; Grayhack *et al.*, 1955). However, Armstrong and King (1971) showed that in the ovariectomised rat treated with progesterone, the injection of prolactin caused a significant increase in the size of the traumatised uteri, even though oestrogens were absent. Injections of oestrone produced a similar effect to progesterone, but prolactin administered together with oestrone inhibited the uterine response to the oestrogen. Oestrone and oestradiol both increased the weight of uterine strips, with consequent increase in the uptake of progesterone by the uterus, and both oestrogens also increased the catabolism of progesterone by the uterine tissues. Prolactin opposed the increase in progesterone catabolism caused by oestrone, but had no effect in the oestradiol-treated rats. This evidence suggests, then, that the actions of prolactin on the uterus are not entirely due to its role as a luteotrophic hormone.

Terranova and Kent (1974) have shown in ovariectomised hamsters that traumatisation of the uterus caused an increase in uterine solids independent of hormone treatment. However, traumatisation also increased the sensitivity of the uterus to prolactin and progesterone, prolactin increasing uterine fluid content and progesterone increasing the uterine solids. The authors suggested that prolactin synergises with progesterone in the formation of deciduomata in this species. In the ovariectomised rat, also, Joseph and Mukabo (1975) have recently shown that traumatisation of the uterus increases its sensitivity to prolactin. Since prostaglandins have been shown to increase uterine growth (Pharris and Hunter, 1971), the authors suggest that prostaglandins might mediate the action of prolactin on the uterus. However more recent work by Joseph and Siwela (1976) has shown that prostaglandin F_{2a} decreases uterine weight, and blocks the stimulatory uterotrophic action of prolactin.

Prolactin has also been shown to affect the growth of the vagina and its secretion of mucus. Kennedy and Armstrong (1972) confirmed earlier reports by Josimovich (1970) that prolactin augmented the action of oestradiol and progesterone in promoting vaginal mucification in the rat, although prolactin alone had no effect. Later work by Kennedy and Armstrong (1973) showed that similar effects could be produced by bovine or ovine growth hormone and by human placental lactogen. However, they showed that the effects of growth hormone and prolactin were additive, whereas those of the placental lactogen and prolactin were not. This suggests that growth hormone acts at a different receptor site from prolactin, whereas placental lactogen acts at the same site.

Mori *et al.* (1974) have also shown that treatment of neonatal mice with oestrogen or prolactin alters the development of the vaginal epithelium. This finding may have physiological implications, since foetal prolactin is secreted prior to parturition (Kohmoto and Bern, 1971; Ingleton *et al.*, 1071). Mori *et al.* (1971) found that both oestrogen and prolactin increased the width of the intracellular spaces between the middle epithelial cells of the vagina. They also point out that neonatal mice bearing hypophysial autografts (Mori and Bern, 1972) show increased ovarian and mammary gland development.

Neonatal treatment of female mice with oestrogen causes persistent vaginal cornification associated with inhibition of the oestrous cycle after puberty. However, conditional upon the dose of oestrogen given, the permanent vaginal cornification may or may not be dependent on the ovary. Neonatal treatment of mice with prolactin increases the uterine levels of nuclear and cytoplasmic oestrogen receptors, and this tends to render the uterine response to oestrogen independent of the ovary (Shymala *et al.*, 1974).

A further role for prolactin may be in regulating the accumulation of uterine fluid in rats (Kennedy and Armstrong, 1972). The increase in uterine fluid content appears to be stimulated by oestrogens (Armstrong, 1968). During early oestrus the cervix relaxes and allows the uterine fluid to drain away (Blandau, 1945; Schwarts, 1964). Hypophysectomised-ovariectomised rats retain fluid in the uterus, whether or not it is ligated. However replacement therapy with prolactin causes a loss of uterine fluid

(Kennedy and Armstrong, 1972), suggesting that prolactin may cause relaxation of the cervix directly rather than stimulating ovarian steroid secretion (Armstrong, 1968).

Like the penis, the clitoris bears preputial glands, and as in the male the rat female preputial glands are stimulated by androgens (Freeman *et al.*, 1964), and hypophysectomy leads to a decrease in preputial gland weight (Huggins *et al.*, 1955). Preputial gland growth can also be stimulated by ACTH (Rennels *et al.*, 1961), growth hormone (Selye, 1951) and prolactin (Bates *et al.*, 1964). Recently Ozegovic and Milkovic (1972) have shown that transplantable pituitary tumours secreting prolactin and ACTH strongly increased the growth of the preputial glands of female rats, elevating the DNA, RNA and protein contents, RNA/DNA ratio and cell content of the glands. This response seems to occur independently of androgenic steroids. However prolactin may act synergistically with androgens on the preputial glands in the normal female, as it does in maintaining the male accessory gland.

In addition to its growth-promoting effects on the mammalian reproductive tract, prolactin also affects the irritability of the myometrium. Ovine prolactin has been shown to exert a marked inhibitory effect on myometrial activity in pregnant guinea pigs (Manku and Horrobin, 1973) and in rats primed with oestrogen and progesterone (Horrobin *et al.*, 1973*a*). However Mati *et al.* (1974) showed that although prolactin inhibited oxytocin-induced contractions in isolated pregnant rabbit myometrium, the effect was less than in other species, the difference perhaps being related to the differing roles of pituitary prolactin in the pregnancy of the various species (pp. 133–136).

The possibility exists that these tranquillising effects on the pregnant myometrium of rabbits and rats may not normally be mediated by pituitary prolactin, but rather by a placental prolactin-like hormone similar to human placental lactogen (HPL). The fact that the human placenta synthesis and secretes a polypeptide hormone with prolactin-like activity (HPL) has been known for some time (Josimovich and MacLaren, 1962; Friesen, 1968) (see Chapter 1). Kaplan and Grumbach (1964) have demonstrated a similar hormone in a monkey, and more recently a hormone with luteotrophic activities has been isolated from the baboon placenta (Josimovich *et al.*, 1973). Comparative studies have revealed the existence of placental lactogens or luteotrophins in the rat (Talamantes, 1973), mouse (Kohomoto and Bern, 1970; Talamantes, 1975), hamster (Matthias, 1974), chinchilla (Talamantes, 1973) and guinea pig (Kelly *et al.*, 1973), and Forsyth (1974) has demonstrated the presence of similar placental hormone molecules in a range of artiodactyls, though none could be detected in the pig. Recently Talamantes (1975) has made a comparative study of the placental lactogenic activity of a range of mammalian species and reported activity in seven species, baboon, chinchilla, guinea pig, rat, mouse, hamster and sheep, but not in the rabbit and the dog. This might help to explain the only slight effectiveness of exogenous prolactin on the pregnant rabbit myometrium (Mati *et al.*, 1974): if the prolactin acts on the pregnant myometrium by occupying receptors for placental lactogen (see Chapter 8), then the lack of, or low secretion rate of, a placental lactogen in the rabbit might as a correlate entail the myometrium having only a few receptor sites. Talamantes (1975) also demonstrated that placental lactogenic activity in at least two species, the rat and the mouse, varied during pregnancy, with peaks of activity at mid-pregnancy and near term (see also chapter 7, pp. 212–214).

From all the evidence it appears that prolactin exerts various direct actions on the female reproductive tract, in addition to those effects mediated *via* elevated progesterone secretion by the corpus luteum. It may also exert a 'quietening' effect on the pregnant myometrium, this latter effect perhaps being a measure of its ability to cross-react with receptors for placental lactogens.

6.3 The mammary gland

6.3.1 *Mammary growth*

The development and secretory activity of the mammary gland are complex processes, and their endocrine control is correspondingly elaborate, involving the concerted action of a host of hormones. Fig. 6.1 illustrates the morphological stages in the maturation of the gland, and a detailed summary of the range of normal growth patterns in different species will be found in the excellent review by Cowie and Tindall (1971). In the present context, only the role of prolactin and certain synergistic hormones will be discussed in any detail.

In maturing female mammals, the development of the mammary gland depends primarily on a functional pituitary-gonadal axis, and the full growth and differentiation of the gland tissues also requires a number of non-gonadal hormones acting synergistically. Reviews of earlier work have been given by Lyons (1958), Lyons *et al.* (1958) and Lyons and Dixon (1966). The earlier studies showed that ovarian and adrenal steroids stimulate mammary growth only in the presence of a functional pituitary. Development of the mammary tissues in rats and mice can be divided into two distinct phases: a) *duct growth*, for which the minimal hormone requirements are oestrogen, corticosteroids and growth hormone; and b)

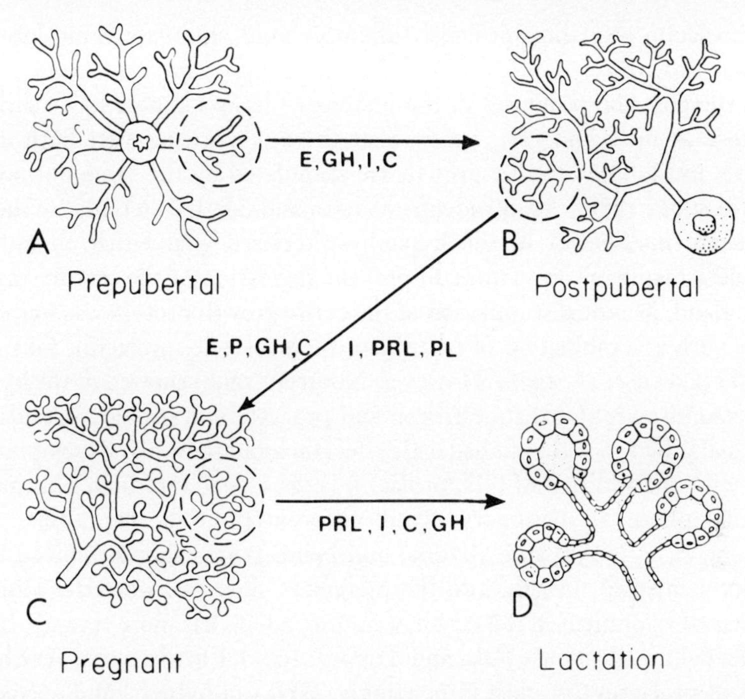

Fig. 6.1 Schematic representation of the hormonal synergism involved in mammary gland development and lactation in rats and mice. C, corticosteroid; E, estrogen; GH, growth hormone (somatotrophin); I, insulin; P, progesterone; PL, placental lactogen; PRL, prolactin. Growth of the duct system from the prepubertal condition (*A*) prior to and after puberty (*B*) requires E+GH+I+C. Development of lobules, as occurs during pregnancy (*C*), involves E+P+GH+I+C; PRL and PL probably participate. Milk secretion (*D*) (indicated by stippling) in alveoli and ducts requires PRL, I, C, and GH. Thyroxine and somatotrophin are beneficial for lactation. (From Nicoll, 1974.)

lobular-alveolar growth, which is promoted by the addition of progesterone and prolactin to the three hormones necessary for phase (a) (Fig. 6.1).

a. *In vivo studies* Evidence from hypophysectomised mice suggests that growth hormone can to some degree substitute for prolactin in stimulating lobulo-alveolar development (Hadfield and Young 1958; Nandi, 1959), but this is true only for certain strains of mice (Nandi and Bern, 1961). Clifton and Furth (1960) showed that some lobulo-alveolar growth occurred in the ovariectomised-adrenalectomised rat bearing an ectopic pituitary transplant, although ovarian steroids are necessary for the normal and full pattern of development. These studies have been confirmed by Talwalker and Meites (1964), who showed in the triple-operated rat (hypophysectomised, ovariectomised and adrenalectomised) that frequent injections of prolactin plus growth hormone (GH), or implantation of GH-

and prolactin-secreting pituitary tumours could stimulate some lobulo-alveolar growth.

In the goat the responses of the mammary tissues appear to be similar to those in mice and rats, except that there is evidence that both duct growth and lobulo-alveolar growth are stimulated by the same hormones (Cowie *et al.*, 1966). As in rodents, ovarian and adrenal steroids by themselves are insufficient in the hypophysectomised goat either to induce complete mammary growth or to prevent regression of the mature mammary gland. Maximal stimulation of the entire growth process was found to occur with a combination of oestrogen, progesterone, prolactin, GH and ACTH (Cowie *et al.*, 1966). However, Norgren (1968) showed in the hypophysectomised rabbit that oestrogen and progesterone strongly stimulated mammary duct growth, but had little effect on lobulo-alveolar development, even with the addition of GH; prolactin is the essential pituitary hormone for this phase of mammary gland differentiation in the rabbit, and Mizumo *et al.* (1955) and Mizumo and Naito (1956) showed indeed that prolactin injected directly into the mammary duct can stimulate lobulo-alveolar development in the rabbit, a finding which has more recently been duplicated for the mouse (Oka and Topper, 1972). Furthermore, very high doses of prolactin (or, most interestingly, GH) can induce lobulo-alveolar growth in the triple-operated rabbit (Denamur, 1969) and rat (Talwalker and Meites, 1964). It should be recognised that, important though prolactin undoubtedly is, none of these findings necessarily precludes physiological roles for steroids and insulin (see below) in *normal* lobulo-alveolar development. This proviso applies, too, in the case of studies showing that ectopic pituitary transplants (secreting mainly prolactin) can stimulate advanced mammary growth in rats (Bardin *et al.*, 1962; Dao and Gawlik, 1963; Meites and Nicoll, 1966) and mice (Browning and White, 1965).

Oestrone and progesterone, singly or in combination, were reported to increase lobulo-alveolar growth in the guinea-pig (Benson *et al.*, 1957), but since oestrogens are now known to release prolactin from the pituitary (p. 178) these results are difficult to interpret. The same difficulty arises in considering the finding that oestrone stimulates mammary growth in the gonadectomised rabbit (Norgren, 1966).

The general ineffectiveness of oestrogens in the absence of the anterior pituitary (Cowie and Tindall, 1971) may be due in part to a decline in the amount of oestrogen-receptor protein in the mammary tissues: the first subcellular action of oestrogen in the mammary gland is to bind to a specific cytoplasmic receptor (Jensen and DeSombre, 1972), and prolactin stimulates the levels of oestradiol-binding protein in the rat mammary gland *in vitro* (Leung and Sasaki, 1973), an effect which is inhibited by progesterone.

The situation is undoubtedly complex: for instance, the limited lobular-alveolar response to oestrogen in the hypophysectomised rabbit can be augmented by cortisone or by thyroidectomy, both factors which by themselves *depress* mammary growth (Norgren, 1968).

Mammary gland responses have been studied with the electron microscope. Either prolactin or reserpine (which releases prolactin) can retard mammary gland involution in the intact rat (Richards and Benson, 1970), but ultrastructural examination showed some degree of involution in the treated animals, associated with the continued presence of macrophages. Other work showed that intraductal injection of prolactin in the pseudo-pregnant rabbit led to a progressive increase in the endoplasmic reticulum and in the number of free ribosomes (Fiddler *et al.*, 1971). However, Mills and Topper (1970) found that insulin and corticosteroids are essential for full development of the endoplasmic reticulum of the mammary cell. Fiddler *et al.* (1971) also reported that prolactin strongly increased cisternal development in the rough endoplasmic reticulum and induced hypertrophy of the Golgi apparatus and of the smooth endoplasmic reticulum. Hallowes and Wang (1971) reported similar changes in mouse mammary gland explants treated with GH or prolactin, both hormones causing growth of the endoplasmic reticulum but prolactin being more effective in producing hypertrophy of the Golgi elements. Following hypophysectomy, the alveolar epithelium displays the classical ultrastructural signs of regression, and prolactin opposes these changes (Hartman *et al.*, 1970).

Tokon *et al.* (1972), like Fiddler *et al.* (1971), showed that prolactin stimulates the endoplasmic reticulum and Golgi apparatus in the mammary cells of pseudo-pregnant rabbits, but in addition they detected a certain amount of duct growth and of the accumulation of fat droplets in the alveolar cells. This might be taken to indicate that prolactin is capable of inducing the full cycle of milk production in the pseudo-pregnant rabbit, but it is obviously not possible to exclude the participation of other hormones in the process in these intact animals. Recently, Mittra (1974) reported similar ultrastructural responses to prolactin in the rat mammary gland, and obtained the same effects by enhancing endogenous prolactin secretion with perphenazine. Thyroidectomy greatly augmented the responses to prolactin, in agreement with earlier reports that mammary growth is greater in hypothyroid rats. Since thyroxine appears to stimulate prolactin secretion (Haigh and Ensor, unpublished observation), these results imply that thyroxine antagonises prolactin at the level of the mammary cell.

b. *In vitro studies*

Studies of the hormonal requirements for development of mammary explants *in vitro* have produced results which in the main agree with the *in vivo* findings, indicating that the synergistic interactions between the various hormones operate largely at the tissue level (Riviera, 1964; Barnawell, 1965; Gadklari *et al.*, 1968; Topper, 1969; Turkington, 1972). In mammary explants from mid-pregnant mice, proliferation of the parenchymal cells and differentiation of alveolar lobes appear to require insulin, corticosterone and prolactin (Topper, 1969; Turkington, 1972). Some development of the alveolar system in explants from virgin mice can be brought about by incubation with insulin, ovarian and adrenal steroids, prolactin and GH, but to obtain full development it is necessary to pretreat the mice with ovarian steroids, prolactin and GH (Ichinose and Nandi, 1966). This pretreatment is not essential for the response in mammary explants of young rats, but again these explants require insulin in the medium to avert regression of the parenchymal cells. Some lobulo-alveolar development can be brought about in these rat explants by insulin and prolactin alone, but the addition of corticosteroids or oestrogen strongly enhances the response (Ichinose and Nandi, 1966).

Insulin appears to act on the mammary epithelium primarily as a mitogenic agent (Topper, 1969, 1970; Turkington, 1972), producing a generation of daughter cells which then respond to prolactin. Turkington (1972) has proposed a scheme whereby various humoral factors (insulin, epithelial growth factor, GH, serum growth factor) can initiate the division of mammary epithelial stem cells during the culture period. These cells then require a period of exposure to cortisol or corticosterone before they will fully respond to prolactin (see Fig. 6.2), although some slight response to prolactin is evinced even in the absence of corticosteroids (Turkington *et al.*, 1967). It has been suggested that insulin and corticosteroids may stimulate the development of the Golgi apparatus and of the endoplasmic reticulum (Mills and Topper, 1970). The degree of development of these organelles would effectively limit the secretory output of the mammary cell, but should not limit the action of prolactin on phosphorylation and levels of RNA and DNA (see Chapter 8).

This idea that insulin is the prime mitogenic hormone has not gone unchallenged. Dilley (1971) showed for the rat mammary gland *in vitro* that insulin could not maintain the initial mitotic activity, but that prolactin could. Alveolar development showed a dose-related response to prolactin in these mammary explants. Oka and Topper (1972) found that prolactin treatment of mature virgin mice converts the non-proliferative mammary epithelium into an actively growing structure *in vivo*, but in agreement

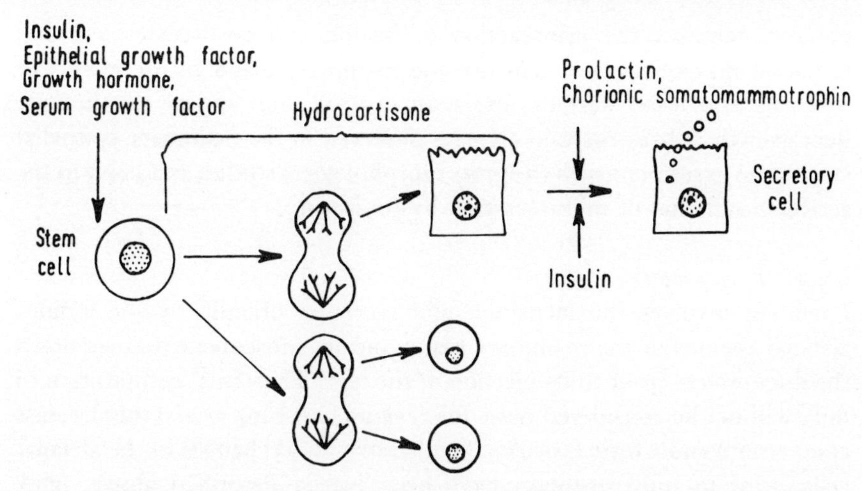

Fig. 6.2 General scheme of the sequence of hormone actions and cellular transitions during mammary epithelial cell differentiation *in vitro*. Chorionic somatomammotrophin ≡ placental lactogen. (From Turkington, 1972.)

with Turkington (1972) they found that prolactin is not mitogenic *in vitro*, although prolactin given to the mice before autopsy did sensitise the cells to the *in vitro* mitogenic action of insulin and mouse serum. This prolactin effect seems to be direct, since injections of prolactin directly into the abdominal mammary glands of adult virgin mice potentiated the mitogenic action of insulin, measured by ^{3}H-thymidine incorporation. Lyons *et al.* (1958) had previously shown that prolactin injected into a single mammary gland of the rat induced proliferation in the injected gland only, the other glands remaining quiescent. Prolactin added to mouse mammary explants did not accelerate the onset of insulin sensitivity, which occurred after 24 hours in culture irrespective of prolactin pre-treatment (Oka and Topper, 1972), but cultured mammary explants from mice with low serum prolactin levels did not readily respond to insulin. This last finding has also been reported for rats (Amenomori *et al.*, 1970; Linkie and Niswender, 1972). In the rat, sustained serum prolactin levels do not rise until late pregnancy, although there is a sharp post-coital surge. Explants of rat mammary gland respond to insulin only if taken in late pregnancy, and do not show the characteristic increase in DNA levels until after the prolactin levels have increased (Oka and Topper, 1972). Cortisol also sensitises mammary epithelial cells to prolactin, but this is unlikely to have physiological significance since the circulating levels of corticosteroids do not rise in pregnancy (Topper and Oka, 1971).

In summary, prolactin appears to stimulate lobulo-alveolar growth to some degree in the mammary gland of various species. The full response possibly requires the prior action of insulin and cortiscosteroids, and prolactin may in its turn sensitise the mammary gland to the effects of insulin and ovarian steroids. Prolactin in itself does appear to stimulate duct growth. Ultrastructural changes observed in the mammary epithelial cells are in agreement with the gross morphological studies and point to the active stimulation of milk secretion by prolactin.

6.3.2 *Lactogenesis*

Lactation involves the formation and secretion of milk by the lobulo-alveolar regions of the mammary gland and its subsequent passage down the duct system prior to its ejection at the teat. The actual composition of milk will not be considered here: for reviews, see Ling *et al.* (1961), Glass *et al.* (1967) and Cowie (1969). The ultrastructural changes in the alveolar cells prior to milk secretion have been briefly described above. Milk secretion entails a complex sequence of events. Following initiation of the synthesis of specific ribosomal, messenger and transport RNA there is an increase in RNA polymerase activity, and then the induction of the enzymes which synthesise milk proteins and the appearance of lactose and casein (Turkington, 1972). Further details of these events are given in Chapter 8.

The effects of prolactin on milk yield and secretion both *in vitro* and *in vivo* have been well documented (see review by Nicoll, 1975). *In vivo* studies have shown that prolactin increases milk production in rats (Grosvenor and Mena, 1973), mice (Nandi and Bern, 1961), rabbits (Fredrickson, 1939; Kilpatrick *et al.*, 1963, 1964) and other species discussed below.

Lactogenesis has been studied primarily in triple-operated rodents and rabbits, in which surgical removal of the pituitary, ovaries and adrenals produces a largely endocrine-deficient animal in which the effects of various hormone combinations can be defined with some precision. In most species a complex of hormones appears to be necessary to induce lactogenesis in the fully developed mammary gland. Lyons *et al.* (1958) showed that combined injections of oestrogen, progesterone, corticosteroid, bovine growth hormone and ovine prolactin would stimulate lactogenesis in the triple-operated rat. Subsequently, the combination of prolactin and corticosteroid is sufficient to maintain lactation (Cowie and Tindall, 1961). Similar results have been obtained for triply operated mice, although in some strains the combination of GH and corticosteroid can maintain lactogenesis, and in others the addition of GH augments the maintenance effects of prolactin with corticosteroid (Nandi, 1959; Nandi and Bern, 1961; Wellings, 1969). Prolactin levels have been shown to increase during lactation in the rat

(Amenomori *et al.*, 1970), in which blood corticosterone and prolactin change in parallel during lactation (Simpson *et al.*, 1973). In the rabbit, prolactin by itself can restore lactation in hypophysectomised (Cowie and Watson, 1966) and triply operated (Kilpatrick *et al.*, 1963, 1964) animals although corticosteroids do improve the milk yield (Cowie, 1966). Bovine GH has little effect in the rabbit, in contrast to its stimulatory action in other rodents (Hartmann *et al.*, 1970).

The physiological signal for the initiation of lactogenesis varies from species to species, and has been reviewed in detail by Cowie and Tindall (1971). The generally accepted view is that the fall in progesterone secretion at the end of pregnancy permits an increased output of oestrogen, which in turn stimulates prolactin release which acts as the immediate trigger. Support for this view comes, for example, from the work of Yokoyama *et al.* (1969), who showed that ovariectomy on day 18 of pregnancy in the rat leads to a rapid accumulation of lactose in the mammary gland, which can be reversed by progesterone. Further evidence comes from work on the goat. Hart (1974) has recently shown that milking causes increased secretion of both prolactin and growth hormone in the goat and that basal prolactin levels rise during late pregnancy and after parturition. Also in the goat, progesterone has been reported as having an inhibitory effect on lactation (Thorburn and Schneider, 1972), and a sharp rise occurs in oestrogen levels immediately prior to parturition (Challis and Linzell, 1971). However, treatment with bromocryptine in the already-lactating goat does not decrease milk yield (Hart, 1973). This anomaly has been reported for other species, in that although serum prolactin levels stay high during lactation, and all the evidence is that prolactin promotes lactogenesis, exogenous prolactin does not necessarily increase the milk yield. However, improved milk yield has been reported in rats treated with prolactin alone or in conjunction with growth hormone or thyroxine (van Berswardt-Wallrabe *et al.*, 1960). These results were not confirmed by Talwalker *et al.* (1960), but more recently Flint and Ensor (1976) have demonstrated that ergocryptine will reduce pup growth rate in the postpartum-mated lactating rat. Similarly controversial results for the rat have been reported with bovine growth hormone (Thatcher and Tucker, 1970). Prolactin does not increase milk yield in the guinea pig (Nagasawa and Naito, 1962, 1963) or in lactating cows (Cotes *et al.*, 1949), although GH will stimulate lactation in cattle (Cowie, 1966). On the other hand, prolactin can increase lactation and elevate lactose levels in the rabbit (Gachev, 1963, 1968).

These anomalies may result from high levels of endogenous prolactin in the blood during lactation. the system may already be operating at maximal capacity and cannot be stimulated further by treatment with exogenous prolactin. This would explain the anti-lactational effects of ergocryptine in rats (Flint and Ensor, 1976), mice (Bartke, 1974) and cows (Smith *et al.*, 1974), although we should note that for the cow (Karg *et al.*, 1972) and the goat (Hart, 1973) some work using ergocryptine suggests that prolactin may not be essential for lactation. Hayman (1972) studied *Bos indiens*, in which lactation is variable and is not automatically maintained by milking, and found that prolactin will maintain lactation, in contrast to its ineffectiveness in the intensely selectively bred strains of Western Cattle.

In summary, although the evidence is somewhat confusing and indicative of marked interspecific differences, it appears in general that the minimally essential combination for initiation and maintenance of lactogenesis is prolactin plus a corticosteroid, although the latter is of only minor importance in the rabbit. Growth hormone may augment the response in some mammals, and there is evidence that thyroid hormone may also be stimulatory in some species (Cowie, 1969; Denamur, 1969; Cowie and Tindall, 1971).

6.3.3 *Suckling and prolactin release*

Undoubtedly the most potent external stimulus for prolactin release is stimulation of the nipple, in suckling or in milking (Tindall and Knaggs, 1970; Cowie and Tindall, 1971). Reece and Turner (1937) were the first to report depletion of pituitary prolactin following suckling during post-partum lactation in rats. Apart from suckling, various stress stimuli have been reported to cause release of prolactin (Johke, 1970; Schams, 1973), and there is evidence of a photoperiodically entrained diurnal cycle of release (Clarke and Baker, 1964; Kocj *et al.*, 1971). These secondary factors will be dealt with later in this chapter. For the present, we shall consider the evidence that suckling promotes prolactin release, a neuroendocrine reflex known as the suckling reflex. Suckling also, and independently of its effect on prolactin, elicits the release of oxytocin, which promotes the discharge of milk from the nipple by inducing contractions of the mammary myoepithelial cells; this is the milk-ejection reflex.

A number of reports have confirmed the early observations of Reece and Turner that prolactin is released in response to suckling (Grosvenor and Turner, 1958; Sar and Meites, 1969), and it now seems that suckling acts by depleting the hypothalamic pool of prolactin inhibiting factor, PIF (Ratner and Meites, 1964; see pp. 162–169). The recent development of homologous radioimmunoassay systems has made it possible to measure circulating prolactin levels in the goat (Johke, 1969, 1970; Bryant *et al.*, 1970), cow (Johke, 1969, 1970; Schams and Karg, 1970), rat (Amenomori

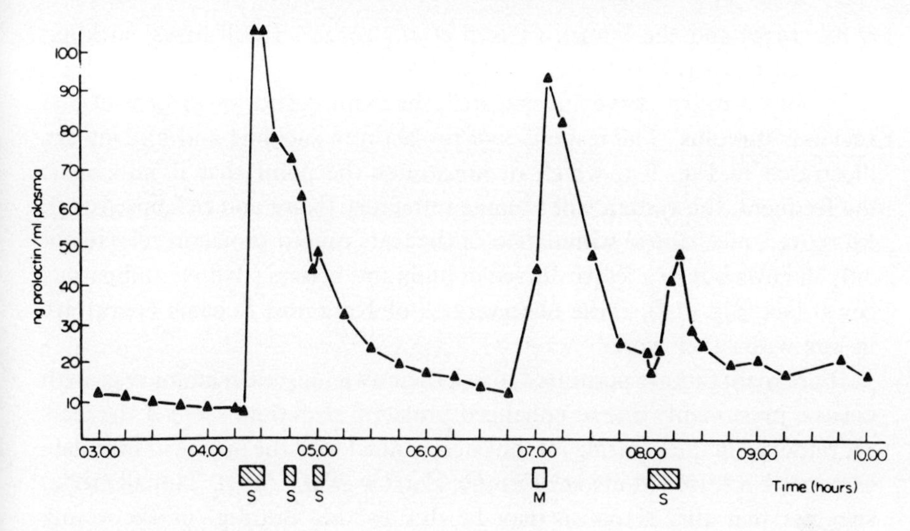

Fig. 6.3 The effect of frequent suckling (S) and milking (M) on prolactin response. (From Karg and Schamms, 1974.)

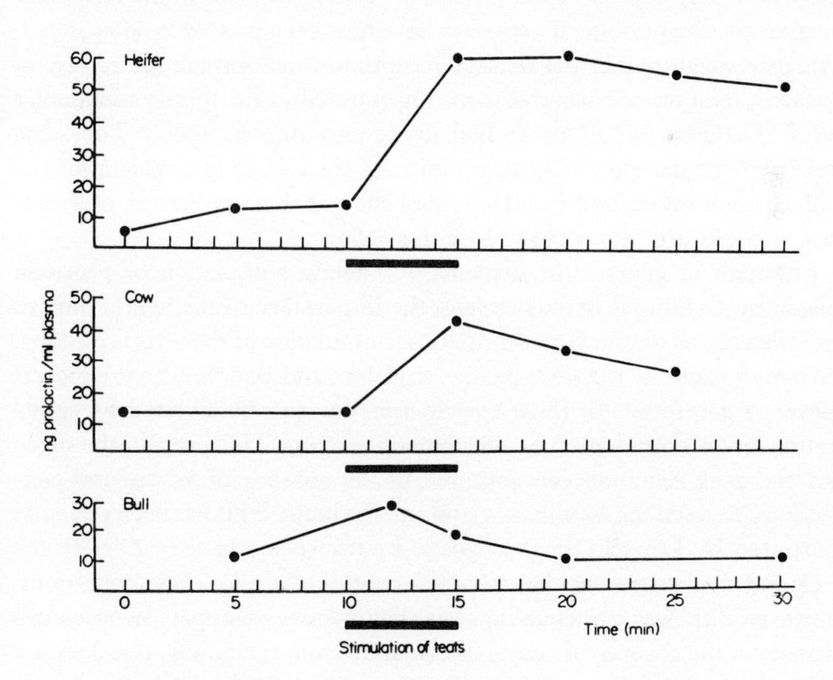

Fig. 6.4 The effect of mechanical stimulation of the teats on prolactin release in a heifer, cow and bull. (From Karg and Schamms, 1974.)

et al., 1970) and the human (Tyson *et al.*, 1972*a*). In all cases, suckling increased prolactin secretion.

Various workers have investigated the nature and specificity of the suckling stimulus. The responses of prolactin to suckling and milking are illustrated in Fig. 6.3, which demonstrates the point that if milking is too frequent, the system will become refractory (Karg and Schams, 1974). Moreover, mechanical stimulation of the teats caused prolactin release not only in cows but to a lesser degree in bulls and heifers (that is, nulliparous cows) (see Fig. 6.4); these observations of Karg and Schams (1974) are in line with other work.

If pregnant rats are permitted to lick their own nipples, mammary growth ensues, presumably due to enhanced prolactin secretion. Even if the teats are removed in the lactating rat, physical contact with the pups will stimulate some milk release (Moltz *et al.*, 1969; Zarrow *et al.*, 1973). Tindall (1974) suggests that this response may be due to the 'butting' of the young stimulating the severed peripheral nerves through the skin.

The actual neural pathways which transmit the suckling stimulus from the nipple to the hypothalamus have been elegantly traced in the rabbit by Tindall and Knaggs (1969, 1972), and are depicted in Figures 6.5 and 6.6. These diagrams show the ascending paths of neurons in the forebrain. In addition, the peripheral pathways have now been plotted in some detail. Evidence suggests that the sensory receptors which initiate the release of prolactin (and other hormones) are concentrated in the nipple and areolar tissue (Cathcart *et al.*, 1948; Ballantyne and Bunch, 1966). The same area (perhaps the same receptors?) initiates the release of oxytocin for the milk ejection reflex, and Findlay (1968) showed that application of a local anaesthetic to the teat would block this reflex.

Although in general the evidence for neural stimulation of prolactin release by suckling is overwhelming, the importance of the neural limb of the reflex seems to vary between species. Stimulation of the teats is essential for maintenance of lactation in the rat (Eayrs and Baddley, 1956) and cat (Beyer *et al.*, 1962). In these species lactation can be arrested by spinal section and is not restored by injection of oxytocin. However, in the sheep and the goat, lactation can continue in the absence of any neural connections between the mammary gland and the brain. This has been elegantly illustrated by Linzell (1963) in goats, by transplanting the udder to the neck and demonstrating that lactation continued under these conditions. However, although evidence suggests that in some mammals lactation can proceed in the absence of neural information from the nipple, this does not eliminate a role of the intact neural pathway in normal lactation, and even more so in milk ejection. Physical stimulation of the ascending pathways in

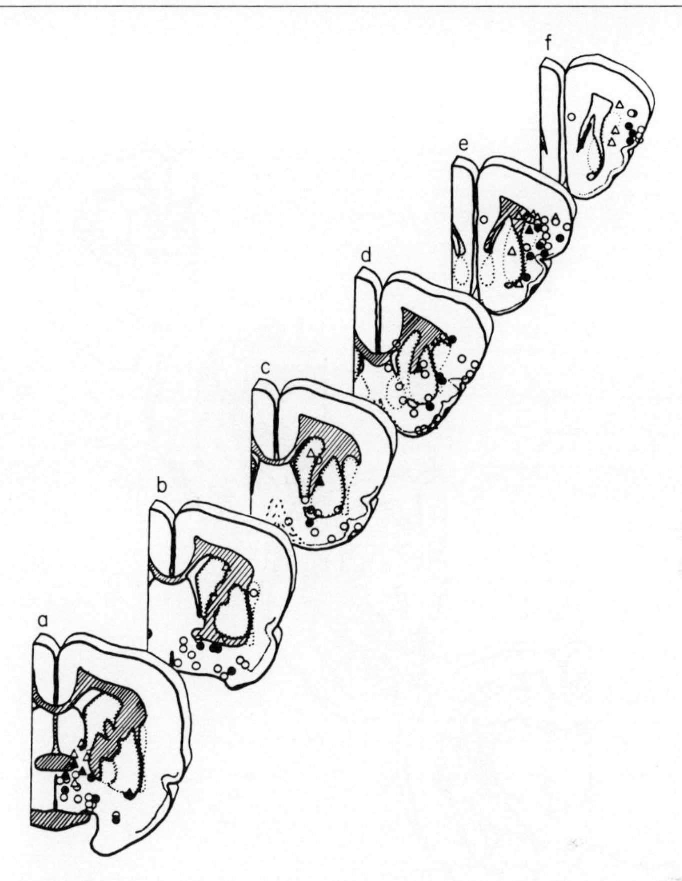

Fig. 6.5 Pathways for release of prolactin in the forebrain of the rabbit (see legend to Fig. 6.6). Note stimulation sites associated with lactogenesis in lateral and medial preoptic area (a), and further forward in, and ventral to, the external capsule (b, c, d), in the claustrum (d, c) and in the adjacent orbito-frontal cortex (e, f). (From Tindal and Knaggs, 1972.)

the spinal cord and brain can evoke prolactin release (Relkin, 1967; Tindall and Knaggs, 1969). Spinal section, even in the goat, does block the milk-ejection reflex, if the section is located in the dorsal funiculi (Popovici, 1963), as well as in the rat and rabbit (Eayrs and Baddley, 1956; Mena *et al.*, 1976*b*).

The pathways in the brain are more clearly delineated. In the rabbit, the ascending pathway described by Tindall and Knaggs (1969, 1970, 1972) is situated in the lateral tegmentum in the midbrain and has been traced forwards to the posterior hypothalamus, passing midway between the third ventricle and the mamillo-thalamic tract; then passing rostrally into the medial forebrain bundle. The tract then appears to pass to the lateral and

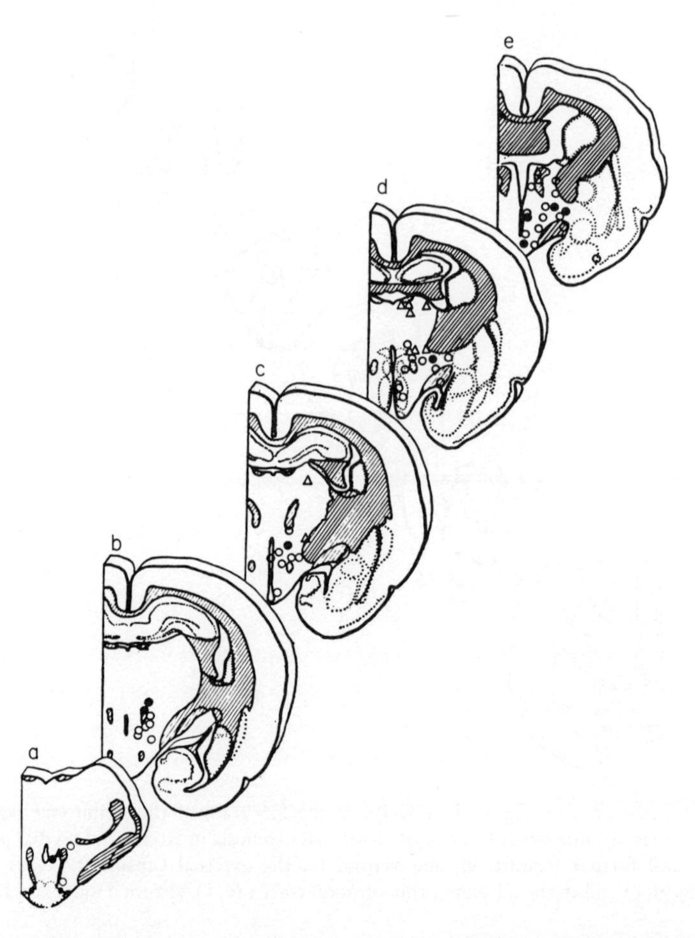

Fig. 6.6 Pathways for release of prolactin in the forebrain of the rabbit showing sites where twice-daily electrical stimulation during 11 days of pseudopregnancy caused the release of prolactin and lactogenesis. It was found that control electrode tips, if situated in a pathway for prolactin release, could also evoke release of prolactin by causing irritation of neurones (see Tindal and Knaggs, 1972). Stereotaxic planes pass rostrally at 1 mm intervals from (a) through to (f). Note stimulation sites associated with lactogenesis close to third ventricle in posterior hypothalamus (a) and further rostrally in medio-dorsal (b, c) and then in lateral hypothalamus (d, e) in association with medial forebrain bundle, also two sites near the third ventricle in the anterior hypothalamus (c). ●, Stimulation followed by lactogenesis; ▲, control electrode, no stimulation followed by lactogenesis; ○, stimulation, no lactogenesis; △, control electrode, no stimulation, no lactogenesis. (From Tindal and Knaggs, 1972.)

medial preoptic areas and thence caudally to the medical anterior hypothalamus close to the third ventricle (Tindall, 1974). There also appears to be a descending pathway capable of stimulating prolactin release passing from the orbito-frontal cortex to the preoptic area (Tindall and Knaggs, 1970, 1972). The hypothalamic sites immediately concerned in prolactin control have been studied in some detail, and are further considered below. Stimulation of sites in the lateral hypothalamus, but not in the medial hypothalamus, causes prolactin release (Shikado, 1966; Weiner *et al.*, 1972), and stimulation of the arcuate nucleus and median eminence also releases prolactin, these apparently being sites at which oestrogens exert their effect in inducing prolactin secretion (Clemens *et al.*, 1971*a*). Stimulation of preoptic sites have been shown to inhibit prolactin secretion (Clemens *et al.*, 1971), but the situation may be rather complicated, since other workers have reported that prolonged electrical stimulation of the preoptic region promotes prolactin release, whereas acute stimulation is inhibitory (Wuttke and Meites, 1972*b*).

Although tactile stimulation of the nipple is probably the main natural signal for prolactin secretion, there is evidence to suggest that prolactin may be released in response to other stimuli supplied by the litter. For example, Grosvenor (1965) has shown that nursing rats separated from their litters, but allowed to see them, exhibit a depletion of pituitary prolactin comparable to that in rats allowed to actually suckle their pups. This response was later shown to depend on olfactory stimuli from the pups, although visual stimuli will suffice if olfaction is blocked (Mena and Grosvenor, 1971). Deis (1968) obtained evidence indicating that auditory stimuli from a mother with pups increases the milk yield in rats, but since the response was mimicked by injections of oxytocin, it may depend on the release of oxytocin and facilitation of milk ejection, rather than on release of prolactin.

6.3.4 *Biochemical changes in the mammary tissue induced by prolactin*

Changes in the specific activity of a number of enzymes in the mammary gland have been reported in the rat (McLean, 1958; Rees and Huggins, 1960; Baldwin and Milligan, 1966) and in the rabbit (Gul and Dils, 1969) during the lactation cycle. These include increases in isocitrate dehydrogenase, pyruvate dehydrogenase and various enzymes associated with the glucose 6-PO_4(G-6-P) pathway. Lipoprotein lipase activity has also been shown to rise in the goat during lactation (McBride and Kern, 1963; Robinson, 1963) and can be stimulated by intraductal prolactin injection in the rabbit (Falconer and Fiddler, 1970). This response was blocked by addition of actinomycin D or cycloheximide, suggesting that it involves

RNA synthesis *de novo*. Similar rises have been reported in galactosyl transferase (Turkington, 1972), G-6-P enzymes (Heitzman, 1969; Hartman *et al.*, 1970) and in ornithine decarboxylase (Oka, 1974).

Chadwick (1960) has shown that prolactin increases the lactose content of the mammary gland, and it also increases the casein levels (Fiddler *et al.*, 1971; Bourne and Bryant, 1974). Fiddler *et al.* (1971) showed that prolactin stimulated the incorporation of amino acid into casein, while Bourne and Bryant (1974) showed that it also promoted the incorporation of P-orthophosphate into individual casein components by rabbit mammary glands. This has been confirmed by Bullough and Wallis (1974) who demonstrated a similar effect in mouse mammary cells, with an increase in the total milk protein content as previously shown by Juergens *et al.* (1965).

The production of the milk sugar lactose from G-6-P has been studied by a number of authors. Heitzman (1969) showed that prolactin increased the titre of certain enzymes associated with the G-6-P pathway in cattle, and a detailed study of enzyme activity in the mammary gland of lactating rabbits (Hartman *et al.*, 1970), showed that prolactin could maintain the levels of 6-phosphogluconate dehydrogenase, phosphofructokinase, and phosphoglucomutase in hypophysectomised animals. However, in most cases the measured levels of the enzymes associated with the G-6-P pathway were in excess of the levels required for milk sugar production and were therefore unlikely to be rate-limiting. On the other hand, the levels of acetyl-coA-carboxylase might be rate-limiting for fatty acid synthesis (see below). In the rabbit, bovine growth hormone was more effective than prolactin or HGH in elevating milk protein and lactose levels (Hartman *et al.*, 1970).

Delouis and Denamur (1972) have shown that prolactin increases the conversion of ^{14}C-glucose to lactose in cultures of rabbit mammary explants. Induction of lactose synthesis by prolactin, or by prolactin and cortisol, appears to take slightly longer *in vitro* than *in vivo*, possibly because of the generally slower rate of protein synthesis in the explants. Lactose synthesis presupposes the production of proteins such as α-lactalbumin, which increases the affinity of UDP-galactosyl transferase for glucose and concomitantly leads to a high level of lactose synthetase activity (Fitzgerald *et al.*, 1970; Klee and Klee, 1970). The fact that prolactin can be shown to act by itself in stimulating lactose synthesis in the pseudopregnant rabbit is unusual, since in the mouse corticosteroids are required together with prolactin to stimulate lactose synthesis (Palmiter, 1969) and UDP-galactosyl transferase and α-lactalbumin levels (Turkington and Hill, 1969). These results have been confirmed by Van der Laar, Owens and Topper (1973) who showed that virgin mice mammary glands exhibit low levels of both UDP-galactosyl transferase (lactose synthetase A protein) and α-lactalbumin (lactose synthetase B protein). Prolactin stimulated the production of both these proteins *in vitro*, when administered with insulin and hydrocortisone. The A-protein activity was shown to reach a maximum level after 72 hours of exposure to the hormones, 24 hours earlier than the peak of B-protein activity. However, priming of mice *in vivo* with either oestradiol or prolactin accelerates the appearance of B-protein activity.

These findings agree with the work of Jones and Cowie (1972) on rabbits who showed *in vivo* that there was a marked decline in lactose synthetase activity following hypophysectomy, correlated with a decline in α-lactalbumin activity. This decline could be prevented by replacement therapy with ovine prolactin but not with bovine growth hormone.

Extensive studies have also been made on the effects of prolactin on mammary fatty acid synthesis and milk fat production. Wang *et al.* (1971) showed that insulin can stimulate the incorporation of ^{14}C-acetate into the fatty acids of mouse mammary gland explants. This increased incorporation was further stimulated by the addition to the medium of prolactin and cortisol, or bovine growth hormone and cortisol. Prolactin was however more effective than bovine growth hormone on a weight basis. Prolactin-treated explants synthesised a higher proportion of ^{12}C and ^{14}C-fatty acids than did explants treated with insulin and cortisol alone. This work confirms the early studies of Abraham *et al.* (1960) who showed that hypophysectomy in the rat reduces fatty acid synthesis by the mammary gland, which can be restored by replacement treatment with prolactin or cortisol. In the rabbit, corticosteroids are not essential for the maintenance of fatty acid metabolism (Bolton, 1971), or for the output of milk fat (Cowie *et al.*, 1969). The rate of lipid synthesis *in vitro* was seven times higher with prolactin, insulin and corticosterone than with insulin and corticosterone (Bolton, 1971). These results have been confirmed in principle by the work of Hallowes *et al.* (1973) who showed that insulin and corticosterone can stimulate fatty acid synthesis in mammary tissue from virgin and pregnant rats, an effect potentiated by prolactin and, to a lesser extent, by growth hormone. However, in this study it was shown that prolactin mainly stimulated the incorporation of ^{14}C-acetate into 8C–10C chain fatty acids, in contrast to the findings for the mouse, where prolactin stimulated synthesis of longer-chain fatty acids (above, Wang *et al.*, 1971).

Further studies of the rabbit mammary gland *in vitro* confirmed that prolactin mainly increased the synthesis of 8C to 10C fatty acids, and that the addition of insulin to the explants further stimulated the production of these medium-chain compounds. The subsequent addition of corticosterone had no effect on these fatty acids, but enhanced the production of long-chain compounds. The addition of all three hormones to the explants resulted in increased synthesis of both types of fatty acids, but with a preponderance of the long-chain molecules (Forsyth *et al.*, 1972). However, normal rabbit milk secreted *in vivo* at about 23 days of pregnancy contains a preponderance of medium-chain fatty acids, similar to those stimulated by prolactin *in vivo* (Strong and Dils, 1972). The composition of milk secreted *in vivo* after corticosterone treatment is not known. It is noteworthy that addition of corticosterone *in vitro* was necessary to maintain the milk yield at *in vivo* levels.

As mentioned earlier, the levels of lipoprotein lipase activity rise in the goat mammary gland during pregnancy (Robinson, 1963), and this enzyme can be stimulated by the intraductal injection of prolactin in the rabbit (Falconer and Fiddler, 1970). The response appears to require the synthesis of new protein, since it can be blocked by the injection of actinomycin D. Similar responses have been observed in rat mammary tissues, in which lipoprotein lipase activity, presumably concerned with increased mobilisation of triglycerides, rises during the two to three days prior to parturition and remains high during lactation (Hamosh *et al.* 1970; Hamosh and Scow, 1971). When suckling stops, the enzyme level rapidly declines (McBride and Korn, 1963). In the rat, hypophysectomy on the fifth or sixth day of lactation reduced lipoprotein lipase activity in the mammary gland, and at the same time increased the activity in adipose tissues (Zinder *et al.*, 1974). Prolactin replacement treatment prevented these changes, and this restorative effect of prolactin was not altered to any extent by injections of dexamethasone, thyroxine or GH. These findings suggest that prolactin may act to divert dietary fatty acids from the body fat stores to the mammary gland prior to the synthesis of milk fats (Zinder *et al.*, 1974).

In addition to changing the activity levels of various enzymes involved in milk production, prolactin also appears to affect transport of polyamines and ions into the mammary tissues. Kano and Oka (1976) have shown that prolactin and insulin augment the transport of the polyamines spermidine, spermine and putrescine, into the mammary tissue of mice. Previous reports had shown that the concentration of spermidine increases during lactation *in vivo* (Russell and McVicker, 1972) and *in vitro* (Oka, 1974; Oka and Perry, 1974*a*). The work of Oka and his co-workers (Oka and Perry, 1974*a,b,c*) has suggested that spermidine may increase the synthesis of milk protein. This is supported by the fact that spermidine can substitute for glucocorticoids in maintaining lactogenesis *in vitro*.

The role of prolactin in osmoregulation in mammals is discussed elsewhere (see p. 189). In 1972 Shulkes and co-workers showed that prolactin produced an increased salt appetite similar to that found in many lactating mammals. Taylor *et al.* (1975) have shown that the aqueous phase of rabbit milk changes in composition during lactation, particularly in late lactation when it resembles colostrum, with high sodium and chloride levels and low potassium (see also Linzell and Peaker, 1971).

Work by Linzell and Peaker (1973, 1974) has suggested that there may be two routes of ion movements in the goat mammary epithelium, ions being transported either by a transcellular route or by a paracellular route *via* the tight junctions between the alveolar cells. Prolactin appears to produce milk similar in composition to that produced by goats in mid-lactation, with low sodium and high potassium (Taylor *et al.*, 1975). These authors suggest that prolactin acts by decreasing paracellular transport, possibly by physically altering the tight junctions.

Falconer and Rowe (1975) have shown that prolactin causes a decrease in mammary sodium levels and stimulates Na^{+}-K^{+}-ATPase activity in the alveolar epithelium, the response being blocked by ouabain. Parallel changes occurred in chloride levels. These findings point to transport across the cell membrane, somewhat contrary to the view held by Linzell and his co-workers (Taylor *et al.*, 1975; Linzell *et al.*, 1975) which holds that prolactin might decrease sodium transport into the milk by affecting the paracellular ion transport. The results of Linzell *et al.* (1975) were obtained, however, with rabbits at the end of lactation, as were those of Cowie (1969) who reported similar changes, whereas the results of Falconer and Rowe (1975) were obtained in mid-lactation 10–20 days postpartum, and therefore perhaps should not be compared with those of Linzell *et al.* (1975).

It can be seen that as well as stimulating mammary growth and milk flow, prolactin appears to be a major hormone stimulating the production of milk proteins, sugars and fats, and in addition is probably important in maintaining the ionic composition of the aqueous phase. The synergistic role of other hormones seems to vary with the species, but as in the development of mammary tissue (see also Chapter 8), insulin and corticosteroids appear to be essential for development of the fully fledged response to prolactin.

II Prolactin in the male

The role of prolactin in male mammals has not yet been clearly established, although there now exists considerable evidence for its involvement in the growth and function of sexual accessory organs, and in testicular steroidogenesis. Prolactin levels are relatively low in male mammals. In the first 20 days of life in the male rat, both prolactin and LH levels are low, but FSH levels are comparable with those in the adult (Negro-Vilar *et al.*,

1973). At 25 days both prolactin and FSH levels rise sharply while the levels of LH remain low. Prolactin levels show two subsequent rises at 60 days and 70 days, before reaching adult levels. In contrast, FSH peaked at 30 days and then declined, while LH increased gradually to reach adult levels at about the same time as prolactin. Basal levels of prolactin in mature males vary with species, being about 20–30 ng/ml in the male rat (Amenori *et al.*, 1970), 50 ng/ml in the ram (Davies *et al.*, 1971), 80–200 ng/ml in the male goat (Hart, 1973; Buttle, 1974), but only 30–40 ng/ml in the bull (Oxender *et al.*, 1972). In male lambs Ravault and Courot (1975) showed that prolactin was detectable in the plasma about one week after birth, and remained relatively constant until the animals were 10–12 weeks old when there was a sharp rise in serum prolactin, which reached a maximum value one to two weeks later. This peak then declined subsequently to basal levels. This pattern is different from that in the rat (Negro-Vilar *et al.*, 1973; Dohler and Wuttke, 1974), although in both species the prolactin rise appears to be independent of LH and testosterone (Cotta *et al.*, 1974). In both species however, the rise in prolactin levels was correlated with testicular growth.

6.4 The testis

An effect of prolactin on the testes was first reported by Bartke (1965, 1966*a,b*) who showed that injections of prolactin caused gonadal development in infertile dwarf mice which have a genetic deficiency limiting the production of growth hormone and prolactin. Prolactin also stimulates spermatogenesis in these dwarf mice (Bartke and Lloyd, 1970). Bartke (1969) has also shown that prolactin affects the cholesterol stores of male mice: dwarf male mice bearing prolactin-secreting adult female pituitary transplants showed a significant increase in esterified cholesterol in the testis, an effect similar to that produced by prolactin in the ovary (see p. 136). In addition prolactin replacement therapy in hypophysectomised normal mice tended to produce an increase in total testicular cholesterol and in the percentage of esterified cholesterol, although these changes were not significant. Recently, Bartke (1976) has demonstrated that prolactin also stimulates spermatogenesis in hypophysectomised normal mice, an action in which prolactin may be synergistic with LH, possibly acting *via* the Leydig cells (Bartke, 1971). Administration of ergocornine or CB154 (bromocryptine) to male mice significantly lowers their total testicular cholesterol levels, most notably the level of esterified cholesterol. LH acts on the Leydig cells to deplete the levels of esterified cholesterol while prolactin blocks this effect and maintains the esterified cholesterol pool as a precursor for steroidogenesis promoted by LH (Bartke, 1974). Bartke *et al.*

(1976) also have shown that prolactin increases the sensitivity of the testis to LH. These results indicate that prolactin acts on the mouse testis in ways analogous to its ovarian action (pp. 136–139).

In the rat, the action of prolactin on the testis is not quite so clearly defined. Woods and Simpson (1961) showed that prolactin slightly increased testicular weight in hypophysectomised rats. Evans (1962, 1964) has reported a stimulatory effect of prolactin on the testicular β-glucuronidase activity, and Musto *et al.* (1972) found that prolactin can also increase 17β-hydroxysteroid dehydrogenase levels in the testes.

Moreover, Negro-Vilar and Saad (1972) have reported that ectopic pituitary transplants in immature hypophysectomised male rats significantly stimulated testicular growth. Since LH and FSH were undetectable in the plasma of these animals, it may be concluded that prolactin alone was responsible for this effect. Later work by Negro-Vilar *et al.* (1973) showed that the sharp rise in FSH and prolactin in the 25-day old rat was correlated with a rapid increase in testicular weight, and a similar correlation was observed in the immature ram by Ravault and Courot (1975).

Hafiez *et al.* (1972) have shown that treatment of hypophysectomised rats with LH and prolactin greatly increases plasma testosterone levels, indicating that prolactin may have similar effects in the rat as in the mouse. This would agree with earlier work by Hafiez *et al.* (1971), showing that prolactin could stimulate 3β-hydroxysteroid dehydrogenase levels in the testes. Boyns *et al.* (1972) demonstrated that blockade of prolactin release by bromocryptine increased serum LH levels in the rat, but reduced serum testosterone levels, suggesting that prolactin may be physiologically important in maintaining normal testosterone secretion in the rat. Confirmatory evidence has been obtained by Johnson (1974). The work of Boyns *et al.* (1972) also indicated that prolactin could stimulate testosterone production by the adrenal, supporting the concept of a widespread general role for prolactin in maintaining steroid hormone precursors (Nicholl, 1974; see p. 136).

6.5 Accessory structures

Prolactin has been shown to have a marked stimulatory effect on the growth and secretory capability of the prostate in both rats and mice (Grayhack *et al.*, 1955; Grayhack, 1963; Grayhack and Lebowitz, 1967; Bartke, 1974), an action which appears to involve synergism with androgens (Thomas and Manandhar, 1974, 1975). The dorsal and lateral lobes of the prostate appear to be more sensitive to the prolactin/androgen combination than the ventral lobes (Moger and Geschwind, 1972).

A number of studies have established that prolactin augments the synthesis of citric acid in response to androgen stimulation in the dorsal lobe and lateral lobes of the prostate in castrated-hypophysectomised rats (Grayhack and Lebowitz, 1969). Reddi (1969) has demonstrated a similar effect in the castrated Indian palm squirrel. Grayhack and Lebowitz (1969) had also shown that prolactin stimulates fructose synthesis by the dorso-lateral lobe in the hypophysectomised-castrated rat, but this effect was not detected by Keenan and Thomas (1975) in castrated mice. However, prolactin does stimulate fructose synthesis in the prostate of the castrated Indian palm squirrel (Reddi, 1969). Prolactin has also been shown to increase ^{65}Zn uptake by the dorso-lateral prostate lobes in the castrated male rat (Moger and Geschwind, 1972), an effect which does not require synergism with androgens.

A major effect of prolactin on the prostate may be to increase its uptake of testosterone. Zepp *et al.* (1973) showed that prolactin increased the uptake of radioactive testosterone and dehydrotestosterone by the anterior and dorsal lobes of the rat prostate, but not by the ventral and lateral lobes. Similar results were obtained by Keenan *et al.* (1973), who found that androgen uptake by the prostate was stimulated by prolactin in castrated rats but not in intact rats, findings which suggest that androgens may be important in maintaining the prolactin-stimulated androgen uptake. This hypothesis has to some extent been confirmed by Tindal and Means (1976) who showed that prolactin in high doses stimulates androgen-binding protein levels in the rat prostate. These authors point out that prolactin also stimulates testosterone production, however, which in turn can increase the level of androgen binding protein. Lloyd *et al.* (1974) confirmed that prolactin can stimulate the uptake of androgens by the anterior and dorsal lobes of the rat prostate *in vitro*. Most interestingly, however, this response was produced by ovine prolactin but not by bovine prolactin, which actually reduced androgen uptake. On the other hand, Keenan and Thomas (1975) recently failed to detect increased prostatic localisation of testosterone in castrated rats injected with ovine prolactin, although the prolactin did augment the classical androgen-stimulated increase in prostate weight. Why this work should in essence disagree with the earlier work of Keenan *et al.* (1973) is not clear. Controversy has existed for some time regarding the effect of prolactin on male sex accessories in the rat. Lostroh and Li (1957) for example, could detect no effect of prolactin on prostate weight, but Woods and Simpson (1961) later demonstrated synergism between androgen and both prolactin and growth hormone.

Golder *et al.* (1972) have reported a stimulation of prostatic adenyl cyclase by low doses of prolactin *in vitro*. Thomas and Manandhar (1974) have shown that both testosterone and prolactin separately reduced cAMP levels in the rat ventral prostate after 30 minutes *in vitro*, but that the two hormones together were ineffective. However, the rat *ventral* prostate is relatively insensitive to prolactin and androgen therapy (Moger and Geschwind, 1972). Thomas and Manandhar (1975) showed that prolactin increased the RNA and DNA content of the ventral prostate of the rat, but did not increase the concentration or uptake rate of ^{3}H-thymidine. This may suggest an increase in cell number, that is a proliferative effect, rather than an increase in protein synthesis. It is clear that all this field requires further investigations.

Not only the prostate is responsive to prolactin. Prolactin also has stimulatory effects on the seminal vesicles of the rat (Pasqualini, 1953), mouse (Bartke, 1974) and guinea pig (Antliff *et al.*, 1960), and stimulates proliferation of rat seminal vesicle cells *in vitro* (Bengmark and Hessebojo, 1963, 1964). Reddi (1969) obtained similar results in the Indian palm squirrel, and also demonstrated that prolactin increases citric acid levels in the seminal vesicle. Moreover, ergocornine- and bromocryptine-blockade of prolactin release in the mouse results in a significant reduction in seminal vesicle weight (Bartke, 1974). Prolactin has also been shown to increase the uptake of androgens in the guinea-pig seminal vesicle and to cause a shift from reductive metabolism of androgens.

In summary, most evidence in this field comes from the effects of exogenous prolactin on male sex accessory glands, but the results of Bartke (1974) with prolactin blockade suggest a truly physiological role of prolactin in maintaining the seminal vesicles in the mouse, and Peyne *et al.* (1968) have shown that reserpine treatment in rats augments seminal vesicle growth in response to androgens. Accordingly, it appears that endogenous prolactin plays a part in the maintenance of the sexual accessory organs, at least in the mouse and the rat. Most intriguing is the finding by Asano (1965) that removal of the prostate led to elevation of prolactin secretion in rats, which was suppressed by a crude prostatic extract. The physiological elucidation of this observation awaits further research.

III Hypothalamic regulation of prolactin secretion

An intact connection between the hypothalamus and the pituitary is essential for the normal regulation of prolactin secretion. Everett (1954) showed that stalk section and ectopic transplantation of the pituitary results in autonomous hypersecretion, and electrolytic lesions in the hypophysiotropic area of the hypothalamus increase serum prolactin concentration

(Chen *et al.*, 1970; Bishop *et al.*, 1971) and the rate of prolactin synthesis (Macleod and Lehmeyer, 1972).

6.6 Inhibition

The overall effect of the hypothalamus on prolactin secretion in mammals is inhibitory. Pasteels (1961) and Talwalker *et al.* (1963) among others have shown that acidic extracts of the hypothalamus contain a factor that inhibits prolactin release from the rat pituitary *in vitro*. The nature of this prolactin inhibiting factor (PIF) is discussed later (pp. 165–169); it is worth noting here that Talwalker *et al.* (1963) were able to show that prolactin secretion by the isolated rat pituitary was not affected by acetylcholine, epinephrine, norepinephrine, serotonin, histamine or oxytocin.

In vivo experiments by Amenomori and Meites (1970) have shown that crude extracts of hypothalamic tissue are able to decrease serum prolactin levels in normal non-lactating rats. A rapid response results from injecting such extracts into the jugular vein, a significant decrease in plasma levels being observed eight minutes after injection (Watson *et al.*, 1971). Other workers have shown that injections of hypothalamic extracts can block the suckling-induced fall in pituitary prolactin levels in lactating rats (Grosvenor *et al.*, 1965; Grosvenor, 1965; Dhariwal *et al.*, 1968), and Kuhn *et al.* (1974) have shown that slightly purified hypothalamic extracts depress the high circulating levels of prolactin characteristics of lactating rats. Kuhn *et al.* (1974) also showed that higher doses of hypothalamic extract were required to suppress prolactin secretion *in vivo* than *in vitro*, and suggested that this might be due to the very short half-life of the PIF in the circulatory system. The failure of Amenomori and Meites (1970) to demonstrate an *in vivo* effect of hypothalamic extracts in lactating rats may have stemmed from this fact, since they measured prolactin levels three hours after injection of the extracts. Kamberi *et al.* (1971) confirmed *in vivo* that hypothalamic PIF acts directly on the pituitary, by infusing hypothalamic extracts directly into the hypophysial portal system and detecting an almost immediate reduction in prolactin release. The response persisted for about 90 minutes after infusion, and was dose-independent.

The work mentioned constitutes only a fraction of the great mass of evidence showing that as in most other vertebrates, prolactin secretion in mammals is under an inhibitory hypothalamic control. Many workers, for example, have studied ectopic pituitary transplants, which secrete prolactin at a high rate (Everett and Nikitowitch-Winer, 1963). Work with these ectopic transplants has raised a striking and important anomaly: Chen *et al.* (1970), among others, reported that multiple pituitary transplants in hypophysectomised rats can never raise plasma prolactin to the levels found during normal lactation. Evidence such as this, and of other kinds as well, has led to the concept of a prolactin releasing factor (PRF), discussed further on pp. 174–178. Nonetheless, prolactin output by ectopic pituitaries is high, and experiments involving lesions of the median eminence in the hypophysectomised-ectopically grafted rat indicate that the pituitary graft might be subject to a degree of hypothalamic inhibition by way of the systemic circulation (Sud *et al.*, 1970). A further complication lies in the fact reported from several laboratories that prolactin is detectable in the serum of hypophysectomised rats (Chen *et al.*, 1970; Lu and Meites, 1972; Niell and Reichert, 1971), and the suggestion has been made that the pars tuberalis, which is left *in situ* at hypophysectomy, may continue to secrete small amounts of prolactin. Even apart from this, completeness of hypophysectomy is critical in all this work: thus, Malven and Portens (1973) have showed that residual fragments even as small as one-eighth of a pituitary can raise serum prolactin levels above those of hypophysectomised rats.

Some discussion still surrounds identification of the site of PIF synthesis. Lesions in the arcuate nucleus, median eminence and the ventral parts of the ventro-medial nucleus can all cause release of prolactin (Haun and Sawyer, 1960; Chen *et al.*, 1970; Welch *et al.*, 1971) and these results could reflect either blockage of the main PIF release pathway or destruction of sites of synthesis. Hejco *et al.* (1972) and Scaraimuzzi *et al.* (1972) have reported that total deafferentation of the hypothalamus causes a marked and permanent fall in serum prolactin levels. Later work by Krulich *et al.* (1975) has shown that postero-lateral deafferentation of the hypothalamus results consistently in low serum prolactin levels, but that deafferentation in the anterior region is relatively ineffective. However if care was taken to eliminate stress at the time of serum collection the difference all but disappeared. Krulich *et al.* (1975) see this as further evidence for a 'stress-stimulated' release mechanism, Terkel *et al.* (1972), and conclude that prolactin is released by stress, such as ether, during blood-sampling and that deafferentation merely prevents this stress response.

The afferent pathways to the hypophysiotrophic regions concerned in prolactin control are not known with any certainty. The ventro-medial nucleus receives a dual innervation *via* the stria terminalis from the amygdala (Heimer and Manta, 1969; Chi, 1970; De Ohnos, 1972). The dual nerve tracts originate in areas of the amygdala which have been shown to bind oestradiol preferentially (Stumpf and Sar, 1971; Pfaff and Keiner, 1972), corresponding to the area in which oestradiol implants cause prolactin release (Tindall *et al.*, 1967). Other evidence suggests that there may be direct neural connections from the preoptic region to the ventromedial nucleus (Heimer and Manta, 1969). These possible local innervatory pathways are shown diagrammatically in Fig. 6.7.

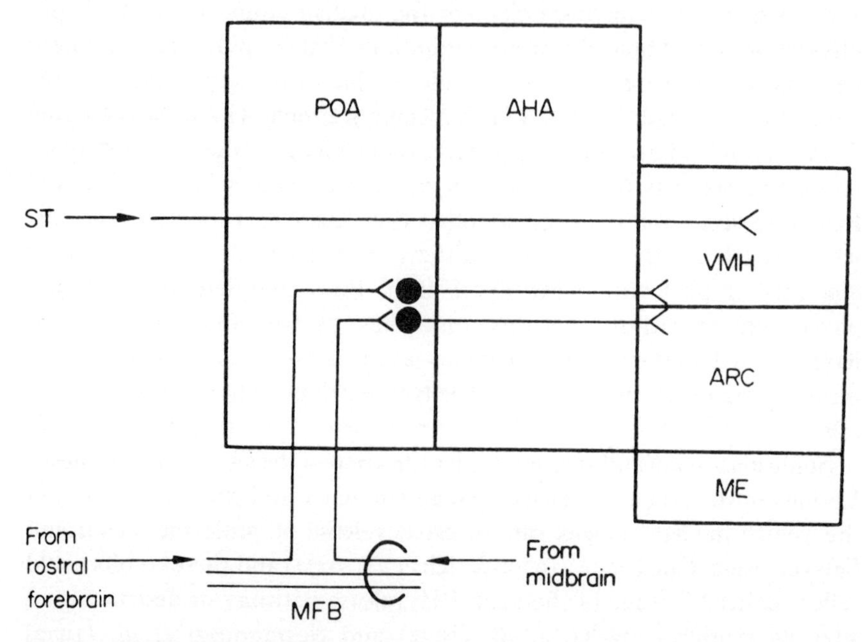

Fig. 6.7 Possible mechanism for major releases of prolactin. Impulses from prolactin-releasing stimuli ascend via the spinothalamic tract and medial forebrain bundle or descend from the orbital cortex, claustrum, external capsule and part of striatum to enter the preoptic area. Stimula could act at the preoptic level on a subsequent neuron or chain of neurons involved in tonic release of PIF by the basal hypothalamus. It is possible that while the somas of such neurons could lie in the arcuate ventromedial area their dendrites might extend rostrally to the preoptic–anterior hypothalamic region. In either case it is proposed that transmission in these final neurons would be inhibited by prolactin-releasing stimuli, causing a temporary inhibition of PIF release and hence a surge of prolactin release. The mechanism could also be modulated by a specific region of the amygdala acting via the stria terminalis on the ventromedial nucleus. AHA = anterior hypothalamic area, ARC = arcuate nucleus, ME = median emminence, MFB = medial forebrain bundle, POA = preoptic area, ST = stria terminalis, VMH = ventromedial nucleus. (From Tindall, 1974.)

The response of the hypothalamus to stimulation is modified by ovarian steroids. Clemens *et al.* (1971*a*) have shown that stimulation of the basal-medial hypothalamus causes increased serum prolactin levels in intact rats. This has been discussed above and could result from either inhibition of PIF or stimulation of PRF, or from the combination of both mechanisms. Both oestrogen (Meites, 1970) and progesterone (Meites and Nicoll, 1966; Sar and Meites, 1968) have been shown to deplete hypothalamic PIF stores. Oestrogen appears to lower the thresholds of certain hypothalamic neurons to specific stimuli (Kawakari and Tevasawa, 1967; Beyer *et al.*, 1970) while progesterone may exert a biphasic response (Sawyer, 1967). More recently Clemens *et al.* (1971*b*) have shown that sham stimulation and electrochemical stimulation of the hypothalamus caused prolactin release in animals treated with oestrogen. Rats treated with both progesterone and oestrogen showed a large rise in serum prolactin levels following electrochemical stimulation but not after sham stimulations. This work illustrates clearly the differential effects of steroids on hypothalamic responses to specific stimulation.

6.6.1 *The nature of PIF*

There has been considerable controversy over the past few years as to whether PIF resembles other hypophysiotrophic hormones in being a peptide, or whether it is a catecholamine. The evidence discussed in this section is by no means encyclopaedic but is representative of the state of present knowledge.

The first indications of a special role for hypothalamic catecholamines in prolactin control came from Kanematsu *et al.* (1963*a*), who showed that reserpine injections induced lactation and decreased pituitary prolactin content in rats. Reserpine is a sedative which probably acts on the brain by depleting catecholamines in adrenergic neurons. Anatomical data (Hökfelt, 1967) has shown that dopaminergic neurones from the tuberoinfundibular region terminate in the median eminence adjacent to the portal blood vessels which feed the anterior pituitary. Further work by Hökfelt and Fuxe (1972) has suggested that prolactin can also modify the activity of these tuberoinfundibular dopaminergic fibres. Using the Falck-Hillarp fluorescent technique, they showed that prolactin, alone of the hormones tested, caused a dose-dependent increase in the turnover of dopamine in these neurons. The response was marked in hypophysectomised male and female rats, and reduced in castrates; the turnover rate was high during lactation, and in this phase turnover was reduced by hypophysectomy or by ergot alkaloids known to reduce prolactin secretion (see p. 170). Hökfelt and Fuxe (1972) concluded that these tuberoinfundibular neurons were part of the normal feedback system for prolactin and that dopamine acted by stimulating the release of PIF.

This conclusion raises the funadamental point in the controversy about the role of dopamine or other catecholamines in the control of prolactin secretion. The two possibilities are, firstly, that dopamine acts directly on the pituitary; and secondly, that dopamine acts on the hypophysiotrophic region of the hypothalamus to promote PIF release (Macleod, 1976). As will be seen, both mechanisms may in fact be operative.

Early work with systemically injected catecholamines was inconclusive. Lu *et al.* (1971) showed that a single systemic injection of epinephrine, norepinephrine or dopamine had no effect on serum prolactin levels, although epinephrine and dopamine caused a small but significant drop in pituitary prolactin levels. However these catecholamines have been shown not to pass readily across the blood-brain barrier, which makes interpretation of these results uncertain (Innes and Nieberson, 1970). In experiments designed to obviate this difficulty, Kamberi *et al.* (1970) showed that a single injection of dopamine into the third ventricle of rats depressed serum prolactin levels and also appeared to increase hypothalamic PIF activity.

Recently the effect of the suckling reflex on hypothalamic monoamine levels has been studied in the rat by Mena *et al.* (1976). These authors took lactating rats which had been separated from their litters for eight hours, and showed that resumption of suckling caused a rapid rise in plasma prolactin levels, associated with a rapid and marked depletion of hypothalamic dopamine and serotonin, with an increase in the levels of 5-hydroxyindoleacetic acid, a metabolite of serotonin. Thus, not only the catecholamine dopamine, but also the indoleamine serotonin, might be involved in prolactin control.

The findings of Mena *et al.* (1976) suggest that dopamine may not be the only hypothalamic monoamine concerned in prolactin control. Lu *et al.* (1971) had previously shown that systemically injected epinephrine slightly reduced pituitary prolactin levels, and Kamberi *et al.* (1970) had reported that intraventricular injections of melatonin can cause prolactin release. Recently, Chen and Meites (1975) showed that 5-hydroxytryptophan, a precursor of serotonin, caused a marked increase in plasma prolactin levels, and that para-chloroamphetamine, which depletes serotonin stores, caused a fall in serum prolactin. Subramanian and Gala (1976) also have shown that serotonin antagonists could block the afternoon surge of prolactin in oestrogen-primed rats, as could several adrenergic blocking drugs. Lawson and Gala (1976) reported that after blockade of prolactin secretion by apomorphine – a dopamine agonist (see p. 172) – there is a marked rise or 'rebound' in serum prolactin levels. This 'rebound' peak of prolactin secretion could be decreased to some extent by giving serotonin-blocking drugs, suggesting that at least part of the rise may be induced by serotonin.

These facts suggest that there are probably two monoamine-based control systems in the hypothalamus, one catecholaminergic, the other serotoninergic. A great deal of evidence indicates not only that inhibitory control is based on catecholamines, but further that dopamine is the specific agent involved. Blake (1976) has basically repeated the earlier work of Lu *et al.* (1973) and shown that of the three possible catecholamines used (dopamine, epinephrine and norepinephrine) only dopamine is a potent inhibitor of prolactin release. Although norepinephrine and epinephrine could partially inhibit the 'rebound' rise in serum prolactin levels following dopamine infusion in the rat, they did not suppress tonic prolactin secretion.

a. *Is PIF a catecholamine?* Evidence for a direct effect of catecholamines on prolactin secretion was first provided by Macleod (1969) and Birge *et al.* (1970) who showed that the release of newly synthesised ^{3}H-leucine-labelled prolactin *in vitro* was inhibited by dopamine (5×10^{-7}M). Radioactive prolactin accumulated in the pituitary cells prior to the arrest of synthesis. Macleod and Lehmeyer (1972) subsequently showed that in rat pituitaries incubated with catecholamines the prolactin cells became loaded with secretory granules and had a greatly hypertrophied endoplasmic reticulum. Shaar *et al.* (1973) have reported *in vitro* inhibition of rat prolactin secretion with 10^{-9}M dopamine, although Macleod *et al.* (1970) were able to show clear-cut inhibition only at a dose level of 10^{-7}M. Wilson and Ensor (1977) have also found that 10^{-7}M is about the lowest effective dose for inhibition.

Shaar and Clemens (1974) showed that treatment of hypothalamic extracts with monoamine oxidase virtually eliminates their prolactin-inhibiting activity. In addition Wilson and Ensor (1977) have recently shown that incubation of cultured male rat pituitaries with dopamine significantly raises cAMP levels and that this action can be mimicked by incubation with dibutyryl cAMP (see Fig. 6.8). This and other evidence suggests that dopamine is capable of exerting a direct inhibitory effect on the rat prolactin cell.

b. *Is PIF a peptide?* Although most workers would now agree that hypothalamic dopamine plays a part in the direct inhibition of prolactin secretion (Chen and Meites, 1974; Macleod, 1976), the possibility exists that dopamine may act to release a peptide PIH from hypothalamic sites into the portal system. Guilleman and his co-workers (Brazeau *et al.*, 1973; Vale *et al.*, 1973) have reported the isolation of a tetradecapeptide capable of inhibiting the release of GH, TSH and prolactin. More recently, Dular *et al.* (1974) have extracted bovine posterior lobes including the stalk and median eminence regions. These extracts were centrifuged, concentrated under vacuum, and passed through Sephadex G-25. Vasopressin contamination was removed by incubation of the active peptide fractions with collidine-pyridine-acetic acid, and the resultant preparations were then shown by

Incubation period mins.	mean p.mols. c.AMP. mg^{-1} AP tissue.			Percent change
	10^{-6} M dopamine	Control	Difference	
0	3.87	4.12	−0.25	− 6
2	2.57	3.02	−0.45	−15
3.5	3.20	3.18	+0.02	+ 0.6
5	3.82	3.41	+0.41	+12
7	1.03	0.98	+0.05	+ 5
10	1.56	1.90	−0.34	−18

Fig. 6.8 The effect of dopamine on cyclic AMP levels in cultured rat pituitary tissue.

bioassays to be free of neurohypophysial peptides. Some of the resulting fractions produced strong inhibition of prolactin secretion by the bovine pituitary *in vitro*. None of these workers, however, treated their preparations with monoamine oxidase to remove any monoamine activity inherent in their preparations. Shaar and Clemens (1974) reported that in their hands monoamine oxidase treatment virtually destroyed the prolactin-inhibiting activity in hypothalamic extracts. However, Dular *et al.* (1974) claim that direct administration of catecholamines is without effect in their *in vitro* system. Further, Greibrokk *et al.* (1974, 1975) have isolated a porcine peptidic PIF, distinct from somatostatin, which clearly merits further investigation.

c. *Is PIF somatostatin?* Another hypothalamic peptide which can alter prolactin secretion is somatostatin (GH-inhibiting hormone, GHIH). This hormone reduces basal levels of prolactin secretion in female rats, and also opposes TRH-stimulated prolactin release in culture (Drouin *et al.*, 1976). However somatostatin appears to have little or no effect *in vivo*, in anaesthetised males treated with oestradiol benzoate, in hypothyroid rats, or in hypothyroid rats treated with oestradiol benzoate.

The fact that somatostatin is a tetradecapeptide could account for some reports, discussed previously (p. 167), which claimed the isolation of a peptidic PIF from hypothalamic extracts. However, Greibrokk *et al.* (1974, 1975) have reported isolation of a peptidic PIF from the pig hypothalamus, and claim that it is distinct from somatostatin.

6.6.2 *Effect of pharmacological agents*

a. *L-dopa* L-dopa (levo-dihydroxy-phenyl alanine) is a precursor in catecholamine synthesis and its administration results in an increase in brain catecholamines, particularly dopamine and norepinephrine. A number of authors have reported effects of L-dopa on serum prolactin levels. Lu and Meites (1972) have shown that L-dopa produces a rapid drop in serum prolactin levels, with a concomitant rise in pituitary content. In addition, a number of workers have reported suppression of serum prolactin by L-dopa within two to three hours in man (Kleinberg and Frantz, 1971; Malarkey *et al.*, 1971; Frantz *et al.*, 1972; Friesen *et al.*, 1972; Turkington, 1972).

Chen *et al.* (1974) showed that L-dopa can block the suckling-induced rise in prolactin levels, and they concluded that L-dopa acts by increasing hypothalamic catecholamine levels and thus releasing PIF (cf Meites and Clemens, 1971). This view had been previously put forward by Lu and Meites (1972), who demonstrated that injection of L-dopa could also inhibit the release of prolactin from the ectopically grafted pituitary. These findings, have been confirmed by Donoso *et al.* (1973, 1974). Kamberi *et al.* (1971) have shown that intraventricular administration of dopamine inhibits prolactin release whereas infusion of saline solutions of dopamine into the hypophysial portal system does not. Donoso *et al.* (1973), on the other hand, found that L-dopa could still stimulatę prolactin release from the ectopic pituitary transplants even when lesions were placed in the median eminence, findings which support the view of Macleod and his co-workers (Macleod, 1969; Macleod *et al.*, 1970) that dopamine has a direct inhibiting action on the prolactin cell. These views have recently been upheld by the results of Takahara *et al.* (1974) who showed that catecholamines infused in glucose solution into the portal blood system were effective in inhibiting prolactin secretion. Glucose protects the catecholamines from monoamine oxidase activity, and the failure of Kamberi *et al.* (1971) to inhibit prolactin release may have been due to oxidation of their infused dopamine by monoamine oxidase, which occurs at high levels in the pituitary.

b. *Dopamine agonists*

Ergot alkaloids Probably the best known pharmacological agents which alter prolactin release are the ergot alkaloids. Most of these substances are based on a tetracyclic ring structure designated as ergoline. Cassady *et al.* (1974) have recently reviewed the efficacy of a wide range of these compounds in inhibiting prolactin release, and have concluded that the basic four-ring ergoline structure is essential for activity. Modification of the D ring (see Fig. 6.9) greatly increased the inhibition, and 8,9-ergolines were more potent inhibitors than the 9-10-derivatives resulting from a shifting of the double bond. Lysergic acid and dihydroxy derivatives also significantly inhibited prolactin release. However, two derivatives of the basic ergoline molecule, ergocryptine and bromocryptine, are now the most widely used pharmacological inhibitors of prolactin release. Another deriva-

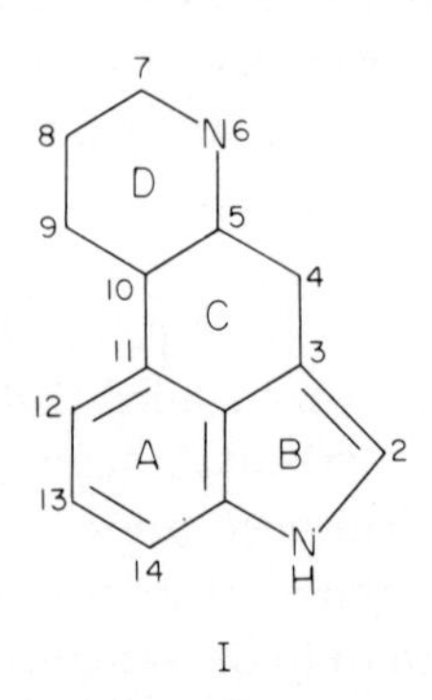

Fig. 6.9 Basic ergoline molecular structure.

tive, ergocornine, also inhibits prolactin release from cultured rat pituitary halves, and it also prevents the oestrogen-stimulated rise in prolactin output (Meites *et al.*, 1972). This inhibition of prolactin release is associated with an increase in the pituitary content (Lu *et al.*, 1971). Ergocornine was also shown to be very effective in reducing the diameter of 7,12-dimethyl-12-benzanthracene (DMBA)-induced mammary tumours in female Sprague-Dawley rats. Surprisingly, in view of its potent prolactin-inhibiting effects in Wistar rats (Flint and Ensor, 1976), ergocryptine was relatively ineffective in inhibiting tumour growth (Meites *et al.*, 1972). However other workers have reported that ergocryptine inhibits DMBA induced mammary tumours in rats (Heuson *et al.*, 1970), that ergot-derived drugs can block the development of hyperplastic alveolar nodules in mice (Yanai and Nagasawa, 1970), and that ergocornine and ergocryptine can suppress growth of spontaneous mammary tumours in old female rats (Quadri and Meites, 1971). As prolactin secretion is known to stimulate the growth of mammary tumours, particularly those induced by DMBA injection (Dao, 1964; Pearson, *et al.*, 1969; Meites, 1972), these results suggest that ergocryptine can induce a long-term suppression of prolactin release.

More recent work has supported this view (Arai *et al.*, 1972; Sinha *et al.*, 1974). Macleod and Lehmeyer (1972, 1973) have confirmed that prolactin release in rats is inhibited by administration of ergocornine and ergocryptine. Wuttke *et al.* (1971) suggested that ergocornine might act by increasing PIF secretion by the hypothalamus, although they did not exclude the possibility that it might inhibit the pituitary gland directly, an effect demonstrated, as already described, by the *in vitro* studies of Meites *et al.* (1972) and other workers (Nicoll *et al.*, 1970; Lu *et al.*, 1971; Pasteels *et al.*, 1971; Macleod and Lehmeyer, 1972). Ergocryptine appears to act as a potent dopamine agonist, with its action being blocked by dopamine antagonists such as haloperidol, pimozide and perphenazine (Macleod and Lehmeyer, 1974).

In non-rodent species, injections of ergocryptine or bromocryptine have been shown to reduce serum prolactin levels in the sheep (Niswender, 1972, 1974), cow (Karg *et al.*, 1972; Smith *et al.*, 1974) and goat (Hart, 1973; McMurty and Malven, 1974). The treatment reduced milk production in the goat, but not in the cow, a species in which the supply of prolactin does not appear to be rate-limiting for lactation (Cotes *et al.*, 1949).

Interestingly, Smith *et al.* (1974) have shown that ergocryptine inhibition of prolactin secretion in rats lasts for about five days following two consecutive daily injections. More recently Flint and Ensor (1976) have shown that ergocryptine in postpartum-mated lactating rats results in reduced litter growth-rates. After a single injection litter growth-rate was depressed for 24 hours and then became normal until day six when it declined again. *In vitro* culture of pituitaries from female rats showed that a single injection of ergocryptine inhibited prolactin release for 24 hours, the autonomous release then recovering but dropping again at six days, and a similar pattern of response was demonstrated *in vivo* by radioimmunoassay of serum prolactin levels (Flint and Ensor, in press). These findings suggest that ergocryptine has a short-term (24-hour) initial effect on prolactin release, and a longer-term suppressive effect on synthesis. By day six, this long-term action becomes so pronounced that prolactin release is again reduced, as a consequence of depressed synthesis.

Horrobin (1974) has pointed out that ergocryptine may have independent effects in its own right which may confuse interpretation of its action on prolactin secretion.

Apomorphine Apomorphine is an alkaloid with pharmacological properties similar to dopamine, and it has been shown to stimulate dopamine receptors with a high degree of specificity (McKenzie, 1971). It was originally considered that apomorphine acts indirectly on prolactin secretion, *via* the hypothalamus, but MacLeod and Lehmeyer (1973) showed that it can in fact act directly on the rat pituitary *in vitro* to cause a 60% decrease in the release of newly synthesised prolactin. This response was abolished if the donor rats were given a prior injection of perphenazine, or if the isolated pituitaries were incubated with perphenazine; perphenazine is known to promote prolactin release, in part at least by blocking inhibition by dopamine at the pituitary level (see pp. 173–175). Smalstig *et al.* (1974) have confirmed that dopamine directly inhibits release of prolactin from the rat pituitary *in vitro*, and they also found that apomorphine can inhibit the suckling-induced rise in serum prolactin levels *in vivo* (see also Lawson and Gala, 1976), as well as inhibiting the surge of prolactin at pro-oestrus (see p. **000**). Apomorphine also significantly attenuated the stimulatory effect of chloropromazine treatment on prolactin secretion (cf Lu *et al.*, 1970).

Thus far, the evidence is entirely consistent with the idea that apomorphine acts directly on the prolactin cell. However, apomorphine also has effects at the hypothalamic level. Ojeda *et al.* (1974), for instance, have shown that intraventricular injections of apomorphine can inhibit prolactin release *in vivo* in oestrogen-treated ovariectomised rats, and their findings suggest that the mechanism of action of apomorphine is to increase PIF secretion by the hypothalamus. This view implicitly assumes the existence of a PIF other than dopamine, a concept to which we shall return later (pp. 165–167).

c. *Tranquillising drugs* A wide range of tranquillising drugs such as reserpine, perphenazine, Chlorpromazine and α-methyl-p-tyrosine have been shown to stimulate prolactin secretion. Early work by Barraclough and Sawyer (1959) and Kanematsu *et al.* (1963) showed that tranquillisers may induce breast enlargement and galactorrhea in humans, and Coppola *et al.* (1965) found that reserpine can induce pseudo-pregnancy, an effect which they ascribed to depletion of hypothalamic norepinephrine stores, and which could be blocked by L-dopa but not by α-methyldopa. In 1965, Ratner *et al.* confirmed that hypothalamic extracts from reserpine-treated rats were less potent in inhibiting prolactin release than extracts from control rats. Further investigations have shown that reserpine strongly increases serum prolactin levels, and that the response can be blocked or depressed by subsequent administration of L-dopa, but can be potentiated by 5-HTP injections (Chen and Meites, 1975). In addition Chen and Meites

(1975) showed that α-methyl-m-tyrosine (MmT) and α-methyl-p-tyrosine (MpT) greatly increased serum prolactin levels. Carr *et al.* (1975) showed that dopamine synthesis from ^{3}H-MpT in the hypothalamus was greatly depressed by the suckling stimulus. On the other hand, injections with MmT significantly reduced the rate of dopamine synthesis, and resulted in an increase in plasma prolactin levels. This suggests that reserpine, MmT and MpT act by decreasing hypothalamic dopamine content.

Phenothiazine derivatives, especially perphenazine, appear to have similar effects to reserpine although they probably act through a different biochemical pathway. Injection of perphenazine into rats causes a rapid increase in serum prolactin levels (Ben-David *et al.*, 1970; Lu *et al.*, 1970; Macleod and Lehmeyer, 1974) and also increases synthesis of the hormone in rat pituitaries *in vitro* (Macleod and Lehmeyer, 1972). When perphenazine-treated pituitaries were incubated with dopamine, the usual inhibition of prolactin release and synthesis was attenuated (Macleod and Lehmeyer, 1974); the longer the interval between perphenazine treatment and incubation, the more resistant the pituitary became to dopamine.

Macleod (1976) has summarised the possible mechanisms of action of perphenazine as follows:

1. Perphenazine acts directly on the hypothalamus to decrease PIF concentration (Ratner *et al.*, 1965; Ben-David *et al.*, 1970).
2. It acts by stimulating catecholamine synthesis in the hypothalamus, thus decreasing the levels of a non-catecholamine PIF (Lu *et al.*, Meites *et al.*, 1972).
3. It can deplete hypothalamic catecholamines and so promote prolactin release.
4. It can act directly on the pituitary to block dopamine action.

Perphenazine does not appear to have a direct effect of its own on the pituitary, but it does seem able to block the effect of dopamine on prolactin secretion. Probably alternatives 3 and 4 above are the most likely in the light of the known effects of other antidopamine drugs.

Haloperidol appears to act as a competitive antagonist of catecholamines (Anden *et al.*, 1966). This drug also increases serum prolactin levels (Dickerman *et al.*, 1972), and hypothalamic extracts from haloperidol-treated rats have reduced prolactin-inhibitory potency, possibly because their catecholamine content is reduced (Dickerman *et al.*, 1974). However Macleod and Lehmeyer (1973) demonstrated a direct stimulatory effect of haloperidol on prolactin synthesis and release by the rat pituitary *in vitro*, and the drug also abolished the inhibitory action of dopamine on these processes. These results have since been confirmed by Quijada *et al.* (1973).

Like perphenazine, haloperidol may have effects both on the hypothalamus and on the pituitary.

d. *Other neuroleptic drugs* In addition to the action of the various tranquillisers and haloperidol a number of other neuroleptic drugs can affect prolactin secretion.

Pimozide is a specific dopamine receptor-blocking drug (Andèn *et al.*, 1970). When injected into rats it causes an increase in serum prolactin levels (Lawson and Gala, 1976) and in the synthesis of prolactin *in vitro* (Macleod and Lehmeyer, 1974). In addition it also opposes the normal *in vitro* action of dopamine (Macleod and Lehmeyer, 1973, 1974). This suggests a biological action similar to perphenazine and haloperidol. Implants of pimozide either in the pituitary or in the region of the median eminence and arcuate nucleus produced a gradual increase in serum prolactin (Ojeda *et al.*, 1974).

Chlorpromazine and *methadone* have also been shown to promote prolactin release (Smalstig *et al.*, 1974; Clemens and Sawyer, 1974). The actions of both chlorpromazine (Smalstig *et al.*, 1974) and methadone (Clemens and Sawyer, 1974) were opposed by apomorphine. Some workers (Carlsen and Lindquist, 1963) have suggested that chlorpromazine may act as a dopamine antagonist, and Sasame *et al.* (1972) have suggested that methadone may also act by blocking dopaminergic receptors.

The accumulated evidence presented above must lead one to the conclusion that dopamine inhibits prolactin secretion, and the *in vitro* studies strongly indicate that dopamine can act directly on the pituitary. However this does not preclude the existence of a non-catecholamine PIF.

6.7 Stimulation

Apart from catecholamines and indoleamines, a number of other naturally occurring substances have been shown to stimulate prolactin secretion, including TRH, somatostatin, oestrogens and thyroid hormones. TRH and somatostatin are hypothalamic neurosecretory products, and their properties relate to the question of the existence of a distinct prolactin-releasing factor (PRF).

6.7.1 *Thyrotrophin releasing hormone (TRH)*

Thyrotrophin releasing hormone (L-pyroglutamyl-L-histidyl-L-proline amide, TRH) has been shown by a number of authors to increase prolactin secretion (Jacobs *et al.*, 1971; Rivier and Vale, 1974; Chen and Meites, 1975). The degree of TRH-stimulation appears to depend upon the endocrine status of the animal. Lu *et al.* (1972) were unable to demonstrate stimulation of prolactin released by male rat pituitaries *in vitro*, but Mueller *et al.* (1973) reported that pretreatment of male rats with oestradiol allowed them to respond to relatively low levels of TRH. Vale *et al.* (1973) found that cultured pituitaries from euthyroid rats are only minimally sensitive to TRH, but that the response can be increased by prior *in vivo* treatment with propylthiouracil. Noel *et al.* (1974) have also shown that the lowest doses of TRH still effective on TSH release also stimulate prolactin release.

Hill-Samli and Macleod (1974, 1975) have suggested that TRH does not stimulate the release of newly synthesised prolactin *in vitro*, but opposes the inhibitory action of dopamine. Blake (1974) found that TRH elevated serum prolactin levels in lactating rats, the response lasting for about 30 minutes. Suckling caused a marked increase in the circulating levels of both prolactin and TSH but the rise in prolactin preceded the rise in TSH by about five minutes. However the time taken by different pituitary cell-types to respond to stimulation might well differ, and this temporal separation of the TSH and prolactin responses to suckling is certainly not sufficient reason for excluding TRH as mediator of prolactin release in this situation. TRH also elevates prolactin levels in rats on the afternoon of pro-oestrus although the accompanying rise in TSH is not normally observed in pro-oestrus (Blake, 1974). TRH has also been shown to stimulate prolactin release *in vivo* in humans (Bowers *et al.*, 1971; Jacobs *et al.*, 1971), sheep (Davies and Berger, 1972) and cattle (Convey *et al.*, 1972).

Tashjian and Hoyt (1972) have shown that TRH increases both release and synthesis of prolactin by cultured rat pituitary cells, and it also stimulates the incorporation of radioactive amino acid into rat prolactin and the rate of *in vitro* prolactin secretion if injected *in vivo* prior to incubation (Dannies and Tashjian, 1973). When TRH was added *in vitro* at zero time, increased synthesis of prolactin did not occur until four hours later, but increased release of prolactin occurred within the first three hours, confirming that TRH can stimulate the release of previously synthesised prolactin (Dannies and Tashjian, 1973). In the GH_3 cell system used by Tashjian and his co-workers (see Ivey and Tashjian, 1975), TRH has been shown to increase intracellular levels of cAMP (Hinkle and Tashjian, 1974). The methyl xanthine *theophylline*, which is known to inhibit the catabolism of cAMP, was shown to potentiate the action of TRH and to cause release of prolactin, although it could not be shown to elevate cAMP levels. The effect of TRH could also be mimicked by addition of dibutyryl cAMP to the medium (Hinkle and Tashjian, 1974). These results have since been confirmed (Dannies *et al.*, 1976), and indicate that TRH stimulates prolactin secretion by using cAMP as second messenger.

6.7.2 *Prolactin releasing factor (PRF) other than TRH*

We have seen (p. 169) that there are good reasons for thinking that in addition to PIF the hypothalamus also produces a factor that stimulates prolactin secretion. Even the highest serum prolactin levels reported in rats bearing ectopic pituitary transplants (that is, with the influence of PIF minimised) are well below the levels found during lactation (Nicoll *et al.*, 1970), even when the rats bear *multiple* pituitary transplants (Chen *et al.*, 1970). Treatment of rat or sheep pituitaries *in vitro* with hypothalamic extracts produces in the short-term a reduction in prolactin secretion (PIF effect), but in the longer-term the same extracts *stimulate* prolactin output, a result ascribable to a prolactin-releasing factor or PRF (Nicoll *et al.*, 1970). Various workers, for instance Valverde *et al.* (1972), have reported the isolation of a PRF from hypothalamic extracts, and Krulich *et al.* (1971) found that PIF, PRF and (significantly) TRH activities are located in different hypothalamic regions in the rat. The fact that circulating levels of prolactin in nursing rats are so much higher than in rats with an ectopic pituitary graft has led to the concept that suckling must simultaneously inhibit the secretion of PIF while promoting the secretion of PRF (Nicoll *et al.*, 1970); on the other hand, some workers feel that the basic suckling-induced rise in prolactin secretion might be ascribed to inhibition of PIF (Tindall, 1974), while trivial stress-induced changes have been ascribed to increased release of PRF into the portal system (Terkel *et al.*, 1972). There are indications that levels of PRF activity in the hypothalamus respond to steroid and other hormones: for example, Meites *et al.* (1970) demonstrated PRF rather than the usual PIF activity when they reported that hypothalamic extracts from postpartum and oestrogen-treated rats *stimulated* lactation when injected into oestrogen-primed rats, and similar findings were reported by Mikinsky *et al.* (1968) using hypothalamic extracts from milking cows.

More direct evidence has been obtained for a PRF. Arimura *et al.* (1969) for example, showed that porcine hypothalamic extracts could elevate serum prolactin levels when injected into sheep and goats. However, since TRH can stimulate prolactin secretion (pp. 174–175), it has been argued that PRF and TRH may be identical. Against this, some workers feel that TRH may have only a pharmacological effect on prolactin secretion, or may only stimulate prolactin release from neoplastic pituitary cells (Ivey and Tashjian, 1975). However, recent evidence seems conclusively to show that TRH can stimulate prolactin secretion by normal pituitary lactotrophs, at least in some species (see Smith and Convey, 1974); infusion of TRH into the portal vessels of the rat stimulates release of prolactin and GH, as well as TSH (Takahara *et al.*, 1974), and cold-stress in rats causes an increase in prolactin secretion which has been explained as a by-product of release of TRH (Jobin *et al.*, 1975). Suckling has generally been reported to cause the simultaneous release of both prolactin and TSH, although in women the suckling-induced rise in serum prolactin is independent of changes in TSH and thyroid hormone (Gautvik *et al.*, 1974).

There can be little doubt that TRH is able to stimulate prolactin secretion (see also, pp. 174–175). The important question is whether or not the hypothalamus produces another factor, distinct from TRH, which is the physiological PRF. On balance, the evidence favours the idea that TRH and PRF are indeed distinct entities. Thus, crude methanol or acid extracts of pig hypothalamus induced a significant rise in serum prolactin levels in male rats primed with oestrogen and progesterone. The response was dose-related, and was not due to contamination with vasopressin, which could be removed chromatographically. TRH, at levels equivalent to those in the extracts, did not stimulate prolactin release in rats primed in this way, and passage of the extracts through G-10 Sephadex removed detectable TRH but left the PRF activity intact. Similar results were obtained with crude rat hypothalamic extracts (Valverde *et al.*, 1972).

Dular *et al.* (1974) obtained similar results using extracts of bovine hypothalamus. They showed that sequential Sephadex chromatography could separate two fractions distinct from TRH, both possessing PRF activity. One of these fractions released prolactin, GH and TSH from bovine pituitaries *in vitro*, in descending order of potency, while the other fraction was specific in releasing only prolactin. Whether such factors are physiologically active *in vivo* is not, of course, known for certain. Many such biologically active substances separated by extraction and chromatography could represent precursor molecules for other active factors. Interest in the TRH-PRF problem continues to run high, and different approaches have been used. Machlin *et al.* (1974) used a rat antiserum to inactivate TRH in hypothalamic extracts, and demonstrated that PRF activity was not impaired, while Frohman and Szako (1975) retained their PRF activity after adsorbing TRH on charcoal. Boyd *et al.* (1976) incubated porcine hypothalamic extracts in rat serum to destroy TRH and vasopressin, and then ran the extract on Sephadex G-25. The final material still contained reduced but detectable levels of PRF activity. The minimal effective dose of TRH that would release prolactin in their system was 10 ng/ml, and the final material contained no more than 0·6 ng/ml of TRH (Boyd *et al.*, 1976). Injections of porcine stalk-median eminence extract (pSME) into rats released both TSH and prolactin. However, the extracts contained more PRF activity than could be accounted for by their TRH content, and when injected together with maximally stimulatory doses of TRH they released prolactin but not TSH. The TRH activity in the pSME was destroyed by incubation with rat plasma, but the PRF activity remained (Szako and Frohman, 1976).

While not all workers have reported concordant findings – Rivier and Vale (1974), for example, concluded that all the PRF activity in their ovine hypothalamic extracts was attributable to TRH – the weight of evidence does seem to favour the existence of a PRF that is distinct and separable from TRH.

PRF activity has been detected in the pineal. Kitay and Altschule (1959) long ago reported that pineal extracts could promote lactogenesis, but since then the pineal has been shown to contain melatonin, serotonin, AVT and TRH (see pp. 174), all of which can stimulate prolactin release. However, Blask *et al.* (1976) have more recently shown that a separate fraction, free of these contaminants, can stimulate prolactin secretion.

In summary, the evidence suggests that prolactin is under complex hypothalamic control. The hypothalamus produces PIF, which itself may be a double entity, dopamine and a peptide; TRH, which under certain conditions, if not all, can release prolactin; and a PRF, separate from TRH but as yet of uncertain nature. The resultant control of prolactin synthesis and release presumably represents a finely adjustable balance in this multifactorial system.

6.8 Sex steroids and hypothalamic control of prolactin

As described previously (pp. 129–130), plasma prolactin levels vary quite markedly during the oestrous cycle, being highest in the afternoon of pro-oestrus in rats (Niswender *et al.*, 1969; Sinha and Tucker, 1969; Yoshinga *et al.*, 1969; Amenomori *et al.*, 1970; Niell, 1970; Smith *et al.*, 1974). A comparable pattern is seen in cattle (Sinha and Tucker, 1969; Swanson and Hafs, 1971; Raud *et al.*, 1971) and sheep (Kahn, 1971; Kahn and Denamur, 1974). Rises in serum prolactin levels are also associated with the later stages of pregnancy in all three species (Linkie and Niswender, 1972; Schams and Karg, 1970; Johke, 1971; Ingalls *et al.*, 1973).

Oestrogens have been shown to cause hypertrophy of the pituitary gland and to increase prolactin secretion in rats (Ratner *et al.*, 1963; Kalra *et al.*, 1970). Conversely, ovariectomy in cycling rats produces a fall in serum prolactin levels, which can be reversed by replacement treatment with either oestrogen or testosterone (Catt and Moffat, 1967; MacLeod *et al.*, 1970). In contrast, ovariectomy of late-term (day 19) pregnant rats elevates serum prolactin levels and initiates lactogenesis (Vermouth and Deis, 1974). A number of workers have reported that plasma progesterone levels gradually decline at term in the rat (Fajer and Barraclough, 1967; Grota and

Eik-Nes, 1967; Weist, 1970), and it is possible that the high blood levels of progesterone in late pregnancy may suppress prolactin secretion, which would explain the effect of ovariectomy in late pregnancy. Yoshinga *et al.* (1969) observed a rapid increase in oestrogen levels in ovarian venous plasma at term, and this may stimulate the high prolactin levels associated with parturition. Pituitary prolactin levels decrease at term (Grosvenor and Turner, 1960), and concomitantly serum levels rise (Amenomori *et al.*, 1970), and Morishige *et al.* (1973) have demonstrated a correlation between progesterone decrease and prolactin increase in the blood on day 21 of pregnancy in the rat.

However, Simpson *et al.* (1973) have claimed that prolactin release in late pregnancy in the rat is independent of ovarian influence. Moreover, Bridges and Goldman (1975) have reported that ovariectomy on day 17 or day 21 in the rat leads not to a rise but to a decrease in serum prolactin levels, in contrast to the findings of Vermouth and Dies (1974) (p. 178). The opposing results cannot at present be reconciled, though Bridges and Goldman (1975) suggest that part of the explanation may lie in the methods of obtaining serum samples, whether by decapitation or by ether anaesthesia (cf Neill, 1974).

The relationship between prolactin and ovarian steroids during pregnancy and pseudo-pregnancy was discussed in detail in relation to the luteotrophic process (pp. 133–136).

In cycling rats, very high doses of oestrogen cause a slight decrease in the peak prolactin levels at pro-oestrus, but still maintain levels higher than in controls (Chen and Meites, 1970) or in male rats (Lu and Meites, 1971). Serum prolactin levels in female rats can also be elevated by androgens, and these androgenised females are then refractory to any further stimulation of prolactin secretion by oestrogens (Mallanipah and Johnson, 1973). However, ovariectomy in such androgenised rats, and in spontaneously constant-oestrus rats, resulted in a fall in serum prolactin levels (Ratner and Peake, 1974). Knudsen and Mahesh (1975) have recently shown that the androgen dehydroepiandrosterone injected into immature female rats causes surges in FSH, LH, and prolactin three days later, with consequent precocious sexual maturation and ovulation.

As described on pp. 129–130, prolactin is released in quantity during the afternoon of pro-oestrus (Kwa and Verhofstad, 1967). This prolactin surge can be eliminated by injection of antiserum to oestrogen on the second day of dioestrus, but the surge occurs normally if the antiserum is accompanied by diethylstilboestrol, a non-steroidal oestrogen (Niell *et al.*, 1971). Non-steroidal anti-oestrogens, such as nafoxidine and tamoxifen, have been used to inhibit oestrogen-induced prolactin release (Heusen *et al.*, 1973; Jarden *et al.*, 1975), and to inhibit the growth of prolactin-dependent chemically induced mammary tumours (Terenius, 1971).

Tashjian and Hoyt (1972) demonstrated that oestrogens will cause a two-fold increase in prolactin secretion from cultured GH_3 cells *in vitro*. However, although oestrogen is certainly capable of stimulating prolactin release *in vitro*, its main site of action is probably hypothalamic (McCann *et al.*, 1968). Ramirez and McCann (1964) have shown that hypothalamic implants of oestrogen stimulate prolactin release, as do implants in the pituitary or median eminence. Ratner and Meites (1964) claimed that the hypothalamic PIF content was lowered by chronic oestrogen administration. Other brain centres have also been reported to be involved. In particular, Tindall *et al.* (1967) found that prolactin release occurred when oestradiol was implanted into the amygdala. All these sites were later found to possess mechanisms for the preferential uptake of oestradiol (see p. 177) (Stumpf and Sar, 1971; Pfaff and Keirer, 1972). Ahren *et al.* (1971) and Hokfelt and Fuxe (1972) have shown that the low prolactin levels during dioestrus are associated with rapid dopamine turnover in the tubero-infundibular region, while the pro-oestrous surge is associated with slow hypothalamic dopamine turnover. Dopamine turnover in the rat is reduced by ovariectomy and restored by treatment with oestrogen or testosterone (Fuxe *et al.*, 1969). Ajika *et al.* (1972) have shown that the subcutaneous injections of oestrogen lowered hypothalamic PIF levels in ovariectomised rats, in agreement with an earlier report that ovariectomy lowered the hypothalamic dopamine content (Stefano and Donoso, 1967). Thus it can be seen that among the events associated with the female reproductive cycle is the stimulation of prolactin secretion at pro-oestrus by ovarian oestrogens. The surges of oestrogens which stimulate this cyclic prolactin release are probably in their turn released by pulsatile release of LH and FSH from the pituitary. A single injection of antiserum to LH induces a long-term hypoprolactinaemia, which may well be the result of abolition of the surges of oestrogen consequent upon the permanently reduced blood levels of biologically active LH (Kerdelhue *et al.*, 1975).

IV Cyclic release of prolactin and the role of the pineal

Prolactin release from the mammalian pituitary can be divided into three distinct categories: suckling-induced release: minor stress-mediated release: and circadian or seasonal rhythmic release patterns. Pituitary prolactin content in female rats varies considerably during the day with a maximum at 16·00 hours and a minimum at 22·00 hours (Clark and Baker, 1964). Diurnal rhythms have been reported in male rats, intact female rats and ovariectomised female rats, in each case with higher values in the afternoon (Koch *et al.*, 1971; Dunn *et al.*, 1972). A circadian pattern of release has also been demonstrated in the cow (Swanson and Hafs, 1971), although to some extent this rhythm was obscured by the rise in serum prolactin induced by milking: in fact, it was not possible to measure prolactin in the post-milking samples because they were too high for the assay (Koprowski *et al.*, 1972). Further work has shown that lactating cattle secrete more prolactin in response to milking in summer than in winter (Koprowski and Tucker, 1973), and indications of a similar seasonal change was observed in non-lactating cattle (Schams, 1973). This suggests a seasonal as well as a diurnal pattern of prolactin release, at least in cattle.

In the goat, the situation appears to be rather different. Hart (1973) showed that in lactating females, in non-lactating virgin females, and in male goats there was a slow rise and fall in the basal prolactin levels throughout the 24-hour period, but not a sharply defined diurnal rhythm. There was evidence of a seasonal rhythm in lactating animals, which was modified by the state of lactation. Hart's findings were later confirmed by Buttle (1974) for male goats. Once again there was no evidence of a strong diurnal rhythm in the animals studied, but there were very marked seasonal variations with lowest values in mid-winter (1·5–4·5 ng/ml) and highest values in summer (40–100 ng/ml, see Fig. 6.10). This difference between winter and summer levels persisted in the castrated male, although prolactin levels in castrates were slightly higher than normal in midwinter.

In male sheep, Pelletier (1973) has demonstrated a photoperiodic control of prolactin release. Blood levels of prolactin are approximately proportional to the length of the light period, prolactin being highest in animals maintained on a 16-hour photoperiod and lowest in those maintained on eight hours. Castration was shown to elevate serum prolactin levels at both photoperiods (16L : 8D, 8L : 16D), but it did not obliterate the light-induced differential. Replacement therapy with testosterone propionate reduced prolactin levels in castrates, but again the photoperiodic differential was maintained. Similar patterns have been demonstrated in female sheep under natural seasonal conditions, although in the female there was a phase-shift in the response, with low levels of prolactin in July, during the long natural photoperiods (Saji, 1966).

Changes in prolactin secretion are also associated with the oestrous cycle in the female rat. A surge of prolactin occurs in the afternoon of pro-oestrus (Kwa and Verhofstad, 1967; Niswender *et al.*, 1969; Wuttke and Meites, 1970), this being stimulated directly by the oestrogen surge occurring on day two of dioestrus (Niell *et al.*, 1971). A similar surge of prolactin release occurs daily during the first two or three days following

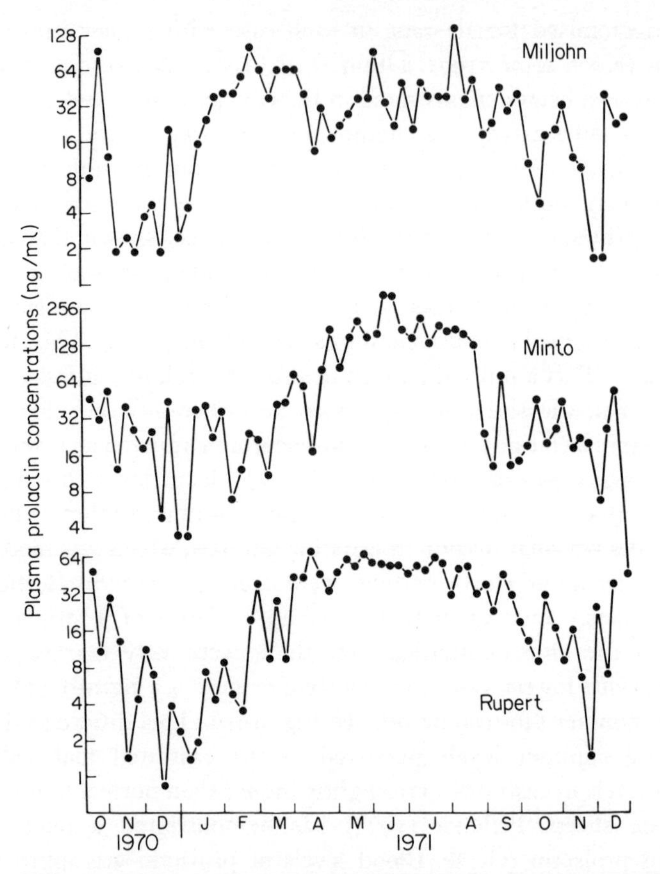

Fig. 6.10 The concentrations of prolactin in three intact male goats over the period October 1970 to December 1971. (From Buttle, 1974.)

the induction of pseudopregnancy (Meites *et al.*, 1972). This daily surge, which appears to be a response to cervical stimulation, peaks at about 05·00 hours (Freeman and Niell, 1972). More recently Smith and Niell (1975*a*) have shown that there is a critical period during which cervical stimulation can produce prolactin release, and that in pseudopregnant and pregnant rats there are in fact two daily peaks of prolactin release, one in the early morning and the other in the afternoon (Smith and Niell, 1976*b*). These peaks always appear at the same time of day no matter when cervical stimulation is applied, and the same pattern occurs in pseudo-pregnancy initiated in the ovariectomised-adrenalectomised rat. In the pregnant female the two daily surges of prolactin persist until about day eight and then disappear. In the pseudopregnant rat the surges disappear on or about day 12 (Smith and Niell, 1976*b*).

In the normal female rat the prolactin surges associated with proestrus and with pregnancy or pseudopregnancy are probably caused by elevated oestrogen in the case of pro-oestrus and by neural mechanisms entrained by cervical stimulation in pregnancy and pseudopregnancy. However, there is increasing evidence that the photoperiodically induced changes in prolactin secretion may be mediated *via* the pineal. The pineal synthesises melatonin (Wurtman *et al.*, 1963*a*), and its synthesis can be inhibited by exposure to constant light (Wurtman *et al.*, 1963*b*). Kamberi *et al.* (1971) have shown that melatonin injections into the third ventricle of male rats resulted in elevated plasma prolactin levels. Relkin (1972) found that exposure of male rats to continuous illumination resulted in increased pituitary prolactin and decreased serum levels. The reverse response was observed in animals kept in constant darkness. In pinealectomised animals the dark-induced elevation of serum prolactin levels was abolished. This work suggests the involvement of pineal melatonin in photoperiodically induced prolactin release, melatonin perhaps acting to inhibit PIF secretion. Confirmatory evidence has since been obtained by Relkin *et al.* (1972). Donofrie and Reiter (1972) have shown that the pineal can depress pituitary prolactin levels in the rat, with a concomitant increase in circulating levels, although more recently Talbot and Reiter (1974) were unable to show an effect of either melatonin or 6-methoxytryptophol (6-MT) on plasma prolactin levels, although 6-MT increased, and melatonin increased, the total amount of prolactin in the pituitary. Both Donofrie and Reiter (1972) and Talbot and Reiter (1974) worked on castrated rats, which have elevated pituitary prolactin levels (Talbot and Reiter, 1974).

More recent work by Reiter (1974) on rats rendered blind and anosmic showed that this dual operation depressed the body weight, reduced the size of the anterior pituitary, and depleted pituitary prolactin. Pinealectomy returned the pituitary prolactin levels to normal. It seems that the pineal became hyperactive as a result of sensory deprivation. There was no increase in plasma prolactin associated with the drop in pituitary content in the blind-anosmic rats; indeed, at 75 days following surgery, plasma prolactin levels were significantly depressed. This is contrary to the earlier results of Relkin (1972) and also to those of Shinino *et al.* (1974) on female rats. These contradictions are difficult to explain, although Reiter (1974) suggests that perhaps the continued automonous secretion of prolactin resulting from pineal hyperactivity exhausted the capacity of the pituitary to secrete prolactin over a long period. However, as Reiter himself points out, ectopically transplanted pituitaries continue to secrete prolactin at a high rate for much longer periods than 75 days (Amenomori and Meites, 1969).

In an attempt to throw some light on this problem, Blask *et al.* (1976) have studied the effect of pineal extracts on prolactin secretion by rat pituitaries *in vitro*. They showed that crude extracts of bovine, rat and human pineal all strongly increase the release of prolactin from the cultured pituitaries; however, further purification of the bovine pineal extracts yielded two fractions (A_I and A_3) both of which *inhibited* prolactin release. The picture is complicated by the finding (Reiter *et al.*, 1976) that arginine vasotocin (AVT), a known pineal constituent, can release prolactin from the pituitary, as can oxytocin (Krulich *et al.*, 1974). What is more, recent studies by Montova *et al.* (1976) have shown that the rat pineal contains significant amounts of thyrotrophin-releasing hormone (TRH) and that TRH levels in the rat pineal show a circadian rhythm with peak production in the dark phase. TRH is known to be a potent prolactin-releasing factor (see pp. 174–175), so it could very well be the active pineal principle detected in these various studies.

In summary, three pineal constituents, melatonin, AVT and TRH have been shown to stimulate the release of prolactin by the pituitary, but the results of *in vivo* studies have not yet clearly defined a physiological role for the pineal in mediating photoperiodic control of cyclic prolactin secretion. Such a role however, is certainly indicated by the evidence. In addition, two partially purified bovine pineal fractions, A_I and A_2, have been shown to *inhibit* prolactin release by the rat pituitary *in vitro*; what this means in physiological terms can be decided only by future investigations.

V Stress and prolactin release

Grosvenor *et al.* (1965) first clearly showed that stress could deplete pituitary prolactin levels. Previously Swingle *et al.* (1951*a* and *b*) had demonstrated that adrenalectomy, injection of tissue extracts, or various drugs, could induce pseudopregnancy in rats, and Nicoll *et al.* (1960) had shown that a range of stresses such as cold, intense light and heat, restraint and the injection of formaldehyde could induce lactation in oestrogen-primed rats, and had concluded that under these conditions both prolactin and ACTH were released from the pituitary. More recently, Niell (1970) has shown that ether anaesthesia causes a three- to eight-fold increase in serum prolactin levels at all stages of the rat oestrus cycle except on the afternoon of pro-oestrus when levels are already very high (see p. 130) and probably could not be further elevated. Ether-laparotomy elevated prolactin levels in ovariectomised female rats, but not in males. It is of interest to note that LH levels were not modified by any of these experimental procedures. Work by Tashjian *et al.* (1970) tends to suggest that prolactin release in stress conditions is a result of direct hypothalamic intervention: they showed that cortisol added to cultured neoplastic pituitary cells caused a 20–40% *drop* in the basal secretion rates, which suggests that prolactin release by stress *in vivo* cannot be a direct pituitary response to high circulating levels of corticosteroids. Furthermore, Copinschi *et al.* (1975) found that glucocorticoids can *suppress* the hypoglycaemia-induced release of prolactin in man, and Jobin *et al.* (1975) have shown that doses of dexamethasone known to block ACTH release also block the release of prolactin in response to cold-stress. Blake (1974), in agreement with Niell (1970), found that ether-stress induces release of both prolactin and ACTH and also *inhibits* TSH secretion, unlike cold-stress which stimulates release of all three hormones (Kraicer *et al.*, 1963; Fortier *et al.*, 1970). Jobin *et al.* (1975) found that ether-stress caused a rise in plasma prolactin corticosterone and thyroxine, but induced a transient *decrease* in serum TSH levels. In their experiments, cold-stress was followed by a sharp but transient rise in prolactin and later by a slower and more sustained release of TSH, findings which led the authors (Jobin *et al.*, 1975) to suggest that ACTH and prolactin under certain conditions may share a common release mechanism which is independent of the mechanism controlling TSH.

Contrary to Niell's (1970) findings, Turpen *et al.* (1975) found that ether-stress can directly affect LH levels, causing a small but transient rise in serum LH, no matter what the time of day. The evidence that ether-stress stimulates prolactin release now seems quite conclusive, having been demonstrated in the dioestrus female rat (Neill, 1970; Morishige and Rothchild, 1974) and in male and female gonaectomised rats (Ajika *et al.*, 1972; Enker *et al.*, 1974), the response being relatively rapid (Wakabayaslu *et al.*, 1974). However results relating to serum LH levels are not so conclusive (Neill, 1970; Euker, 1973; Morishige and Rothchild, 1974).

Stress has also been shown to alter the reproductive capacity of mice and rats (Ogle, 1934; Barnett and Coleman, 1959; Pennycuik, 1966; Hensleigh and Johnson, 1971). Pennycuik (1966) showed that heat-stress impairs lactation, and Paris and Ramaley (1972) showed that heat-stress in prepubertal mice caused loss of body weight and delayed vaginal opening.

Paris *et al.* (1973) also studied the effect of short-term stress on the fertility of female postpubertal mice. The response seems to differ with age: younger mice showed a depressed fertility while fertility was enhanced in fully cycling females. The response in the younger mice may be mediated by ACTH (Christian, 1964; Hagino *et al.*, 1969). However, in the adult ovulation appears to continue independently of the adrenal

(Gorski and Lawton, 1972), and in adult rats adrenalectomy appears to have no effect on reproductive function (Thoman *et al.*, 1970). Luteinisation of the ovaries has been shown to occur in severely-stressed rats under conditions which may be expected to produce hyperprolactinemia, and there are also reports that prolactin may alter gonadal function in humans (Archer, 1977). Prolactin has been shown to have an inhibitory effect on LH secretion in man (Villalobes *et al.*, 1976). These authors showed that in lactating women there was a gradual increase in FSH during the first two weeks of lactation while LH levels remained low. When pituitary prolactin release was blocked by administration of ergocryptine, it was found that the FSH increase was inhibited but that there was a marked rise in serum LH levels. Rolland *et al.* (1975) have also shown that high levels of circulating prolactin in women can suppress synthesis of oestradiol 17-β. Recently Lu *et al.* (1976) have invoked an inverse relationship between the prolactin and the LH responses to suckling, suggesting that suckling may inhibit LH secretion while stimulating prolactin, rather than that prolactin itself suppresses gonadotrophin.

Obviously a more detailed study is needed of the effect of stress-induced prolactin release on gonadal function. However, the evidence suggests that in mature animals, levels of corticosteroids, which are always elevated by stress, are relatively unimportant in the control of ovarian function and that stress-induced alterations of gonadal function may be mediated by short-term elevations in prolactin secretion.

The mechanism by which stress can bring about prolactin release is not clearly understood. The ability of either to induce prolactin release has been shown to vary with the condition of the animal (Neill, 1970; Wuttke and Meites, 1970; Euker *et al.*, 1975). The prolactin-release response to ether-stress can be modified by sex steroids (Reier *et al.*, 1974), being increased by oestrogen treatment and decreased by progesterone. However these hormones may act on the pituitary directly to limit its ability to secrete prolactin. As described above, Jobin *et al.* (1975) have suggested a common pathway of prolactin and ACTH control in both ether-stress and cold-stress, and Marchlewska-Koj and Krulich (1975) have suggested that a serotoninergic release mechanism might mediate ether-induced prolactin release. The dopamine inhibitory system however, does not appear to be involved (Reier *et al.*, 1975). Recently Grosvenor and Whitworth (1976) have shown that frequency of suckling caused a subsequent increase in the size of the prolactin surge. As a central serotoninergic mechanism has also been implicated in the suckling reflex (Kordon *et al.*, 1972), this may indicate that the two processes share a common pathway for inducing prolactin release.

Some controversy remains about the prolactin response to cold-stress. We have seen that Jobin *et al.* (1975), and others, have reported that cold increases prolactin secretion. On the other hand Mueller *et al.* (1974) reported that cold causes a *decrease* in blood prolactin, which then increases when the temperature is raised, while TSH levels show the opposite response. Whether TRH is involved in mediating cold-stressed release of prolactin is uncertain, though TRH is certainly secreted in response to cold. Even those authors who have demonstrated stimulation of prolactin secretion in response to cold stress have been doubtful about the part played by TRH. What is, however, certain is that a wide variety of stresses can alter the circulating levels of prolactin and other pituitary hormones. Whether stress-induced hyperprolactinaemia has any major physiological influence on reproductive processes remains for further investigation.

VI Prolactin in parental behaviour

The first reports of stimulation of maternal behaviour by prolactin in mammals came from Riddle *et al.* (1934, 1935), who found that injections of prolactin into virgin female or male rats increased the incidence of pup-retrieval and intensified the care of rat pups placed with them. However, later workers (Beach and Wilson, 1963; Lo and Fuchs, 1962) have failed to demonstrate increased retrieval in rats following prolactin administration. One of the main problems with research into this area is that parentally inexperienced male and female rats may suddenly develop a high incidence of retrieval quite spontaneously (Rowell, 1961; Rosenblatt, 1967).

Recent work by Moltz *et al.* (1970) with virgin female rats suggests that ovarian steroids may also be involved in parental behaviour, and Zarrow *et al.* (1971) have reviewed the field. They confirmed the finding of Terkel and Rosenblatt (1968) that plasma taken from lactating rats could initiate retrieval in virgin female rats. Terkel (1970) has also obtained retrieval by blood exchange between two unrestrained rats. However these results, while indicating some humoral control of maternal behaviour, do not identify the hormone(s) involved. Denenberg *et al.* (1969) have shown that injections of oestradiol and progesterone into gonadectomised female rats are insufficient to produce maternal behaviour, but addition of prolactin or cortisol to the oestradiol and progesterone did reduce the latency time for pup-retrieval (Zarrow *et al.*, 1971). Moltz *et al.* (1970) have reported similar findings.

Terkel and Rosenblatt (1978), in confirmation of their earlier results, have shown that a humoral factor in the blood of rats at parturition will initiate maternal behaviour in virgin females. In the light of previous evidence this is likely to be a mixture of prolactin with progesterone or oestrogen, gonadotrophin levels being low at parturition. More recently Bridges *et al.* (1974) have tried to correlate serum prolactin levels with maternal behaviour in the rat. Using ovariectomised pregnant rats, sham-ovariectomised pregnant rats, and virgin ovariectomised rats, they showed that there was no difference in prolactin levels or maternal behaviour between the three groups and also that high levels of prolactin did not in themselves initiate maternal behaviour. However, ovariectomy on day 17 of pregnancy may reduce the terminal oestrogen surge and may interfere with the normal interactions of prolactin and sex steroids. These findings raise some doubts about the role of prolactin in *initiating* maternal behaviour.

Early work on the mouse suggested that nest-building behaviour can be induced by progesterone (Koller, 1952, 1956; Lisk *et al.*, 1969). Recently Voci and Carlson (1971) have shown that implantation of 70 μg of prolactin either into the hypothalamus or subcutaneously significantly increased various components of maternal behaviour. Mice with progesterone implants in the hypothalamus built superior nests, but were no different from controls in pup-retrieval. This separation of the nest building and pup-retrieval responses to progesterone treatment is very interesting, and may explain the evidence that prolactin stimulates pup-retrieval but not nest building in rats. This work has since been repeated and confirmed (Voci and Carlson, 1973).

The plucking of hair from the body to line the nest is a characteristic element of material behaviour in the pregnant rabbit (Sawin *et al.*, 1960), and treatment with oestradiol, progesterone and prolactin significantly stimulated hair-plucking (Farooq *et al.*, 1969). Unfortunately, these authors used a treatment time of 56 days, far longer than the duration of pregnancy. Zarrow *et al.* (1963, 1965) have shown that oestrogen and progesterone can induce nest building in ovariectomised rabbits. However these results were obtained with animals which were not hypophysectomised. In hypophysectomised-ovariectomised rabbits the steroid treatment was ineffective (Anderson *et al.*, 1971), suggesting that the earlier results could have been due to steroid-stimulation of prolactin release. In confirmation, Zarrow *et al.* (1971) have shown that injection of ergocornine maleate in rabbits prevents nest building and that this blockade can be overcome by the subsequent injection of prolactin.

VII Prolactin and osmoregulation

Prolactin has been shown to be an important osmo- and iono-regulatory hormone in the lower vertebrates. Evidence has been growing during the past few years that it may also play a part in mammalian osmoregulation. The first indications that prolactin may affect mammalian kidney function came from work by Lockett and Nail (1965), who showed that both bovine and ovine prolactin reduced the urinary excretion of sodium in the rat. If water diuresis was induced in rats treated with propylthiouracil, the action of prolactin was reversed and it promoted sodium excretion; these workers found no effect of prolactin on renal water retention in rats. Urinary sodium excretion was also reduced in rats by ovine, bovine and hGH, the hGH being by far the most effective. Lockett (1964) also found that hGH could promote sodium retention by the perfused cat kidney. To what extent such renal effects of hGH can be attributed to its intrinsic prolactin-like properties (see Chapter 1) is unknown; it has also been reported by several laboratories to cause increased renal plasma flow, increased clearances of insulin, creatine and PAH, and increased tubular reabsorption of sodium (Gerahberg, 1960; Bergenstal and Lipsett, 1960; Biglieri *et al.*, 1961; Corvilain and Abramow, 1962; Beck *et al.*, 1964).

More recently, Ensor *et al.* (1972*a*) found that pituitary prolactin levels fell in dehydrated rats, and that lactating rats resist dehydration better than non-lactating females. Injected ovine prolactin induces a dose-dependent antidiuresis, and also appears to stimulate water intake. Thus prolactin during lactation acts as an antidiuretic hormone, although it cannot maintain lactation in extremely dehydrated rats (Ensor *et al.*, 1972*a*). This antidiuretic effect of prolactin has recently been confirmed by Burstyn *et al.* (1975). Smith *et al.* (1974) showed that water-restriction caused prolongation of the oestrous cycle in rats, and that the rats became sexually receptive earlier than usual during pro-oestrus. Smith and co-workers interpreted these changes as being due to enhanced secretion of progesterone and/or corticosterone from the adrenal cortex, but in the light of present knowledge, increased prolactin secretion, stimulated by dehydration, could have prolonged dioestrus in these rats (cf Ensor *et al.*, 1972*a*). The antidiuretic response to prolactin is not shown by male rats, and only be females at certain stages of the oestrus cycle (Beynon and Ensor, unpublished), which suggests that the effect of prolactin is not due to vasopressin contamination, but may involve a synergism with sex steroids. In these respects, the action of prolactin would be comparable to its effect on the frog skin (Howard and Ensor, 1976), which shows a sexually dimorphic response (see Chapter 3).

Relkin and Adachi (1973) have shown that pituitary prolactin levels increase in rats maintained on a low sodium diet. Furthermore, Richardson (1973) reported that prolonged treatment with bromocryptine caused increased urinary sodium and potassium excretion in rats, and also reduced the incidence of spontaneously occurring kidney lesions. This 'protective' effect of prolactin on the kidney has also been recorded by Köhnlein *et al.* (1966) who showed that prolactin shortened the recovery time of the kidney after damage by ischemia.

Work by Burstyn *et al.* (1972) suggests that prolactin may act synergistically with aldosterone in female merino sheep. As in most mammals, aldosterone stimulated sodium retention when given to sheep on a low sodium diet. However when the salt-intake was increased, aldosterone promoted sodium excretion. Prolactin when administered with aldosterone to sheep on a high salt diet restored the sodium-retaining action of aldosterone. Sheep on this regime also showed a significant weight increase, which disappeared when the prolactin injections were discontinued, an observation which suggests that prolactin might also have a water-retaining effect under conditions of high salt loading. More recent work by Horrobin *et al.* (1973*c*) indicates that these effects are not due to contamination of the prolactin preparations with arginine vasopressin (AVP) (see Chapter 3). The water-retaining effect of AVP on the ovine kidney could be abolished by cortisol, and diuresis then occurred in response to further cortisol injections. The subsequent administration of prolactin or oxytocin, however, restored the antidiuretic effect of AVP. Horrobin, *et al.* (1973*a*) have further shown that cortisol can also reverse the sodium-retaining effect of aldosterone, and that prolactin reversed this effect of cortisol and restored the sodium-retaining action of aldosterone.

This evidence would suggest that prolactin has a sodium-retaining and antidiuretic action in mammals. As Nicoll (1974) points out, if a major role of prolactin is to retain sodium, then pituitary prolactin levels would be expected to show a direct relationship with plasma osmolality. However, Relkin and Adachi (1973) showed that pituitary levels in rats were high when plasma sodium levels were low, and that increasing the plasma osmotic pressure (OP) by dehydration led to a significant fall in pituitary prolactin levels. This suggests that prolactin secretion is triggered primarily by high plasma OP, and that prolactin is mainly an antidiuretic agent, perhaps producing its effect by promoting a temporary increase in sodium reabsorption. However, prolactin secretion in man may not respond to wide changes in serum OP, and nor does prolonged hyperprolactinaemia necessarily affect renal water metabolism in the human (Baumann *et al.*, 1977).

To what extent synergism with aldosterone is important for the renal actions of prolactin is not clear, although the findings of Horrobin and his colleagues cited above are suggestive. There are several reports indicative of a relationship between prolactin and adrenal cortex. Thus, human placental lactogen stimulates secretion of aldosterone in man (Melby *et al.*, 1966), and prolactin has been shown to inhibit 5α-reductase activity in the rat adrenal, which would lead to an increase in the corticosterone pool (Witorsch and Kitay, 1972). It was early shown that adrenalectomy in the rat results in pseudopregnancy, presumably in consequence of enhanced prolactin secretion (Swingle *et al.*, 1951), and Ben-David *et al.* (1971) have since more directly shown that adrenalectomy in male rats leads to increased prolactin secretion. These data suggest some form of reciprocal relationship between prolactin and the adrenal cortex, such that prolactin stimulates aldosterone output either directly or indirectly by enhancing the aldosterone-precursor pool, and so 'priming' the adrenal to respond to specific aldosterone-releasing stimuli. This concept obviously has important consequences for our understanding of the total function of prolactin, but it must remain hypothetical pending further research.

Lucci *et al.* (1975) have recently studied the renal effects of prolactin in the rat in some detail. They found that prolactin had no effect on the urinary parameters of normal male rats. However, in the male rat in which volume-expansion had been induced by oral water-loading, prolactin inhibited the concomitant diuresis and renal salt-loss. This response appeared to be dependent on haemodilution, since when volume expansion was induced by infusing whole blood, prolactin was without effect. Although this action of prolactin in the male rat may well be non-physiological, it does not preclude the possibility of a physiological role in the lactating female (Ensor *et al.*, 1972).

Prolactin and water-balance in mammals are also related by the effects of the hormone on the uptake of ions and other solutes across the wall of the intestine (Ramsey and Bern, 1972; Mainoya *et al.*, 1974). Prolactin stimulates the transport of fluid, sodium, potassium, calcium magnesium and chloride across inverted jejunal sacs from rats (Mainoya *et al.*, 1974), and hypophysectomy significantly decreases the jejunal fluid and sodium absorption. The action of prolactin can be inhibited by the simultaneous administration of vasopressin, which by itself is without effect. Stimulation of intestinal transport by prolactin seems to be a fairly generalised action, in that prolactin also increases the jejunal uptake of sugars and certain amino acids (Mainoya, 1975).

In summary, the evidence suggests that prolactin has an osmoregulatory role in mammals, which may be especially important in the female during lactation. However, it is obvious that further work needs to be done to clarify the renal actions of prolactin, and to define more clearly the relationship between prolactin and the adrenal cortex.

7 Prolactin and the primates

The literature on the physiology of primate (particularly human) prolactin is now vast and it is not proposed that the present review should be encyclopaedic. For detailed studies of the subject the reader is referred to the reviews by Frantz *et al.* (1972*a*), Van der Laar (1973), Pasteels (1976) and Archer (1977), among many others. In this chapter an attempt has been made to present an overview of the more recent developments, to enable the reader to place primate lactogenic hormones in the context of the comparative physiology of prolactin. Human placental lactogen (hPL) will be considered, as well as human pituitary prolactin (hPRL).

7.1 Identification of human prolactin

The problems surrounding the identification of human prolactin (hPRL), as distinct from human growth hormone (hGH), were discussed briefly in Chapter 1 (see p. 4). Prior to 1970 there were a number of reports which suggested that hGH possesses all the properties of prolactin, most significantly that it exhibits strong lactogenic potency (Chadwick *et al.*, 1961; Ferguson and Wallace, 1961; Hartree *et al.*, 1965). Concurrently, the failure to isolate distinct PRL activity from human pituitaries collected at postmortem (Wilhelmi, 1961; Tashjian *et al.*, 1965) led workers to question the existence of prolactin as a separate entity in humans (Bewley and Li, 1970).

On the other hand, there did exist a substantial body of circumstantial evidence to suggest the presence of two hormonal moieties. Lactation had been shown to exist in dwarfs with isolated hGH deficiency (Rimoin *et al.*, 1961) and galactorrhoea had been shown to occur in the presence of normal or even subnormal hGH levels (Benjamin *et al.*, 1969). Normal or low levels of hGH had also been found during pregnancy and during post-partum lactation (Spellacy and Buhi, 1969; Spellacy *et al.*, 1970*a*,*b*; Varma *et al.*, 1971), and it was also known that prolactin-rich pituitary tumours existed (Takanati *et al.*, 1967) which did not increase serum hGH activity (Peake *et al.*, 1969).

The application of specialised staining techniques identified separate and distinct lactotrophs and somatotrophs in the human pituitary (Herlant and Pasteels, 1967). Golubuff and Ezrin (1969) later confirmed these identifications and showed that the lactotrophs were extremely scarce at all times apart from pregnancy and lactation, which would explain the earlier failure to detect prolactin activity in postmortem pituitaries not taken from pregnant or lactating females.

Probably the major breakthrough in the isolation and characterisation of human prolactin came from the observation by Pasteels (1962) that *in vitro* cultures of human pituitaries secreted increasing amounts of a substance possessing pigeon crop-sac stimulating activity; later, Brauman *et al.* (1964) demonstrated that under the same conditions the level of GH activity in the medium decreased. A protein extract from this culture medium, rich in pigeon crop-sac activity (but low in GH activity) was used to produce antisera which were capable of neutralising lactogenic activity in the sera of lactating women (Pasteels *et al.*, 1965). These reagents were later used by Greenwood *et al.* (1971) to develop a radioimmunoassay system for human prolactin (Bryant *et al.*, 1971). In 1972, Frantz *et al.* (1972*b*) were able to show that prolactin activity in the serum of patients suffering from galactorrhoea was not neutralised by anti-hGH serum, but could be inactivated by antisera to ovine prolactin. At about the same time Herbert and Hayashida (1970) were able to show that antisera to ovine prolactin localised in the so-called 'pregnancy cells' in a monkey, confirming that ovine, human and monkey prolactins had the same histological background. Later studies by Pasteels (1971) showed that this property was not possessed by antisera to hGH or hPL.

At about the same time, Friesen and his co-workers, working with human (Friesen *et al.*, 1970) and monkey (Friesen and Guyda, 1971) pituitary fragments incubated *in vitro* with ^{3}H-leucine, demonstrated directly the biosynthesis of primate prolactin. Separation of the radioactive protein gave a single large peak which cross-reacted with antisera to ovine prolactin but not with hGH or hPL antisera (Friesen, Guyda and Hwang, 1971). It has also been shown that tumour fragments from patients suffering from galactorrhoea synthesised prolactin exclusively with no trace of hGH (Hwang *et al.*, 1971; Friesen *et al.*, 1972*b*). Using the knowledge that human prolactin shows similar antigenic properties to sheep prolactin but not to hGH and hPL, Friesen and his co-workers were able to use hPL antisera attached to Sepharose in order to remove GH from extracts of monkey and human pituitary glands, so producing prolactin free from

GH (Guyda and Friesen, 1971; Guyda *et al.*, 1971; Hwang *et al.*, 1971, 1972).

Probably one of the main reasons why human prolactin has been so difficult to separate from hGH is that while hGH forms about 5% of the dried pituitary powder, hPRL represents only about one hundredth of this fraction (Friesen and Hwang, 1973). Furthermore, hGH does have inherent prolactin activity (Li, 1972), so that even after careful purification the molecule retains activity equivalent to about 10–20% of that of the best ovine prolactin (Friesen and Hwang, 1973). This obviously means that the bioassays used to monitor pituitary separation procedures will inevitably tend to track hGH rather than hPRL. As previously stated, the concentration of prolactin in the human pituitary increases during pregnancy. In consequence, Lewis *et al.* (1971) were able to separate electrophoretically a second band from the pituitary of a woman who died in childbirth, and this band was subsequently shown to be hPRL, a finding which led to a method for the purification of hPRL, (Lewis, Singh and Seavey, 1971), subsequently improved by Hwang *et al.* (1972) and Friesen *et al.*, (1972) to give yields of 40–50 μg prolactin/mg pituitary.

The isolation of satisfactory yields of hPRL has not only led to the development of radioimmunoassay systems for the hormone (Friesen *et al.*, 1972; Sinha *et al.*, 1973) but also has permitted some study of its structural properties. Initial studies (Niall, 1972) have shown that there is considerable homology between human and ovine prolactin and between hGH and hPL, but not between hGH and hPRL. Lewis *et al.* (1972) have shown that 75% of the tryptic peptide maps of human and ovine prolactin are in fact identical.

Recent work has shown that like other peptide hormones, hPRL exhibits a size heterogeneity, being separable into 'big' and 'little' components. This property has been detected both by electrophoresis (Turkington, 1973) and by gel chromatography (Rogol and Rosen, 1974). More recently von Werder and Clemm (1974) confirmed the presence of two molecular species, a 'big' prolactin with a molecular weight of approximately 44 000, and a 'little' component of about 22 000. The larger component makes up 23–24% of the total immunoreactivity but both molecular species cross-react with antisera to hPRL.

7.2 Prolactin in the female

7.2.1 *Prolactin and the normal ovulatory cycle in primates*

While it appears from the evidence cited above that prolactin plays a part in the genesis of certain female reproductive disorders, its role in the control of normal cycling in female primates is far from clear. A number of workers have reported measurements of prolactin levels in human subjects (Hwang *et al.*, 1971; Jacobs *et al.*, 1972; Noel *et al.*, 1972; L'Hermite *et al.*, 1972). Mean basal levels have been found to be slightly higher in females than in males (9–14 ng/ml v. 6·2–13 ng/ml), but this difference has not always proved to be statistically significant (L'Hermite, 1973). Different workers hold conflicting views as to the role of prolactin in the menstrual cycle. A number of authors have reported rises in circulating levels at mid-cycle, with higher levels persisting during the luteal phase (L'Hermite *et al.*, 1972; Robyn *et al.*, 1973). On the contrary, other workers (Hwang *et al.*, 1971; Tyson and Friesen, 1973; Jaffe *et al.*, 1973; McNeilly *et al.*, 1973; Ehara *et al.*, 1973; McNeilly and Chard, 1974) have failed to show any consistent changes in prolactin levels during the menstrual cycle. McNeilly (1975) reports that although occasionally there was coincidence between mid-cycle peaks of prolactin and LH, the prolactin levels showed marked irregular variation. No consistent relationship could be detected between the levels of prolactin and those of LH, FSH, oestrogen or progesterone (McNeilly *et al.*, 1973; Ehara *et al.*, 1973; Jaffe *et al.*, 1974) or between prolactin and menstruation (McNeilly and Chard, 1974).

In addition, the exogenous administration of hPL to human volunteers, though slightly prolonging the menstrual cycle, failed to prolong luteal function as measured by urinary oestrogen and pregnanedial excretion, (Josimovich *et al.*, 1974). However, a more recent paper (Corenblum *et al.*, 1976) has shown that elevated hPRL levels in women can lead to short-term luteal defects.

The possibility of a luteal role of prolactin has been investigated in non-human primates. Reduction of serum prolactin levels in rhesus monkeys by 2-Br-α-ergocryptine caused a reduction in the menstrual interval and decreased serum oestradiol and progesterone to below normal luteal levels (Espinosa-Campos *et al.*, 1975). Lactating rhesus females have also been shown to have increased levels of progesterone and more active corpora lutea than non-lactating post-partum controls (Weiss *et al.*, 1973). On the other hand, Reyes *et al.* (1975), working with chimpanzees reported that as in the human, prolactin levels showed no definite pattern during the menstrual cycle, varying within the normal human range. During pregnancy, prolactin levels gradually rose to a peak level of about 120 ng/ml (Reyes *et al.*, 1975), a pattern similar to that found in humans (L'Hermite and Robyn, 1972; Tyson *et al.*, 1972*a*) but not to that found in the rhesus monkey and baboon (Hodgen *et al.*, 1972), in which prolactin levels remain low throughout pregnancy. This difference between the higher primates

(Hominoidea) and the monkeys may relate to the corresponding differences in oestrogen metabolism (Friesen, 1972; Reyes *et al.*, 1975). Sassin *et al.* (1972) showed that prolactin levels begin to rise 60–90 minutes after the onset of sleep, and then rise gradually through the sleep period to peak between 05·00 and 07·00. The levels then fell rapidly after waking. Vanhaelst *et al.* (1973) have questioned these results, and have shown that although serum prolactin levels do vary through the 24-hour period, this variation does not necessarily follow the cycle of sleep and wakefulness, although they agreed that the lowest levels did occur between 10·00 and 12·00. Sexual differences in prolactin secretion patterns were reported by Nokin *et al.* (1972), who showed that in women peak serum levels occur slightly earlier (01·00–05·00) than in men (05·00). Other work by Ehara *et al.* (1973), and McNeilly and his co-workers (McNeilly and Chard, 1974; McNeilly *et al.*, 1974) has suggested, however, that there is in fact very little difference between male and female prolactin release patterns, with maximal secretion in both sexes occurring as a spike usually in the early hours of the morning.

7.2.2 *Prolactin and ovarian steroidogenesis*

Although the bulk of the work so far reviewed tends to suggest that prolactin in women and anthropoid apes is not luteotrophic, there is some evidence that prolactin can affect ovarian steriodogenesis. McNathy *et al.* (1974) have shown that prolactin reduces progesterone biosynthesis by human granulosa cells cultured *in vitro*. More recent studies by McNathy *et al.* (1975) have also shown an inverse relationship between prolactin and progesterone in human follicular cyst fluid, elevation of hPRL blood levels being associated with reduced progesterone titres.

High circulating levels of prolactin, either in puerperal lactation or in galactorrhoea, reduces the ovarian steroidogenic response to exogenous prolactin, which suggests that prolactin promotes ovarian refractoriness to gonadotrophin (Zarate *et al.*, 1972; Thorner *et al.*, 1974). In galactorrhoea, reduction of prolactin levels by bromocryptine treatment restores the ovarian steroidogenic response to the normal range (Thorner *et al.*, 1974). These indications of an antigonadal action of prolactin in humans cannot readily be harmonised with the evidence from non-primate mammals, in which prolactin has been shown to enhance the production of a gonadal LH-binding protein (see p. 132). McNeilly (1975) has further suggested that prolactin may directly reduce steroidogenesis by the human ovary which in turn will affect the hypothalamus and pituitary, leading to the breakdown in cyclic gonadotrophin release and hence to amenorrhoea.

However, as we have seen, other workers locate the primary lesion in amenorrhoea in the hypothalamus (Archer, 1977).

7.2.3 *Prolactin before puberty*

There have been very few reports of circulating prolactin levels in prepubertal primates, but preliminary observations indicate that the levels are similar to those found in adults (Hwang *et al.*, 1971; Daughaday *et al.*, 1971). More detailed investigations related to the mechanisms controlling the onset of puberty (Lee *et al.*, 1974; Ehara *et al.*, 1975) show that blood prolactin levels do not vary markedly during puberty in boys, and remain at about the adult male level. However, Ehara *et al.* (1975) showed that in girls, prolactin levels rise significantly in late puberty, usually at about 14–15 years of age. This latter finding has more recently been questioned by Lee (1976), who detected no significant changes in prolactin levels in prepubertal girls. A slight drop in prolactin levels between the ages of nine and ten was not thought to be related to breast development. Lee (1976) mentions that following the menarche, plasma prolactin levels show the adult female pattern, but give no further details.

Francks and Brooks (1976) studied the response of the prepubertal pituitary to a variety of treatments designed to assess the pituitary prolactin reserve. There was no significant difference between the circulating prolactin levels of girls and boys, the levels in both sexes lying within the normal adult female range (3–15 ng/ml). Insulin hypoglycaemia induces prolactin release in adults (Frantz *et al.*, 1972), and in children, too, insulin hyperglycaemia caused an elevation of serum prolactin levels, even in children with isolated hGH deficiency. However, children suffering from panhypopituitarism failed to respond to hypoglycaemia in the normal manner (Francks and Brooks, 1976).

A similar pattern was observed in children subjected to TRH-induced prolactin release. As in adults, normal and hGH-deficient children showed a rise in serum prolactin levels (Jacobs *et al.*, 1976), but children with panhypopituitarism did not respond (Franks, *et al.*, 1975). Thus the pattern of release and responsiveness of the prepubertal pituitary seems to be very similar to that in adults.

7.2.4 *Pathology of prolactin action in women*

a. *Amenorrhoea galactorrhoea* For obvious reasons the primary interest of workers studying the role of prolactin in primate reproductive cycles has related to the part played by hPRL in certain pathological conditions in women. These disorders are mainly manifest as the amenorrhoea-galactorrhoea syndrome (Thorner *et al.*, 1974), which can be subdivided into three types on clinical grounds:

1. The Chiari–Frommel syndrome, the persistence of amenorrhoea and galactorrhoea for at least a 12-month period after parturition, not coupled with evidence of a pituitary tumour.
2. The Ahumada–del Castillo syndrome, the spontaneous appearance of amenorrhoea and galactorrhoea without pituitary adenoma.
3. The Forbes–Albright syndrome, which has similar clinical symptoms to the above but with evidence of pituitary adenoma, (Lloyd *et al.*, 1975).

Abnormal milk secretion and hyperprolactinemia have been found in association with hypothyroidism, and with galactorrhoea, precocious puberty and hypothyroidism, in all cases the symptoms being corrected by thyroid treatment (Ross and Mussynowitz, 1968; Baylis and Va'nt Hoff, 1969, Edwards *et al.*, 1971). Excessive post-partum lactation has also been reported in association with Cushing's syndrome (Mahesh, *et al.*, 1969).

As we have seen, prolactin stimulates post-partum lactation in the human as in other mammals. In addition, the hormone is involved in the normal maintenance of the menstrual cycle. Normal menstrual cycles resume earlier in women who do not breast-feed, that is, when post-partum lactation is suppressed (El-Minairi and Foda, 1971). In many cases of galactorrhoea (Thorner *et al.*, 1974) and normal puerperal lactation (Reyes *et al.*, 1972; Jaffe *et al.*, 1973) there is a consistent pattern of high circulating prolactin levels and low oestrogen levels, with levels of LH and FSH within the normal range. Most patients with galactorrhoea also show enhanced LH and FSH responses to treatment with gonadotrophin-releasing hormone. In contrast, amenorrhoea results from failure of cyclical release of both gonadotrophins, which arrests follicular development and ovulation. Mortimer *et al.* (1973) and McNeilly (1975) have suggested that this failure of gonadotrophin cyclicity could result from reduced production of gonadotrophin releasing hormone (Gn-RH).

Blockade of prolactin release by bromocryptine in patients with amenorrhoea-galactorrhoea (Besser *et al.*, 1972; Thorner *et al.*, 1975) arrests the abnormal milk secretion and in most cases returns ovarian function to normal in response to the resumption of cyclic release of FSH and LH.

Rolland *et al.* (1975*a*, *b*) studied normal lactating women, and showed that prolactin levels are high in late pregnancy and in the early post-partum period, and then decline during the period of breast-feeding. FSH shows an inverse relationship with prolactin, being low during late pregnancy and gradually rising to normal levels between days 7–18 post-partum. Following the drop in human chorionic gonadotrophin (hCG) levels in early pregnancy, LH remains relatively low for at least 28 days, and a similar pattern is shown by serum oestrogen levels once placental oestrogen has been cleared. Treatment of normal lactating women with bromocryptine quickly stops lactation and causes a marked drop in plasma prolactin levels. FSH levels show the same pattern as in controls, but the LH and oestrogen levels return more rapidly to normal after bromocryptine, within 7–14 days post-partum. Bromocryptine-treated women also show an early resumption of the normal menstrual cycle, and endometrial biopsis show a normal uterine secretory pattern, often absent at this stage in the non-induced return to menstruation (Rolland *et al.*, 1975*b*).

In the linked amenorrhoea-galactorrhoea syndrome, as stated earlier, high prolactin levels are usually associated with normal to low gonadotrophin titres and with low circulating oestrogens (Archer *et al.*, 1974; Francks *et al.*, 1975; Sparke *et al.*, 1976). Studies have suggested that this pattern may result from ovarian refractoriness to gonadotrophins, and that this ovarian failure may be a consequence of the high levels of circulating prolactin (Zarate *et al.*, 1972, 1974). However, in a more recent study Nakano *et al.* (1975) have shown in puerperal women that high circulating prolactin does not prevent oestrogen secretion in response to exogenous gonadotrophins. This would appear to confirm the view expressed earlier by Mortimer *et al.* (1973) that in amenorrhoea-galactorrhoea, the primary lesion is hypothalamic. Women with amenorrhoea and elevated levels of prolactin can be induced to ovulate, with subsequent normal luteal function, when treated with hCG or menopausal gonadotrophins (Archer and Josimovich, 1976). In addition, the elevation of serum prolactin levels by exogenous TRH in normal menstruating women does not result in amenorrhoea, and prolactin levels, although elevated, remain within the normal range (Zarate *et al.*, 1974*c*; Jewelewicz *et al.*, 1974).

This evidence has led Archer (1977) to conclude that the increased hPRL secretion of amenorrhoea-galactorrhoea is probably a result of reduced PIF secretion by the hypothalamus. The evidence so far available suggests that prolactin in primates may well be under dopaminergic inhibition, as in other mammals. This view is based on the reduction in serum prolactin levels by 2 Br-α-ergocryptine or L-dopa (Rocenziweig *et al.*, 1973), and by the increase in circulating prolactin levels following chlorpromazine treatment (Apostolakis *et al.*, 1972; Turkington, 1972). Archer (1977) points out that reduced dopaminergic activity in the hypothalamus could result in lowered Gn-RH (gonadotrophin releasing hormone) activity. Secretion of Gn-RH has been shown to be stimulated by dopamine and its agonists (McCann *et al.*, 1973). Recent work has demonstrated increased serum FSH and LH levels and elimination of galactopoiesis following L-dopa treatment in humans (Zarate *et al.*, 1973; Ayalan *et al.*, 1974). L-dopa also reduces serum prolactin levels, so the problem remains with regard to the part played by prolactin in amenorrhoea.

Further evidence shows that women suffering from hyperprolactinaemia fail to show the normal positive-feedback LH surge in response to exogenous oestrogen and, in addition, the serum gonadotrophin levels do not respond normally to treatment with clomiphene citrate (Thorner *et al.*, 1974; Glass *et al.*, 1975). Again this suggests a dysfunction at the hypothalamic level.

From the literature it appears that there is a sharp distinction between the hypothalamic function of patients suffering from amenorrhoea without adenoma and those cases in which a pituitary adenoma can be demonstrated. In patients without a pituitary adenoma, treatment with exogenous Gn-RH leads to increased secretion of FSH and LH. This response has been reported as being in the normal-to-low normal range (Zarate *et al.*, 1974) or greatly enhanced (Wents *et al.*, 1975; Archer *et al.*, 1976). In contrast, this normal or enhanced response is not shown by patients suffering from pituitary adenoma (Zarate *et al.*, 1974).

It appears to be generally accepted that prolactin plays a part in amenorrhoea and galactorrhoea. It is not clear, however, whether prolactin is responsible for the suppression of gonadotrophin levels, either directly or by reducing the ovarian oestrogenic response, or whether this is a consequence of hypothalamic dysfunction independent of the hypersecretion of hPRL.

b. *Breast cancer* In rats and mice, which have been studied extensively, there now appears to be conclusive evidence that PRL and oestrogens act synergistically to cause the development of a high percentage of mammary tumours (Meites *et al.*, 1972). Sterental *et al.* (1963) have shown in the rat that oestrogen does not produce mammary tumours in the absence of the pituitary. Long-term injection of PRL or the transplantation of extra pituitaries into mice, causes an increased incidence of mammary tumours (Muhlbock and Boot, 1967; Boot, 1969), and bilateral lesions in the median eminence, which have been shown to increase PRL secretion in rats (Chen *et al.*, 1970) also increase the incidence of mammary tumours (Welsch *et al.*, 1971). For detailed reviews of the situation in rodents see Meites (1972, 1973; Pearson *et al.*, 1968).

There have been numerous reports that drugs which are known to elevate serum PRL levels, such as reserpine, increase the incidence of human breast cancer (Ettigi *et al.*, 1973; Armstrong *et al.*, 1974; Heinonen *et al.*, 1974), yet the role of PRL in breast cancer development remains unclear (Heuson, 1973). On the one hand, surgical hypophysectomy, which should abolish PRL secretion, is regarded as a successful way of inducing breast cancer regression (Hayward *et al.*, 1970), while injection of ovine PRL has been said to increase breast cancer growth (McAllister and Wellboun, 1962).

On the other hand, stalk section, which results in elevated serum PRL (Turkington *et al.*, 1971), can induce regression of some breast cancers (Ehni and Eckles, 1959). Again, some workers have found high circulating prolactin levels in post-menopausal women with metastatic breast cancer (Murray *et al.*, 1972) but this has not been confirmed by others (L'Hermite *et al.*, 1973; Sheth *et al.*, 1976). Mittra *et al.* (1974) were unable to demonstrate that the PRL response to TRH in women with breast cancer was different from the normal response. A possible explanation of this anomaly may be found in the work of Kwa *et al.* (1974) who showed that serum PRL levels in breast cancer patients were elevated only in women with a familial history of the disease. It seems that possibly only about 20% of breast cancers in women are in fact prolactin-dependent (Salih *et al.*, 1972). However, suppression of PRL secretion by the use of drugs such as CB154 (Schulz *et al.*, 1973) and L-dopa (Frantz *et al.*, 1973) can cause regression of mammary tumours in some cases.

In summary, it seems that in some cases, mammary tumours are prolactin-dependent, and that tumour incidence is increased by elevated prolactin levels. The risk of prolactin-dependent tumours appears to be greater in women with a familial history of breast cancer (Kwa *et al.*, 1974) and in certain cases treatment with prolactin-inhibiting drugs may prove successful. Many of the prolactin-dependent tumours appear to be also oestrogen dependent, and a synergism between the two hormones may be a more important factor in stimulating the development of tumours than high prolactin levels *per se*.

c. *Pre-eclampsia* The relationship between prolactin and pre-eclampsia has been suggested primarily by Horrobin (1975), and the following deductions are based largely upon this article. Horrobin's hypothesis is based mainly upon the fact that prolactin has been shown to cause renal retention of sodium, potassium and water in man (Horrobin, 1974) while also raising arterial blood pressure in rabbits (Horrobin *et al.*, 1973*b*).

This earlier point has, however, been questioned recently by a number of authors (see p. 173) and it has also recently been suggested that prolactin preparations may be contaminated with vasopressin (Horrobin *et al.*, 1973a). On the other hand, Horrobin and his co-workers (Manku *et al.*, 1973) have shown that prolactin at levels comparable to those found in human pregnancy can potentiate the effects of norepinephrine and angiotensin on rat arterioles. Such levels can also produce a nephrotic type condition in rats which can be prevented by the administration of CB154 (Furth *et al.*, 1956; Richardson, 1973).

Clinical examination of patients with pre-eclampsia and from non-pre-eclamptic patients, both of whom produced still-born infants, shows a remarkably high degree of follicular development with some luteinisation in the patients suffering from pre-eclampsia (Govan and Mukherjee, 1950). This fact could suggest the presence of a luteotrophic factor in pregnancy. These facts, together with the excitatory effects on heart (Nasser *et al.*, 1974) and vascular smooth muscle (Manku *et al.*, 1973), suggest that it is possible for prolactin to be involved in the development of the pre-eclamptic syndrome.

7.3 Prolactin in the male

7.3.1 *Prolactin and testicular function*

As in women, hyperprolactinaemia in men is usually associated with hypogonadism, suggesting a direct effect of prolactin on testicular function (Friesen and Hwang, 1973; Boyer *et al.*, 1974; Thorner *et al.*, 1974; Besser and Thorher, 1975; Francks, Jacobs and Nabarro, 1976). The evidence regarding the precise action of prolactin in the human male, however, is still unclear. In their study of four men with hyperprolactinaemia and galactorrhoea, Thorner *et al.* (1974) found that plasma 17β-hydroxysteroid levels (testosterone and dihydrotestosterone) were within the normal range. However in another patient with marked hyperprolactinaemia (100 ng/ml), Besser and Thorner (1975) showed that hCG stimulation of the testes resulted in *decreased* 17β-hyroxysteroid production. After treatment with bromocryptine, the testicular response to hCG was restored to normal.

Magnini *et al.* (1976) treated six normal men with Sulpiride to elevate plasma prolactin levels. Although the treatment did not block hCG-induced rise in testosterone synthesis, it did impair the conversion of testosterone to dihydrotestosterone in response to hCG. This impaired conversion of testosterone to the more active dihydro-derivative was returned to normal by pretreatment with bromocryptine. Although it is remotely possible that bromocryptine could act directly on the testis,

Magrini *et al.* (1976) interpret their findings as indicating a direct inhibitory action of prolactin on testicular 5α-reductase activity.

Somewhat in contrast to these findings, Rubin *et al.* (1975) demonstrated nocturnal rises in blood testosterone levels in men, associated with elevated prolactin levels. Similar results have been reported for nonprimate mammals, in particular the rat (Hafiez *et al.*, 1972), in which prolactin has been shown either to synergise with LH or to increase the levels of LH-receptor protein. Furthermore, prolactin has also been shown to stimulate androgen formation by the human adrenal (Boyns *et al.*, 1972). Seppala and Hirronen (1975) reported elevated serum prolactin levels in women with secondary amenorrhoea and hirsutism, including at least one case in which hyperprolactinaemia was associated with high circulating testosterone levels. In a more recent paper (Seppala *et al.*, 1976), these authors found that although high circulating levels of prolactin and testosterone can occur simultaneously in women, there was no definite correlation between high testosterone levels and hyperprolactinaemia. Of 24 women with amenorrhoea, 12 showed hyperprolactinaemia and 12 had normal prolactin levels, but there was no significant difference between the two groups in blood levels of testosterone and dihydrotestosterone. Nor did blood testosterone levels change in any consistent way when the hyperprolactinaemia of the first group was suppressed by bromocryptine treatment.

Thus, there is some evidence that prolactin can induce symptoms of testicular deficiency in men, probably by reducing the rate of conversion of testosterone to its more active metabolite dihydrotestosterone. However, a comparable relationship does not appear to hold for androgen secretion in women by the adrenal cortex or the ovary. Opposing the evidence that prolactin induces hypogonadism in men is the evidence cited above that nocturnally elevated prolactin is associated with high levels of serum testosterone (Rubin *et al.*, 1975). Cunningham and Huckins (1977) recently reported the case of a man with a long-term chromophobe pituitary adenoma which secreted large amounts of FSH and prolactin, while blood levels of ACTH, TSH, LH and GH were relatively low. The testicular Leydig cells were regressed, in correlation with the low LH levels, but nevertheless sufficient testosterone was secreted to maintain normal sexual development. The authors concluded that this testosterone must have originated in the seminiferous tubules, and that this *tubular* production of the androgen, therefore, was not depressed by the persistent hyperprolactinaemia. However, an alternative view would be that the testosterone was of adrenal origin, and this would be consonant with the work quoted above on hirsute women with amenorrhoea, which indicates that adrenal secretion of testosterone is not depressed by prolactin.

7.4 Prolactin and osmoregulation in man

A number of conflicting reports relating prolactin to water and electrolyte balance in man have recently appeared in the literature. The evidence for an osmoregulatory role of hPRL is based primarily on the observation that prolactin treatment can cause sodium retention in man (Horrobin *et al.*, 1971). Furthermore, it has been reported (Buckman and Peake, 1973*a*) that patients suffering from hyperprolactinaemia are unable to excrete water load, which suggests an antidiuretic role for prolactin in man, which would be in line with its actions in other vertebrates (see p. 12). In a later paper Buckman and Peake (1973*b*) showed that water loading not only suppresses prolactin secretion in normal men but also suppresses the 'functional' hyperprolactinaemia associated with pituitary adenoma. Further evidence of a causal relationship between high plasma prolactin levels and decreased sodium excretion comes from two sources. Firstly, two groups of workers have demonstrated a circadian rhythm of prolactin secretion, with high levels at night when sodium retention is highest and low levels during the day when there is maximal sodium excretion (Nokin *et al.*, 1972; Simpson *et al.*, 1975). Secondly, Auty *et al.* (1976) showed that in subjects undergoing frusemide diuresis there was a strong negative correlation between plasma prolactin levels and sodium excretion. This finding is in agreement with the hypothesis that prolactin directly stimulates renal sodium reabsorbtion. However, Auty *et al.* (1976) report that the picture can be complicated by the action of mineralocorticoids, particularly aldosterone, and that the circadian rhythm of prolactin is not modified by treatment with various drugs. Thus, although prolactin may respond to alterations in sodium levels, it probably does not form a major part of the acute endocrine response to osmotic stress, prolactin responses being more gradual than those of other hormones.

These differences in response to prolactin, and the possible interaction with aldosterone (Solyom *et al.*, 1971), may help explain the failure of various authors to detect a close correlation between prolactin and human renal function. For instance, Baumann *et al.* (1977) found that patients suffering from hyperprolactinaemia show normal responses to a water load and to hypertonic saline loading, findings in agreement with other work indicating that endogenous prolactin does not affect renal sodium or water excretion (Baumann and Loriaux, 1976; Adler *et al.*, 1975). In addition, del Pozo and Ohnhaus (1976) have questioned the relationship between nocturnal prolactin surges and sodium retention: elimination of the diurnal rhythm of prolactin secretion by bromocryptine produced no significant change in either water or sodium retention in comparisons with normal control periods in the same subjects. At present, there is insufficient evidence to reconcile these conflicting findings. It is interesting to note, however, that Cole *et al.* (1975) have used bromocryptine treatment with great success to alleviate water retention associated with premenstrual tension. Similarly Beynon and Ensor (unpublished observations) have shown that the efficacy of the renal action of prolactin in female rats varies with the stages of the oestrus cycle. This might indicate that the prolactin action on the human kidney may not be primary, as at first supposed, but may depend on the hormonal state of the patient.

7·5 Prolactin in amniotic fluid

A number of authors have detected high levels of prolactin in human and monkey amniotic fluid (Friesen *et al.*, 1972; Josimovich, *et al.*, 1974*b*; Ben-David *et al.*, 1974; Chochinov *et al.*, 1976), the levels being particularly high during the second trimester of pregnancy (Tyson *et al.*, 1972*b*; Schenber *et al.*, 1976; Freeman *et al.*, 1976; Chochinov *et al.*, 1976; McNeilly *et al.*, 1976). The source of this prolactin is not yet clear, whether foetal (Fang and Kim, 1975; Chochinov *et al.*, 1976), maternal (Schenber *et al.*, 1975; Josimovich *et al.*, 1974) or placental (Friesen *et al.*, 1972; McNeilly *et al.*, 1976). However, Clements *et al.* (1977) have recently assessed the secretory capacity of the human foetal pituitary, and have shown fairly convincingly that it is unlikely that all the prolactin in the amniotic fluid is of foetal origin, although it remains possible that a high percentage does emanate from this source.

Not only does prolactin occur in the amniotic fluid, but also Josimovich *et al.* (1977) have shown that cell membrane fractions from the foetal rhesus monkey are capable of binding amniotic prolactin, and that these receptors occur in the foetal placenta, liver, lung, myocardium and brain, a distribution similar to that found in adult mammals (see Chapter 8). A number of speculative hypotheses have been put forward regarding the role of amniotic prolactin, including suppression of foetal testicular testosterone secretion in the latter half of pregnancy (Aubert *et al.*, 1975), the control of foetal adrenal growth (Winters *et al.*, 1975), and regulation of salt and water homeostasis in the amniotic fluid and the foetus (Friesen *et al.*, 1972; Josimovich and Merisko, 1975). This last concept is supported by a number of recent reports. Josimovich *et al.* (1977) showed that injection of ovine prolactin into the amniotic fluid of rhesus monkeys results in a decrease in amniotic fluid volume, a response not produced by vasopressin. Work on isolated amniotic membranes of other mammalian species has yielded similar results (Vizsolyi and Perks, 1974). Experiments with the isolated guinea pig amnion show that prolactin can stimulate transport of water into the maternal compartment (Manku *et al.*, 1975), a finding which agrees with the *in vivo* results of Josimovich *et al.* (1977). This recent work suggests that an important role of the prolactin found in amniotic fluid is to regulate water balance in the amniotic cavity and so to protect the foetus.

7.6 Human placental lactogen

Human placental lactogen (hPL) was discovered and named by Josimovich and MacLaren (1962) and as discussed previously, it displays a much stronger immunological cross-reaction with hGH than with hPRL, presumably in consequence of the fact that its amino acid sequence more closely resembles that of hGH than hPRL (Sherwood *et al.*, 1971; 1972). As shown in Fig. 1.4. hPL is a single chain polypeptide of 190 amino acids with two intrachain disulphide bonds, and a total of 163 out of the 190 amino acids are common to both hPL and hGH (Niall, 1971; Niall *et al.*, 1971). According to Sherwood *et al.* (1972), differences which do occur between the two molecules are usually of a minor nature and generally involve pairs of amino acid residues with similar chemical properties. The only major difference is that hPL contains a larger number of methionine residues than hGH (6:3) which results in a larger number of peptide fragments when the hPL molecule is treated with cyanogen bromide (Sherwood *et al.*, 1972). As a result of this marked structural similarity the two hormones share many physiochemical properties. Both have a tendency to aggregate in solution and both exhibit similar mobilities in various electrophoretic procedures (Sherwood, 1967; Floriani *et al.*, 1966). After incubation in alkaline solutions both molecules undergo deamidation (Lewis and Cleaver, 1965; Sherwood and Hardwerger, 1969). The tertiary structure of both hPL and hGH appears to be similar, but is not essential for lactogenic activity (Dixon and Li, 1966; Bewley *et al.*, 1969; Hardwerger *et al.*, 1971; Sherwood *et al.*, 1971), nor for the growth-promoting properties of both molecules as measured by the rat tibia test (Sherwood *et al.*, 1969; Hardwerger *et al.*, 1971). However, Breur (1969) has suggested that complete reduction and alkylation of the hPL molecule does result in loss of ability to stimulate synthesis in costal cartilage.

7.6.1 *Synthesis and production of hPL*

Many aspects of the physiology of hPL have been ably reviewed by Friesen (1974). Synthesis of hPL appears to be localised within the syncytiotrophoblast (Beck and Currie, 1967; Sciarra *et al.*, 1963; Kaspi *et al.*, 1976; Chatterjee, 1976), as demonstrated by immunofluorescent techniques. Gua and Chard (1975) have used an immunoprecipitation assay on sections of freshly frozen term placentae, and have shown that the levels of hPL are closely correlated with the amount of trophoblast tissue in the section. Ectopic production of hPL by tumour tissue has been reported in males with choriocarcinoma (Frantz *et al.*, 1965), and in women, hPL production has also been linked with both trophoblastic (Saaman *et al.*, 1966) and non-trophoblastic (Weintraub and Rosen, 1971) tumours.

Normal production of hPL during pregnancy was first measured by Kaplan and Grumbach (1965), who showed that hPL rose progressively to term in maternal serum, but remained relatively low in foetal serum. As a result of further studies it was calculated that the syncytiotrophoblast produced on average 1 gm of hPL/day (Grumbach *et al.*, 1968; Kaplan *et al.*, 1968), and this figure was later confirmed by Friesen *et al.* (1969) and by Suwa and Friesen (1969*a*,*b*). These authors showed that up to 60% of all protein synthesised by the placenta could be hPL, although lower figures of around 10% were more common. As with other peptide hormones, two molecular species appear to be synthesised, with different molecular weights.

Placental production of hPL has been shown in some studies to be correlated with placental weight (Sciarra *et al.*, 1963; Seppala and Ruoslatiti, 1970; Boire, 1976), although in other studies this relationship has not been found (Saaman *et al.*, 1966). In addition, Tyson *et al.* (1971) have shown that fasting in the mother can cause an elevation in serum hPL levels.

In clinical work, serum hPL levels have been used in attempts to measure placental function. Although normal values vary over a wide range at all stages of gestation, it is possible to correlate hPL levels with some clinical disorders. For instance, some patients with diabetes mellitus and patients with rhesus-immunisation show elevated serum hPL levels (Saaman *et al.*, 1968; Saxena *et al.*, 1969; Seppala and Ruoslathi, 1970; Singer *et al.*, 1970). In addition Spellacy *et al.* (1971) showed in women suffering from hypertension, who eventually gave birth to low-weight or still-born infants, that the serum hPL levels were below normal, in agreement with preliminary observations by Genazzani *et al.* (1969). hPL measurements are now widely used clinically, and numerous reports indicate that hPL levels in maternal serum give a reasonable indication of foetal growth (Macmillan *et al.*, 1976) and of the likelihood of premature abortion (Edwards *et al.*, 1976; Kunz and Keller, 1976; Gartside and Tindall, 1975; Stroobarts *et al.*, 1975). In a recent study it has also been shown that there is a positive correlation between heavy smoking by the mother, low circulating hPL levels and, usually, low birth weight of the baby (Boyce *et al.*, 1975).

7.6.2 *Biological effects of hPL*

Although hPL possesses a wide range of biological activities, overlapping extensively with both hPRL and hGH, no physiological role has been ascribed with certainty to this hormone (Friesen, 1974). Primary growth-promoting effects have been reported in hypophysectomised rats (Florini *et al.*, 1966) and in hypopituitary dwarfs (Grumbach *et al.*, 1968) as well as in the rat tibia assay (Friesen, 1965; Kaplan and Grumbach, 1964), although this latter observation has been questioned by other authors who failed to demonstrate tibial growth stimulation with hPL (Florini *et al.*, 1966; Josimovich and Mintz, 1968). hPL certainly appears to be consistently less potent than hGH in stimulating growth in these various test systems (Josimovich, 1966; Kaplan and Grumbach, 1964; Murakawa and Raben, 1968). However, there is evidence to suggest that the dimer of the molecule is more effective than the monomer (Breur, 1969) and that hPL is particularly effective in acting synergistically with a minimally effective dose of hGH (Josimovich, 1966).

As well as quantitative differences between the responses to hGH and hPL, there is also some evidence that qualitative differences may occur. Early work comparing the activity of the two hormones in hypopituitary dwarfs suggested that hPL produced a full range of hGH-like effects, including nitrogen retention, increase in serum free fatty acids, increased hydroxyproline excretion and hypercalciuria (Kaplan and Grumbach, 1964). However Schultz and Blizzard (1966) reported subsequently that they were unable to demonstrate nitrogen retention in hypopituitary dwarfs undergoing short-term treatment with hPL. Similarly in a recent study of the effects of hGH, hPL and hPRL on serum fatty acids in nonmenstruating women, it was found that hPL did not produce the characteristic rise in acetoacetate and 3-D-B-hydroxybutyrate induced by hGH, and was in fact less effective than hPRL (Berle, *et al.*, 1974). Similar results had previously been observed in men and women undergoing continuous hPL infusion (Beck and Daughaday, 1967). Conflicting results have also been reported concerning the effects of hPL on fat metabolism in other species. In the rat, hPL produced a rise in the fat fraction of the ketone bodies (Friesen, 1967) and a lipolytic effect has been reported on the isolated fat pad and fat pad cells (Genazzini, *et al.*, 1969; Felker *et al.*, 1972). However, Friesen (1967) could find no effect of hPL on serum free fatty acids in the rat, whereas Riggi *et al.* (1966) had previously shown quite large increases in serum free fatty acids of both rabbits and monkeys following hPL treatment. Thus there remains some doubt regarding the possible GH-like effects of hPL, and it is interesting to note that in humans at least, hPL at levels comparable to those found at term and early post-partum, do not have an effect on lipid metabolism (Beck and Daughaday, 1967; Berle *et al.*, 1974).

These findings tend to argue against the early hypothesis of Grumbach *et al.* (1968) who suggested a lipolytic role for hPL during pregnancy. They envisaged that the increased free fatty acids would furnish energy for maternal metabolism and that elevated intracellular fatty acids in adipose tissue would inhibit glucose metabolism, acting therefore in a glucose-sparing capacity (see Friesen, 1974). Friesen (1974) suggests instead that the role of hPL may mainly be amino-acid sparing, due to its anabolic and anticatabolic effects on protein metabolism, acting in concert with insulin. In this connection, it is significant that pretreatment of hypophysectomised rats with hPL does induce enhanced synthesis and secretion of insulin by isolated pancreatic islets (Malaise *et al.*, 1969; Martin and Friesen, 1969). In addition, infusion of hPL into normal subjects causes decreased glucose tolerance even though plasma insulin levels are raised (Beck and Daughaday, 1967; Kuhl *et al.*, 1975) and hPL causes ketosis and hyperglycaemia in diabetics (Saaman *et al.*, 1968). These observations do not necessarily mean that hPL has a direct effect on insulin production, and Beck (1970) has shown that it is more likely that a complex interaction between hPL and progesterone controls insulin synthesis during pregnancy.

The other possible major role of hPL is as a lactogenic hormone stimulating the development of the mammary gland prior to lactation. Friesen (1966) has claimed that hPL stimulates mammary development in the rabbit to an even greater extent than prolactin. Turkington and Topper (1966) have also shown that hPL will stimulate alveolar growth and casein synthesis in explanted mouse mammary gland tissue which has been pretreated with insulin and cortisol. Frantz (1976) however, claims that prolactin is more potent in the mouse mammary assay than hPL, the hPL potencies ranging from 30–60% of either human or ovine prolactin which are equipotent on a weight-for-weight basis (Frantz *et al.*, 1972). However, Doneen (1976), using the same assay system, was unable to differentiate the lactogenic potencies of the three human hormones. This suggests that, although there may exist quantitative differences in effects in different assay systems, both hPL and human prolactin are potent lactogenic hormones. As an interesting sideline, Doneen (1976) also assayed hPL activity in a teleost assay (water permeability in *Gillicthys* bladder), and showed that in this assay hPL was much less potent than hGH and ovine prolactin. hPL has also been shown to have a very low potency (1–4 iu/mg) in the pigeon crop-sac assay (Forsythe, 1967; Friesen, 1965), in which it behaves very like hGH.

The main question raised is why two potent lactogenic hormones should co-exist in the pregnant female. Figures for the levels of hPRL and hPL are variable, but reliable measurements suggest that at term hPRL blood levels vary between 50–500 ng/ml (Hwang, *et al.*, 1971). Postpartum levels fluctuate markedly (L'Hermite and Midgeley, 1971), but tend to rise after each suckling period (Bryant *et al.*, 1971; Hwang *et al.*, 1971). On the other hand Kaplan and Grumbach (1964) have demonstrated blood hPL levels at term of 25 μg/ml, and blood levels during pregnancy below 4 μg/ml are regarded as indicative of placental insufficiency (Spellacy *et al.*, 1971). Thus, blood levels of hPL appear to be always much higher than those of hPRL. Even taking the highest reported biological potency for hPRL and the lowest for hPL, the implication remains that in the pregnant woman about 75% of the total prolactin-like activity resides in the hPL molecule. Thus it is difficult to comprehend the part played by hPRL in pregnancy and mammary gland development.

hPL has been reported as being luteotrophic in the rat (Josimovich and MacLaren, 1962); whether it is luteotrophic in the human is uncertain, since doubt exists as to whether hPRL plays a luteotrophic role. Recent evidence indicates that a placental lactogen exists in the rat and in other mammals (see below), and it is possible that the physiological luteotrophic action of the native rat placental lactogen may be mimicked by hPL. hPL has also been shown to stimulate erythropoiesis in pregnant women (Tyson and Friesen, 1968), and to increase aldosterone production by the maternal adrenal cortex, assessed by its urinary metabolities (Melby *et al.*, 1966). In a recent review on the physiology of hPL, Ong (1976) concludes that while no definite role can yet be assigned to hPL, probably its major effect is to stimulate maternal steroidogenesis.

7.7 Placental lactogens in nonhuman primates

It now appears certain that placental lactogens (see below) are present in a range of mammalian species. Investigations in monkeys have identified a substance in the placenta which cross-reacts with antisera to hGH and hPL (Kaplan and Grumbach, 1964; Josimovich and Mintz, 1968; Forsythe, 1974; Talamantes, 1975). This substance, designated monkey placental lactogen (mPL) by Friesen (1974) has been purified by Grant *et al.* (1970) and by Shome and Friesen (1971). The yield of mPL from the monkey placenta appears to be considerably less than the equivalent yield from humans (Shome and Friesen, 1971). As with many of the other peptide hormones, mPL exhibits a molecular heterogeneity when separated by polyacrylamide gel electrophoresis. The crude extract separates into two molecular fractions, mPL1 with an Rf of 0·52 and mPL2 (Rf 0·8). These figures suggest that mPL2 is more like hPL in its structure whereas mPL1 may be more similar to hGH. Both molecular entities have been found to possess four half-cysteine residues, again comparable to hPL and hGH. However, tryptic hydrolysis of the mPL molecule suggests a greater degree of similarity between hPL and hGH than between either of these two molecules and mPL. Work by Belanger *et al.* (1971) has shown that both mPL1 and mPL2 are synthesised *in vivo* and that the mPLs of the four primates studied (chimpanzee, African green monkey, squirrel monkey and macaque) showed considerable mutual immunological cross-reactions.

mPL appears to possess the same biological properties as hPL. Talamantes (1975) has shown that mPL from the Hamadryas baboon is a powerful stimulator of lactogenesis in cultured BALB/cCrgl mouse mammary tissue. A similar preparation was found to be a potent luteotrophic agent in mice (Josimovich *et al.*, 1973) and mPL has also been shown to be more potent than hPL in its growth promoting effects (Friesen, 1968).

7.8 Placental lactogens in other mammals

The existence of placental lactogenic hormones in mammals other than primates has been known for some time (Ray *et al.*, 1955), but only recently, with the development of co-incubation techniques (Kohmote and Bern, 1970), have detailed studies been made of these hormones. A list of species in which a placental lactogen has been detected is shown in Table 7.1. Some discussion of the similarities between these mammalian placental lactogens and those of the higher primate is germane, because of the ready potentialities for model experimental studies offered by the nonprimate mammals. Placental lactogens have now been purified and at least partially characterised from cattle (Belander, 1976), sheep (Chan *et al.*, 1976) and rat (Robertson and Friesen, 1975). These molecules bear a close resemblance to primate placental lactogens, although the bovine hormone in particular shows only very limited cross-reaction, failing to cross-react with antisera to both bovine PRL and bovine GH. Ovine PL does not cross-react with antisera to ovine PRL (although it can be assayed using radioreceptor assays to lactogenic hormones), but it does cross-react with antisera to hGH. This high degree of molecular specificity is perhaps common: for example, Forsyth *et al.* (1976) have shown that although rat placental lactogen can stimulate lactogenesis in mammary tissue from other rodents, it does not cross-react with antisera to rat PRL. Some controversy still surrounds the question of whether all mammals secrete a placental lactogen. Talamantes (1975) failed to demonstrate lactogenic activity in the placentae of rabbit and dog, although Gusden (1970) had previously isolated from these species placental proteins immunologically and electrophor-

Table 7.1 Non-primate mammals in which a placental lactogen has been detected

Species	*Authority*
Rat	(Mathias, 1967, 1974; Linkie and Niswender, 1973; Talamantes, 1973; Shiu *et al.*, 1973; Kelly *et al.*, 1975).
Mouse	(Cerrute and Lyons, 1960; Kohmote and Bern, 1970; Kohmote, 1975; Talamantes, 1975.
Vole (bank and field)	(Forsyth *et al.*, 1976).
Hamster	(Talamantes, 1973; Kelly *et al.*, 1973; Mathias, 1974).
Chinchilla	(Talamantes, 1973).
Guinea pig	(Kelly *et al.*, 1973).
Goat	(Bulte *et al.*, 1972).
Cow	(Forsyth, 1974; Buttle and Forsyth, 1976).
Sheep	(Kelly *et al.*, 1974).
Fallow deer	(Forsyth, 1974).
Pig	(Gudsen, 1970; not confirmed by Forsyth, 1974).
Cat	(Gudsen, 1970).
Dog	(Gudsen, 1970; Talamantes, 1975).
Horse	(Gudsen, 1970).
Rabbit	(Gudsen, 1970; not confirmed by Talamantes, 1975).

etically related to hPL. A similar situation exists in the pig, Forsyth (1974) being unable to detect biological activity in the placenta, although Gusden (1970) had previously isolated a porcine hPL-like protein. Thus, placental lactogens may well prove to be highly species-specific, both immunologically and in terms of biological activity (see Amoroso, 1955).

Few physiological studies of non-primate placental lactogens have been reported. Most have been shown to possess lactogenic activity in the Kohmote and Bern (1970) cross-culture assay with mouse mammary explants, and in addition rat and mouse placental lactogens are known to be luteotrophic (Josimovich and MacLaren, 1966). Hardwerger *et al.* (1971) have made a detailed study of the metabolic effects of ovine PL in an attempt to produce a model system for the study of human PL. They showed that there was a marked similarity in the effects of the two preparations, both inducing in the rat (or the sheep) a decrease in serum free fatty acid, glucose and amino nitrogen. This was correlated with an initial decrease followed by a significant increase in plasma insulin levels. The authors conclude that the main role of placental lactogens may be to stimulate maternal intermediary metabolism, but more work is required to substantiate this idea, and particularly to determine how far it may be applicable to human physiology.

7.9 Hypothalamic control of prolactin secretion in primates

As in other mammals (Chapter 6), studies on PRL control in primates have pointed to the probable existence of a dual mechanism involving TRH as a prolactin-releasing agent, and dopamine as an inhibitory factor. Other neurotransmitters may also be involved, and a peptidic PIF could also exist.

7.9.1 *Stimulation of prolactin secretion by TRH*

The original observations that administration of physiological doses of TRH in humans (Bowers *et al.*, 1971; Jacobs *et al.*, 1971) and monkeys (Bowers *et al.*, 1972) cause a dramatic increase in prolactin secretion were unexpected, since TRH had previously been considered to be specific for the control of TSH. Subsequent work has shown that the minimal effective dose of TRH (5 to 10 mg in man) will stimulate the secretion of both prolactin and TSH, and that the effective dose-range appears to be identical for both hormones (Friesen and Hwang, 1973). Along with various psychotrophic agents, such as the phenothiazines, TRH is now regularly used to evaluate the status of the hypothalamo-PRL system in man (Friesen *et al.*, 1972).

The effectiveness of TRH as a prolactin releasing hormone varies with the condition of the subject, and TRH has been shown to produce greater effects in women than in men (Bowers *et al.*, 1972). As discussed earlier in this chapter, PRL levels are usually high in conditions of amenorrhoea/galactorrhoea (Hwang *et al.*, 1971; Friesen *et al.*, 1972; Frantz *et al.*, 1972) and in amenorrhoea without inappropriate lactation (Zarate *et al.*, 1973; Seppala *et al.*, 1976). In hypothyroidism, which is often linked with secondary amenorrhoea (Hirvonen *et al.*, 1976), serum PRL levels are commonly high and the TRH-PRL and TRH-TSH responses are often exaggerated (Bowers *et al.*, 1971; Friesen and Hwang, 1973). However, it is possible to separate the two responses to TRH, for instance treatment of hypothyroidism with triiodothyronine or thyroxine is relatively more inhibitory to the TSH response to TRH than to the PRL response (Bowers *et al.*, 1973). In other conditions, such as breast feeding, surgical stress or exercise, high serum PRL levels are recorded in the presence of normal levels of TSH (Frantz *et al.*, 1972; Noel *et al.*, 1972; Bowers *et al.*, 1973). Furthermore, recent studies with bromocryptine have shown that this drug can block TRH-induced prolactin release without altering circulating TSH levels in euthyroid women with secondary amenorrhoea (Hirvonen *et al.*, 1976). Interestingly, in women suffering from normo-prolactinaemic amenorrhoea, the response to TRH was attenuated in those patients who already had relatively low serum prolactin levels, while women with hyperprolactinaemia also showed a reduced TRH response (Hirvonen *et al.*, 1976).

This 'blunted' response in women with hyperprolactinaemia has been observed by other workers (Glass *et al.*, 1976). Some of these patients had clinically manifest pituitary tumours, which may possibly explain the 'blunted' response to TRH stimulation. Other patients have been reported with blood prolactin levels as high as 120 ng/ml and no detectable pituitary adenoma and who show little or no response to exogenous TRH (Hirvonen *et al.*, 1976). This level of blood prolactin is in excess of the functional pituitary reserve of normal women under maximal TRH stimulation (Jacobs *et al.*, 1972). It has been suggested that the development of pituitary adenoma may be a very slow process, and that a functional disorder may not necessarily be manifested clinically (Jacobs and Daughaday, 1973). Findings similar to the above have been reported by del Pozo *et al.* (1972) and Tyson and Zacur (1975) in patients without pituitary tumours, and by Schwinn *et al.* (1975) and Zarate *et al.* (1975) in patients with pituitary tumours. As a result of this work, del Pozo *et al.* (1972) have suggested that elevated prolactin secretion may be associated with pituitary microadenoma not readily detectable by standard clinical assessment, and that, in patients with an adenoma or a developing adenoma, the TRH control of PRL secretion is impaired.

There is also evidence that abnormal thyroid function in women may affect pituitary PRL secretion. Keye *et al.* (1976) report the case of a 22 year-old multiparous woman who was diagnosed as suffering from primary amenorrhoea, primary hypothyroidism and hyperprolactinaemia, and who presented radiological evidence of pituitary enlargement. In this patient the circulating levels of both TSH and prolactin were elevated, but both hormones were further elevated by TRH and subsequently reduced by triiodothyronine. However, in contrast to the cases discussed previously, all clinical symptoms disappeared with administration of triiodothyronine, and the authors concluded that in this patient hyperprolactinaemia was a secondary condition dependent upon primary thyroid failure.

It is apparent that the pattern of response to TRH in hyperprolactinaemic patients is variable. The reduced PRL response to TRH in amenorrhoeic hypoprolactinaemia has already been mentioned. A similar syndrome has been observed in children suffering from protein-calorie malnutrition, a condition characterised by high serum GH and TSH levels, and low prolactin levels. In these children the response to TRH is low or within the normal range. Correction of the nutritional intake results in an elevation of prolactin levels and increases the prolactin response to TRH, suggesting that the magnitude of the response to TRH may depend on the functional state of the prolactin cell (Becker *et al.*, 1975). This observation on patients with high serum TSH levels also suggests that although the effective doses of TRH required to release both prolactin and TSL are similar, the response mechanisms may well be complex, and that independent physiological control of the two hormones does exist. That the response of the prolactin cell to TRH may be modulated by other factors is also suggested by work on men suffering from Klinefelter's syndrome (decreased androgenicity). As well as hypogonadism such patients also have depressed thyroid function with impaired thyroidal responsiveness to TSH, and an enhanced prolactin response to TRL. Following treatment with testosterone the prolactin response is reduced, though remaining higher than normal. It is not clear whether testosterone modifies the prolactin response to TRH directly or inhibits prolactin release through some other mechanism. Gonadal steroids have been shown to modify prolactin secretion (Frantz *et al.*, 1972), an action that will be discussed later.

What does seem clear is that peripheral factors can modify TRH-prolactin interactions independently of the TRH-TSH axis (for example, Wartofsky *et al.*, 1976), which raises the possible existence of a prolactin-stimulating agent other than TRH. Gautvik *et al.* (1974) showed that suckling elevated serum prolactin levels in post-partum women, without affecting circulating TSH, triiodothyronine or thyroxine. Their studies also indicated that the prolactin elevation was not due to TRH release, since in the patients studied both prolactin and TSH responded to synthetic TRH. Thus, although it is reasonable to suppose that endogenous TRH can stimulate prolactin release in humans, it is probably not the only physiologically active stimulating agent. Indeed, there is evidence from the work of Marlarky and Parkratz (1974) and from Hagen *et al.* (1976) that heat-stable factors occur in human plasma which are capable of initiating prolactin release without altering TSH levels. Sephadex filtration indicates that the activity belongs to small peptide molecule(s), immunoreactively distinct from TRH (see also Blask *et al.*, 1976).

In summary, there is considerable evidence that TRH can increase prolactin secretion in man and other primates. This response, although often linked to release of TSH, can occur independently, and may be modified by peripheral factors such as gonadal steroids and thyroid hormone. Also, TRH does not appear to be the only stimulatory factor for prolactin secretion, although positive identification of other releasing factor(s) is not yet available.

The mechanism by which TRH stimulates prolactin release is not yet understood. Evidence from the work of Dussault *et al.* (1976) suggests that the release of TSH by TRH is mediated by prostaglandins, and in consequence it is blocked by acetylsalicyclic acid. However, Dussault and his co-workers were unable to demonstrate any effect of acetylsalicyclic acid on prolactin release, in agreement with the evidence that prostaglandins have little effect on prolactin release in non-primates (Harms *et al.*, 1974; Ojeda *et al.*, 1974; Sato *et al.*, 1974; Labrie *et al.*, 1974). Again, we see the clear separation of two distinct release mechanisms, a separation further emphasised by the fact that prostaglandins appear generally to act via cyclic GMP (guanosine monophosphate), which seems to affect prolactin release only slightly in rats (Wilson and Ensor, 1977). There may also be long-term changes in response to TRH, as Tashjian (1972) has shown that TRH induces morphological alterations in cultured rat prolactin cells.

7.9.2 *Inhibition of prolactin secretion by dopamine*

As for other mammals, there is uncertainty as to whether in inhibiting prolactin secretion in primates, dopamine acts directly on the pituitary, or on the hypothalamus to stimulate production of a prolactin-inhibiting factor (PIF). The main evidence that dopamine is involved in the control of prolactin secretion comes from the widespread use of neuroleptic and other drugs in treating various reproductive disorders. Both L-dopa and ergot alkaloids (mainly bromocryptine, CB154) have been shown to lower serum prolactin levels in normal subjects (Friesen *et al.*, 1971, del Pozo *et al.*, 1972) and in patients suffering from galactorrhoea (Malavky *et al.*, 1971; Lutterbeck *et al.*, 1971). This response has been demonstrated by a number of workers (Besser *et al.*, 1972; Vorga *et al.*, 1972; Rollards and Schellebaus, 1973; del Re *et al.*, 1973; Thorner *et al.*, 1974; Utien *et al.*, 1975; Nader *et al.*, 1975), all of whom have concluded that CB154 can suppress prolactin levels in both galactorrhoea and puerperal lactation. In most cases, bromocryptine was found to be equally or more effective than oestrogens in suppressing lactation (Vorga *et al.*, 1972; Utien *et al.*, 1975). Bromocryptine also appears to be more effective than L-dopa, in that it suppresses prolactin release for a longer period (Friesen and Hwang, 1973).

Further evidence for the involvement of dopamine in control of PRL comes from studies on the effects of apomorphine in normal men. Like the ergot alkaloids, apomorphine is considered to be a specific dopamine agonist (see Chapter 6 for a more detailed discussion). The action of apomorphine in man appears to be slightly more complex than in other mammals, since although it failed to lower serum prolactin levels in the male, it did suppress TRH-induced prolactin release (Nilsen *et al.*, 1975). This evidence suggests that in man, dopamine may act as a counter-balance to stimulation, rather than directly as an inhibitory factor. This view is to some extent confirmed by the work of Martin *et al.* (1974), who showed that apomorphine (0·75 mg) could lower serum prolactin in patients with abnormally high initial levels, although the same dose of apomorphine has no effect in normal subjects (Lal *et al.*, 1973; Nilsen *et al.*, 1975). However, a higher dose (1·5 mg) did somewhat reduce serum prolactin levels (Lal *et al.*, 1973), which complicates the interpretation. Similar results were obtained with L-dopa by Noel *et al.* (1973).

These dopaminergic drugs readily cross the blood-brain barrier in contrast to dopamine, which, as a molecule with a high polarity, does not readily pass from blood to brain. Thus, whereas many of the effects of dopaminergic drugs could be attributed to their action on hypothalamic dopamine receptors, this explanation is unlikely to apply to the effects of dopamine itself when directly infused into the blood system (Camanni, *et al.*, 1975*a*; Camanni, *et al.*, 1975*b*). It seems likely that the inhibition of prolactin secretion by dopamine infusion in acromegalic patients represents a direct effect on the pituitary (Camanni *et al.*, 1977), an interpretation supported by the work of Besses *et al.* (1975), who showed that systematically infused dopamine could block both the TSH and prolactin responses to TRH. Lack of concordance between the effects of infused dopamine and apomorphine in the human may be attributable to the fact that whereas apomorphine stimulates both hypothalamic and pituitary receptors, dopamine, not reaching the hypothalamus, stimulates only the latter. The evidence suggests that both sets of receptors mediate inhibition of PRL secretion. The physiological relationship between the two sets of receptors is unclear.

7.9.3 *Effects of acetylcholine and serotonin on prolactin secretion*

Apart from dopamine, two other neurotransmitters have been shown to affect prolactin secretion in primates. These are acetylcholine and serotonin. The serotonin precursors, L-tryptophan and 5-hydroxytryptophan have been reported as stimulating prolactin release, (MacIndoe and Turkington, 1973; Kato *et al.*, 1974). In addition methysergide, a competitive antagonist of serotonin, has been shown to diminish sleep-related prolactin secretion in man (Mendelson *et al.*, 1975). Similar studies with methergoline, a potent peripheral and central inhibitor of serotonin, have also demonstrated a significant decrease in serum prolactin levels in acromegalic subjects (Chiodini *et al.*, 1976). However, these authors argue that the reduction in prolactin levels by methergoline is probably due to stimulation of dopamine receptors rather than to any blockade of a serotoninergic mechanism. They support their interpretation by citing the marked similarity between the effects of methergoline and CB154, and by demonstrating that cyproheptadine, a specific serotonin blocker, was not capable of depressing PRL levels in acromegalic patients. On the other hand, similar observations on acromegalics have led Delitalia *et al.* (1976) to favour the hypothesis that methergoline inhibits serotonin release. These latter authors admit the possibility that the drug might stimulate dopamine receptors, but conclude that this is not yet proven.

More recent work supports the views of Delitalia *et al.* Lal *et al.* (1977) showed for GH, which alters in parallel with PRL in the studies cited above, that pretreatment with methysergide did not block apomorphine-induced GH release. This suggests that methysergide, at least, does not act through a dopamine-receptor mechanism and points to a serotoninergic prolactin-stimulating mechanism. Delitalia *et al.* (1977) have suggested that serotonin stimulates prolactin release during suckling concurrently with an inhibition of dopamine release. A serotoninergic stimulating mechanism has also been demonstrated in nonhuman primates by Gala *et al.* (1976), who found that serotonin in rhesus and crab-eating monkeys caused a rapid and significant increase in serum prolactin levels.

Acetylcholine may also be concerned in prolactin control. High doses of perphenazine stimulate a pituitary muscarinic component which inhibits prolactin release, an action potentiated by atropine; this observation led Gala *et al.* (1976) to postulate a cholinergic mechanism which inhibits prolactin release, similar to the mechanism proposed for the rat (Subramanian and Gala, 1976).

Thus it seems that in primates a complex multifactorial control of prolactin secretion is exerted by the hypothalamus, involving dopaminergic, serotoninergic, cholinergic components, and perhaps others. There are also indications that as in non-primates, dopamine may act directly on the pituitary to inhibit prolactin secretion.

7.10 Effects of steroid hormones on prolactin secretion

7.10.1 *Sex steroids*

The most striking influence on human prolactin (hPRL) secretion by sex steroids is seen in women, although there are reports that testosterone alters the secretion of hPRL in response to TRH in men (see p. 214), and oestrogens have been shown to cause an elevation of serum prolactin in men (Frantz *et al.*, (1972).

The levels of circulating hPRL are higher in the female than in the male at all ages except before puberty and after the menopause (Jacobs *et al.*, 1973; Ehara *et al.*, 1975). There is controversy regarding variations in hPRL levels during the menstrual cycle. Studies on the urinary excretion of hPRL, by bioassay, suggested that levels are higher in the luteal than in the follicular phase of the cycle (Copedge and Sekaloff, 1951). This finding has been confirmed by RIA (radioimmunoassy) measurements of serum hPRL in women by Vekemans *et al.* (1972), who showed that there was a modest peak of hPRL release at midcycle and that titres of the hormone were significantly increased during the luteal phase. On the other hand Hwang *et al.* (1971) failed to find any correlation between hPRL levels and the stage of the cycle. More recently, Franchimont *et al.* (1976) showed that serum hPRL levels are correlated with serum oestradiol during the cycle.

In addition to possessing higher basal secretion levels than men, women show a more pronounced response to chlorpromazine and TRH. The response in men can be enhanced to the female level by oestrogen treatment (Jacobs *et al.*, 1973; Buckman and Peake, 1973; Carlson *et al.* 1973).

Recently the widespread use of oestrogen-based oral contraceptives has prompted numerous reports that oestrogens affect hPRL secretion. It is now known that hPRL levels in hypogonadal and post-menopausal women are relatively low. This appears to be due to reduced oestrogen secretion by the ovary rather than to any inherent failure of the prolactin secretion mechanism. Yen *et al.* (1974) showed that injections of low doses of ethinyloestradiol significantly increased serum hPRL levels in both hypogonadal and post-menopausal women within one week. With continued treatment hPRL levels then rose and plateaued three to four weeks after treatment had commenced. These results were recently confirmed by Robyn and Vekemaus (1976) with lower doses of the same steroid. However, this response is not elicited by all contraceptive steroids: for example, Spellacy *et al.* (1975) found that the progestagen contraceptive medroxyprogesterone, although producing breast nodules in dogs and monkeys (Rosenfield, 1974) does not increase hPRL secretion. This finding is in line with the work of Meites (1966), who showed that progesterones, corticosteroids and testosterone do not affect the secretion of prolactin from human pituitary fragments *in vitro*, but that oestrogens strongly enhance the secretion.

It seems from mammalian and human studies that the action of oestrogens on prolactin secretion is twofold. Studies in the rat suggest that oestrogens act by lowering the concentration of PIF in the hypothalamus (Nicoll *et al.*, 1970) while human studies have suggested that the steroids act directly on the pituitary to cause an increase in the activity of the lactotrophs (Abu-Fadil *et al.*, 1976; Raboff *et al.*, 1973).

7.10.2 *Corticosteroids*

Although hPRL is released in response to stress, there seems little evidence that glucocorticoids modify hPRL secretion. Several reports have suggested that glucocorticoids can reduce the TSH response to exogenous TRH (Faglia *et al.*, 1973), but not the prolactin response (Re *et al.*, 1976). In agreement, Meites (1966) could show no direct effect of glucocorticoids on the human pituitary *in vitro*.

7.11 Prolactin secretion by pituitary tumours

Prolactin-secreting adenomata are probably the most common of all pituitary tumours (Friesen *et al.*, 1972). Most such tumours do not produce symptomatic galactorrhoea or gynecomastia, and as a result are generally classified as non-functional chromophobe adenomata. Friesen and Hwang (1973) report that of 15 consecutive patients suffering from pituitary tumours, five had serum prolactin levels higher than 200 ng/ml. These authors emphasize that serum prolactin levels in the range 30–200 ng/ml cannot be considered as diagnostic of a prolactin-secreting tumour, since it is possible for these levels to stem from other disorders, such as tumour extension into the hypothalamic region, sarcoidosis, histiocytosis X, metastases etc. While prolactin levels in patients with chromophobe adenomata fluctuate during the day, Friesen and Hwang (see above) claim that the levels rarely fall to within the normal range.

Many disorders such as amenorrhoea and galactorrhoea have been shown to be caused by functional prolactin-secreting chromophobe adenomata (Canfield and Bates, 1965; Pasteels, 1967; Takatoni *et al.*, 1967; Peake *et al.*, 1969; Friesen *et al.*, 1972; Nasr *et al.*, 1972; Turkington, 1972) or by other pituitary or suprasellar tumours causing high prolactin levels, such as craniopharyngiomata (Kleinberg and Frantz, 1971), ectopic pinealomata (Canfield and Bates, 1965; Turkington, 1972) and Nelson's syndrome (Jacobs *et al.*, 1972). These latter conditions probably result from physical or mechanical or morphological interruption of the hypophysial portal system interfering with the normal prolactin-inhibitory control mechanism.

Francks *et al.* (1975) have reported hyperprolactinemia in 73% of patients with pituitary tumours, the remaining patients having non-functional chromophobe adenomata. They concluded that the greater part of the excessive prolactin secretion came from the tumour itself, since serum prolactin levels returned to normal following hypophysectomy. Saaman *et al.* (1977) studied serum prolactin levels before and after hypophysectomy in patients with non-functional chromophobe adenoma, and again demonstrated a drop in levels after surgery. Of the 11 patients in this study, only four had markedly elevated serum prolactin levels (see also Assies *et al.*, 1974). This point was also raised by Friesen and Hwang (1973), who showed that there is no strict correlation between high serum prolactin and clinical conditions such as galactorrhoea, so that even patients with serum levels in excess of 500 ng/ml do not necessarily present clinically detectable symptoms. Friesen and Hwang suggested that such anomalies may be due to lack of the oestrogens or corticoids which are necessary for development of galactorrhoea.

Similar functional normality in the presence of high circulating prolactin levels has also been demonstrated by Child *et al.* (1975). Although amenorrhoea and inhibited ovulation are usual consequences of sustained prolactin elevation (see p. 198), Child *et al.* (1975) found nine cases of pregnancy in women with radiologically demonstrable pituitary tumours. In one case it was necessary to terminate the pregnancy, and one pregnancy spontaneously aborted. However, both Child *et al.* (1975) and Schewchuck (quoted in Archer, 1977) have shown that pregnancy may fail even if the presence of microadenoma is not detected, primarily because of hypertrophy of pituitary tumours during pregnancy. Appreciable enlargement of the normal pituitary occurs during pregnancy, with a rapid but not necessarily complete return to normal after delivery (Erdheim and Stumme, 1909). This size change has also been recorded in pituitary tumours. In the cases cited by Child and his co-workers, pregnancy was maintained either by radiological treatment involving implants of yttrum-90 into the pituitary or by chemotherapy with CB154 or L-dopa. Even so, some of the patients showed side-effects due to growth of the pituitary tumours.

It is interesting to note that a number of authors beside Child *et al.* (1975) have reported successful treatment of pituitary adenomata with drugs such as CB154 or L-dopa. Corenblum *et al.* (1975) reported successful CB154 treatment of tumours in two patients with relatively large adenomata, while Assies *et al.* (1974) have shown that prolactin-hypersecretion by a chromophobe adenoma can be inhibited or reduced by L-dopa, CB154 or hypoglycaemia. This suggests that hypersecretion is not due to abnormality of the lactrotroph secretory physiology, but rather that the hypothalamic inhibitory mechanism may be impaired. Very high levels of serum prolactin can be induced by TRH or chlorpromazine in patients with pituitary adenomata, and Friesen and Hwang (1973) found that in a number of patients, rapid increases in serum prolactin levels followed surgical removal of the tumour, perhaps as a result of hyperproduction of a PRH by the hypothalamus.

7.12 Miscellaneous conditions affecting prolactin levels in humans

A number of other factors, some physiological and some pathological have been reported as increasing pituitary prolactin secretion and/or causing elevated serum levels. These factors are listed in Table 7.2.

Friesen and Hwang (1973) reported that 20% of patients with advanced renal failure have elevated serum prolactin, and also that the clinical picture is often complicated by treatment with antihypertensives (for example, methyldopa) which may secondarily stimulate prolactin secretion.

Table 7.2 Factors causing an elevation of plasma prolactin levels.

Physiological
Pregnancy
Lactation
Exercise
Stress (surgical and other)
Sexual intercourse
Diving
Pathological
Pituitary adenoma
Pituitary stalk section, or impaired portal blood flow
Hypothyroidism (especially in juveniles)
Renal failure
Ectopic production of prolactin by malignant tumours
Hypoglycaemia
Pharmacological
Insulin induced hypoglycaemia
High oestrogen therapy
Psychotrophic drugs (phenothiazines, butyrophenones, sulpiride)
Reserpine and α-methyldopa
TRH
General anaesthesia

Noel *et al.* (1972) have demonstrated elevation of serum prolactin in a variety of stressful situations, including physical exercise, anaesthesia, surgery and in women after sexual intercourse. Daughaday and Jacobs (1972) reported a modest elevation of serum prolactin in men following sexual intercourse, but the work of Noel *et al.* (1972) suggests a much closer relationship in women between orgasm and serum prolactin levels. In general, factors which elevate serum prolactin have been shown to be more effective in women than in men, and as discussed previously the female basal secretion rate also appears to be higher (Noel *et al.*, 1972). Karmali and Horrobin (1976) have reported an increase in prolactin secretion during a simulated dive, another situation that could be characterised as 'stressful'. However, as L'Hermite (1973) has emphasised, to attribute these increases in serum prolactin levels to stress *per se* is merely to underline our ignorance of the physiology of this unusual hormone, particularly since most of this information was obtained from subjects which were suffering no apparent pain or anxiety.

In addition to pituitary adenomata, large amounts of prolactin are secreted by some ectopic hormone-producing tumours. Turkington (1971*c*) reported ectopic prolactin production in a patient with an undifferentiated bronchial carcinoma and in another subject with a hypernephroma. This second case may have involved some impairment of kidney function. Increased serum PRL levels have also been observed in a patient with a medullary thyroid carcinoma which secreted substances with ACTH-like and CRF (corticotrophin releasing factor)-like activity (Birkenhager *et al.*, 1976). However, in this case the prolactin was probably of pituitary origin, in response to a prolactin-secretion-stimulating substance secreted by the carcinoma.

8 Prolactin receptors and the subcellular mechanism of hormone action

Until recently very little was known about the mechanism of action of prolactin, but during the last six years there has been a considerable expansion in our knowledge about the nature of the prolactin receptor and the subsequent expression of its effects. Our knowledge is entirely limited to mammalian tissues although Snart and Debnam (1974) have shown that specific binding sites exist in the toad bladder which are capable of high affinity binding of ovine prolactin.

Many of the physiological processes which have been discussed in previous chapters have depended on synergism with steroid hormones for the full expression of the prolactin effect. Only by knowing the mechanisms of action of prolactin and of steroids is it possible to understand how this synergism might operate. Of the mammalian tissues which have been studied, the response of the mammary gland cells to prolactin is the most fully understood. For this reason this chapter will concentrate upon work on the mammary gland, but reference to other prolactin receptors will be made for comparison.

8.1 Action of prolactin on the mammary gland

The nature of the mammary gland response to prolactin has been discussed in some detail in the previous two chapters. It would seem, according to Turkington (1972) (see Fig. 8.1), that the main effect of prolactin is to stimulate specific protein synthesis, at least in culture, and that the development of this response requires a period of pretreatment with cortisol or some other active adrenal corticosteroid (Turkington *et al.*, 1967). However prolactin may also effect the differentiation of alveolar cells and may even be mitogenic under some conditions (see p. 144).

8.1.1 *Initial action of prolactin binding to cell membrane receptors*

Turkington *et al.* (1973) have shown in an elegant series of experiments

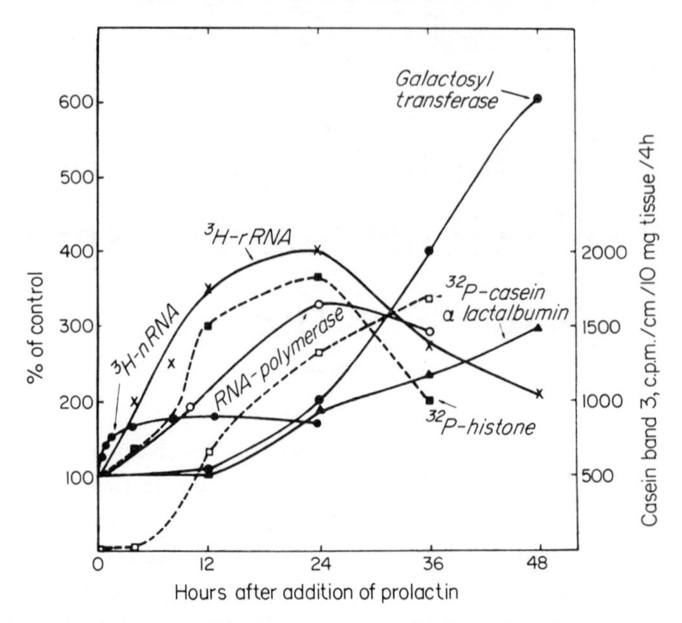

Fig. 8.1 Sequence of molecular events in the induction of milk proteins by prolactin. Mouse mammary explants were incubated with insulin and hydrocortisone for 96 hours prior to the addition of prolactin to the medium.

that the initial action of prolactin on the mammary epithelial cell is to bind to a specific receptor site. In order to pursue these studies, an iodinated prolactin had to be provided. It was possible using lacto-peroxidase oxidation of the iodine to produce prolactin of high specific radioactivity which retained its biological effectiveness when assayed with mouse mammary epithelial cell particles (Turkington, 1971).

Incubation of this prolactin preparation with a pellet produced by low speed centrifugation of mouse mammary gland homogenate showed that there were two possible sites of prolactin action (Turkington and Frantz, 1972; Frantz and Turkington, 1972): i) a low affinity binding site from which the iodinated prolactin could be removed by repeated washings in hormone free medium, and ii) a high-affinity site which bound iodinated prolactin even in the face of several centrifugal washings. It is now generally accepted that the high affinity site is the true prolactin receptor. Turkington (1970) has also demonstrated that prolactin bound to Sepharose beads can still show a high affinity for mammary cell preparations. This indicates that there is a membrane-based site to which prolactin is covalently linked, since the prolactin–bead complex would be unable to pass through the cell membrane.

Studies with the mouse mammary gland pellet have enabled Turkington *et al.* (1973) to make estimates of two important parameters:

1. The total binding activity of the preparation, as measured by the amount of radio-iodinated prolactin bound to the cells after repeated washings with hormone free medium.

2. The specific saturable binding activity, that is the fraction of the total bound iodinated prolactin which can be removed by subsequent incubation with 'cold' prolactin but which is not displaced by other hormones (See Fig. 8.2).

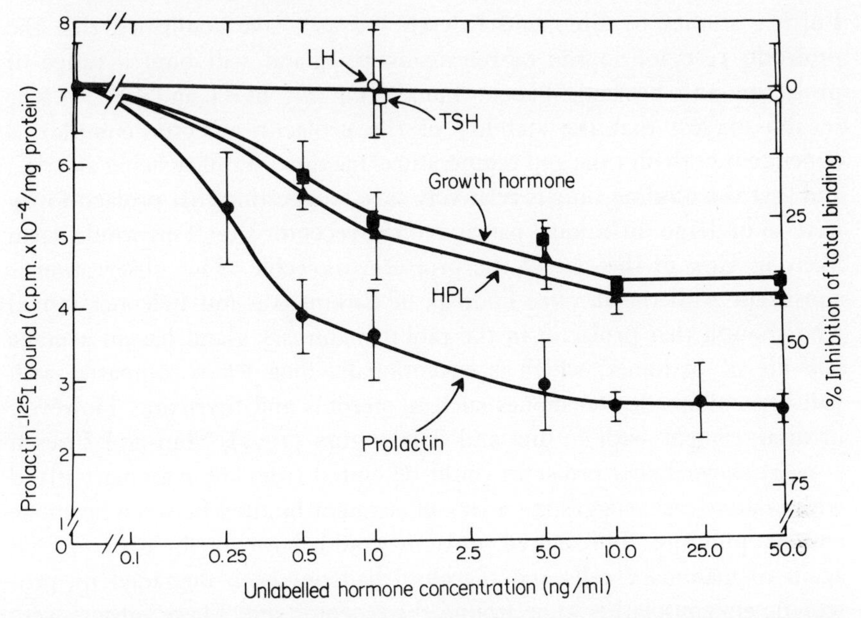

Fig. 8.2 Competitive displacement of [^{125}I]prolactin (ovine) from mouse mammary epithelial plasma membranes by prolactin (ovine). Bovine TSH, ovine LH and Ovine FSH failed to compete with [^{125}I]prolactin for binding at these concentrations.

Under these conditions cold prolactin was able to displace about 60% of the iodinated membrane bound hormone. Bovine growth hormone and human placental lactogen (hPL) also showed significant competitive displacement but were not as effective as cold prolactin. TSH, LH and ACTH were ineffective, although a crude preparation of FSH did possess some competitive activity, which could, however, have been due to prolactin contamination. Turkington *et al.* (1973) demonstrated that saturable and specific protein binding activity can also be found in the liver, kidney, brain, ovary, testis and seminal vesicles, all sites of prolactin action in mammals. However, attempts to bind prolactin to other tissues, that is, heart, lung, spleen and cerebral cortex, were unsuccessful. The total binding affinity of the mammary tissue increased during lactation, suggesting the possible production of new receptor protein.

Falconer (1972) and Birkinshaw and Falconer (1972) have confirmed the work of Turkington *et al.*, and shown that similar receptor characteristics are possessed by rabbit mammary tissue. Using autoradiographic techniques they were able to show that the iodinated prolactin was bound to the epithelial cell surface adjacent to the blood capilliaries.

8.1.2 *Properties and possible nature of the prolactin receptor*

Further studies by Shiu and Friesen (1974*a*) have confirmed that the prolactin receptor in the rabbit mammary gland will bind a range of prolactins and prolactin-like hormones, notably hGH and hPL. Their results suggest that the stability of the prolactin-receptor complex is dependent both on time and temperature, maximal stability being at 37°C, and that the binding time is relatively slow, suggesting that prolactin may have to undergo diffusional passage to the receptor site. This would seem likely in view of the size of the prolactin molecule. This observation is consistent with the *in vitro* findings of Birkinshaw and Falconer (1972) who showed that prolactin in the rabbit mammary gland has an average half-life of 52 hours, which is exceptionally long when compared with half-lives of smaller hormones such as steroids and thyroxine. However, in disagreement with Frantz and Turkington (1972), Shiu and Friesen (1974*a*) showed that prolactin could be eluted from the mammary gland with relative ease, suggesting a lack of covalent binding between hormone and receptor site. This eluted prolactin could, however, be used to bind again to mammary cells and therefore had not been degraded by proteolytic enzymes either at or around the receptor site. These authors were also able to show that the activity of the receptor site could be destroyed by trypsin and phospholipase C. Turkington *et al.* (1973) have also shown that receptor activity can be destroyed by trypsin in the mouse, but that collaginase, neuraminidase, RNAase and DNAase were ineffective. Shiu and Friesen (1974*a*) also showed that treatment with steroid hormones did not effect the prolactin binding activity of the membrane preparations. From this evidence the receptors are probably protein-phospholipid complexes, and the mediation of prolactin action by steroids is not brought about by modification of the binding process or by an increased titre of prolactin-binding protein.

More recently Shiu and Friesen (1974*b*) have managed to solubilise a prolactin receptor molecule from the crude particulate membrane fraction of the pregnant rabbit mammary gland, using Triton X 100 (1%). Concentrations of Triton X 100 greater than 0·01% (v/v) were shown to cause alteration of iodinated prolactin, its molecular weight increasing to 80 000 with loss of biological activity. However, hGH which also binds to the receptor was not altered noticeably by this treatment and could therefore be used in affinity chromatography to purify the receptor protein. The solubilised receptor showed a higher binding affinity for prolactin ($Ka = 16 \times 10^9 m^{-1}$) than did the particulate fraction ($Ka = 3 \times 10^9 m^{-1}$). However in respect to its other properties, and to the time-temperature dependent nature of the binding process, it was identical to the particulate receptor.

The receptor appears to be a protein molecule, with a molecular weight of 220 000 measured by gel filtration on Sepharose 6B. Subsequent analysis of the partially purified receptor by acrylamide gel electrophoresis revealed several protein bands, but prolactin-binding activity was linked specifically with one (or possibly two) of these bands with an R_f of 0·12.

8.2 Prolactin receptor sites in tissues other than the mammary gland

Turkington *et al.* (1973) reported prolactin-binding activity from a number of organs including the ovary, testes, seminal vesicles and liver. Recently the nature of some of these receptors has been studied in more detail.

Shiu and Friesen (1974) confirmed the presence of prolactin receptors in the rabbit ovary and have shown that they have a specific binding affinity for prolactin. Midgeley (1973), using an autoradiographic technique, has found that ^{131}I-labelled prolactin was bound to all corpora lutea of the pseudopregnant rat ovary. This has recently been confirmed for the porcine corpora lutea by Rolland *et al.* (1976), and Carlsen *et al.* (1972) have shown that ^{131}I-labelled prolactin shows essentially the same distribution as hCG in rat ovarian slices. Rolland *et al.* (1976) however, demonstrated that the receptor sites for the two hormones were quite separate in the porcine corpora lutea, although prolactin was capable of inducing the synthesis of LH-binding receptor proteins.

A recent study by Saito and Saxena (1975) has shown that the prolactin receptor in the ovaries of rats, cows and humans showed several properties in common with the mammary gland receptor. Binding was a saturable phenomenon and was dependent upon the binding-protein concentration. Maximal binding was obtained at a pH 7·0 and an incubation temperature of 37°C. The binding capacity could be inhibited by pH 10·0, and was irreversibly destroyed at pH 3·0. Bound ^{125}I-hPRL was displaced competitively by 'cold' human, ovine and bovine prolactins, but not by LH or FSH. In addition, as might be expected hPL and hGH also displaced the iodinated prolactin. Turkington *et al.* (1973) reported a change in the total binding affinity of the mouse mammary gland preparation with oestrogen treatment but Shiu and Friesen (1974*a*) reported that steroid hormones did not alter the binding efficiency. This latter finding has since been questioned by some authors including Friesen (1974) who showed oestrogen increased the number of prolactin binding sites in the rabbit mammary gland. This is also true of DMBA-induced mammary tumours. Saito and Saxena (1975) in their studies of ovarian binding sites have shown that the binding affinity for prolactin changes during the rat oestrus cycle. The lowest levels of prolactin binding sites/mg protein were found at metoestrus ($0{\cdot}8 \times 10^{-12}$m), these increased during dioestrus (11×10^{-12}m) and reached a peak at pro-oestrus ($24{\cdot}6 \times 10^{-12}$m). It will be recalled that oestrogens have been shown to stimulate the pro-oestrus peak release of prolactin (Niell *et al.*, 1971). The number of binding sites in the ovary correlates well with serum oestrogen levels at the different stages of the oestrus cycle and Saito and Saxena (1975) suggest that oestrogen may stimulate the level of prolactin receptor proteins, an idea that has since been confirmed by de Sembre *et al.* (1976). In recent studies, Moodbiri *et al.* (1975) have shown that specific prolactin receptors exist with similar properties in the myometrium and uterine myomata, and in the rat ventral prostate respectively.

Posner (1974) has shown that mouse liver homogenates are also capable of binding ovine prolactin and bovine growth hormone. Pregnancy appeared to increase the specific binding affinity for growth hormone but not for prolactin. hGH showed binding affinities similar to those of bovine GH but could be displaced more readily by ovine prolactin. At the same time, Posner (1974) was able to show that human lymphocytes have sites which specifically bind hGH and that neither non-primate GH or prolactin competitively displaced hGH from the sites. In contrast the hepatocyte binding-sites bound bovine GH and hGH with equal affinity, and hGH was displaced preferentially by ovine prolactin. This suggests that structures of the three hormones are related (see p. 4) but that the different receptors may bind to different parts of the molecules, indicating that these molecules possess more than one active site. In another study, Posner *et al.* (1974) have confirmed that rat liver membranes, as well as binding growth hormones will bind a wide range of lactogens, and that the properties of these receptors are similar to those characterised for the rabbit mammary receptor (Shiu and Friesen 1974*a*). Hepatic lactogen receptor sites were numerous in adult females and were to some extent augmented by pregnancy, contrary to Posner's (1974) earlier findings. They were scarce or virtually absent in adult male rats (Kelly *et al.*, 1974), but could be induced at high levels in male rats by oestrogen treatment (Posner *et al.*, 1975). Hypophysectomy of female rats resulted in a significant decrease in the total binding affinity of the liver, while hypophysectomy in males resulted in their becoming refractory to oestrogen treatment (Kelly *et al.*, 1975). Ectopic pituitary transplants induced an increase in hepatic prolactin binding in males and restored oestrogen sensitivity (Posner *et al.*, 1975). These findings have since been confirmed by Posner (1976) who showed that increased binding was due to an increase in the number of prolactin receptor molecules and not to increased receptor affinity. He also suggests that the increase in prolactin receptor levels is induced by prolactin itself and that oestrogen stimulation is indirect, being mediated by increased prolactin release from the pituitary. This would be different from the rabbit mammary receptor for which Friesen (1974) showed that an increase in receptor concentration could be brought about by oestrogen treatment *in vitro*. The level of prolactin receptors are also modified by growth hormone-secreting tumours (Aragona *et al.*, 1976) and in the hypophysectomised female rat (Bohnet *et al.*, 1976). Both these studies support Posner's (1976) contention that the increase in prolactin receptor sites in the liver is brought about primarily by prolactin induction.

However, Sherman *et al.* (1976) have recently shown that prolactin binding ability of a $100\,000 \times g$ particle from rat liver was significantly reduced by injections of cortisol, testosterone and the progestin, medroxy-progesterone acetate into both intact and castrated female rats. These hormones did not appear to lower serum prolactin levels and were also effective in oestrogen-treated rats with hyperprolactinaemia. It would appear that the steroids have a direct effect on the liver, which in the case of testosterone and cortisol was very rapid and complete desensitisation of the liver to prolactin occurred after three days. This action was specific for the lactogen receptors and did not alter insulin binding to the liver cells, and therefore did not result from membrane-induced alteration of the hormone molecules or changes in the membrane of a general nature. Preincubation of liver cells with steroids *in vitro* did not alter subsequent binding of prolactin. This confirms that the steroid hormones are not altering the liver cell membranes in any permanent way, but are probably inhibiting the synthesis or increasing the catabolism of the receptor protein, blocking the prolactin-induced rise after the membrane stage of the effect. These findings confirm those of Aragona and Friesen (1975) who showed that hepatic prolactin binding was increased by castration in male rats. Testosterone administration has also been shown to decrease the binding of prolactin to kidney membranes, while castration raised binding levels in both the kidney and adrenal (Marshall *et al.*, 1976). A full interpretation of these results awaits further study of the role of prolactin in liver metabolism in the rat.

In its turn, prolactin appears to be able to regulate the levels of receptor protein for other hormones. Rolland *et al.* (1976) have demonstrated that prolactin can induce LH receptor-proteins in porcine corpora lutea. Prolactin also stimulates the levels of oestradiol-binding protein in experimentally induced mammary cancers of rats (Sasaki and Leung, 1975) and injection of prolactin *in vivo* increases oestrogen binding in the liver (Chamness *et al.*, 1975).

In summary, it would seem that the prolactin receptors from a range of mammalian tissues have similar physico-chemical properties, showing maximal binding capacity at pH 7·0 and 37°C. The evidence suggests that the prolactin receptor is a protein with a molecular weight of around 200 000, although a phospholipid element may be incorporated in the molecule, since its binding capacity is degraded by phospholipase C. Its relationship with steroid hormones varies from tissue to tissue, but the evidence suggests that modifications of prolactin action by steroids, and reverse interactions, may be mediated at the cellular level by alterations in the concentrations of specific binding proteins.

8.3 Subcellular actions of prolactin

We have seen that there is now general agreement that the first event in the actions of prolactin is its binding to a specific receptor on the outer cell membrane of its target cells. However, the subsequent changes induced by the formation of this receptor/hormone complex are not so clear. In general terms, most polypeptide hormones have been shown to act through the stimulation of some second messenger, generally cyclic AMP (cAMP), which then generally activates a specific protein kinase to mediate the final response (see Fig. 8.3; Butcher, Robertson and Sutherland, 1971).

There is some evidence that prolactin can stimulate cAMP levels in certain target tissues. Golder *et al.* (1972) have demonstrated elevated cAMP levels in the rat prostate after prolactin treatment. However these results have been questioned by Thomas and Maudaler (1974) who found that high doses of prolactin decreased cAMP levels after 30 minutes incubation, while low doses were totally ineffective. Recently, however, Shani (1976) has obtained evidence that cAMP might act as a second messenger in the pigeon crop-sac response.

On the other hand, prolactin has been shown to stimulate nucleic acid synthesis and nucleotide incorporation in a variety of target organs, the pigeon crop-sac (Nicoll and Bern, 1968), the mammary gland (Turkington, 1972), and the prostate (Thomas and Manardhov, 1975), suggesting that

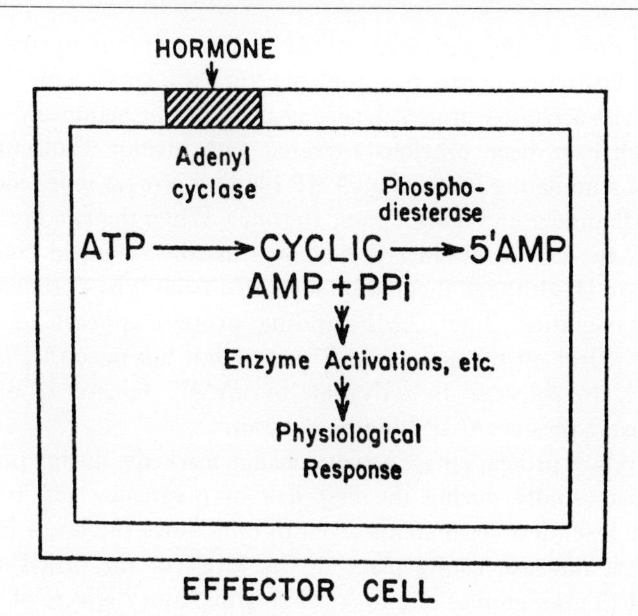

Fig. 8.3 The components of the cAMP mechanism. (From Butcher, Robertson and Sutherland, 1971.)

the primary action may be at the level of the genome. However it is certain, as described previously, that prolactin bound to Sepharose beads can be biologically active, although it cannot cross the cell wall (Turkington, 1970). It is therefore necessary to postulate a second messenger of some kind if prolactin can act at the genome level.

Turkington *et al.* (1973) have proposed a model of prolactin action at the subcellular level which involves a primary prolactin action on nucleic acids, leading to the synthesis of a protein kinase. The protein kinase in turn is activated by cAMP, the production of which is stimulated by an adenylate cyclase receptor activated by a second modifying hormone. The evidence for Turkington's model is outlined below. It should be emphasised that this type of system has only really been demonstrated for the mammary gland, and may not apply to other organs.

8.3.1 *Induction of cyclic AMP-dependent protein kinase*

Early work by Turkington and Riddle (1969) and Turkington (1971) showed that prolactin stimulated ^{32}P-uptake into mammary cells and that this was incorporated into specific histones and into non-histone nuclear proteins. Phosphorylation of these nuclear proteins was shown to occur at the same time as the activation of RNA transcription by insulin and prolactin. Prolactin appears to stimulate a catalytic kinase unit (Kinase I) and a cAMP-activated protein kinase (Kinase II) in mammary epithelial cells which have been previously treated with insulin. Prolactin is also capable of stimulating levels of a cAMP binding protein independently of insulin (Majumder and Turkington, 1971*a,b*). When the two kinases were separated, the cAMP-binding protein was usually found in conjunction with Kinase II, although it could be separated from it by chromatography on hydroxyapatite. The cAMP-binding protein appears to have an inhibitory effect on the activity of Kinase I but not on cAMP-activated Kinase II. In response to activation by cAMP, Kinase II dissociates into Kinase I plus a cAMP-binding subunit.

The levels of protein kinase activity change markedly during pregnancy, rising most rapidly during the first half of pregnancy and reaching a maximum level near term about seven to nine times the levels in virgins, the greatest increase taking place in the levels of the cAMP-activated Kinase II (Turkington *et al.*, 1973). The stimulatory effects of prolactin on these enzymes, and their increase during pregnancy could be prevented by actinomycin D and cyclohexamide suggesting that induction of the kinase enzymes requires synthesis of RNA and protein (Majumder and Turkington, 1971*b*).

8.3.2 *Induction of milk proteins – sequence of events*

In cultured mammary explants, insulin acid and cortisol stimulate a single cell division which ceases after 72 hours (Lockwood, Stockdale and Topper, 1967; Turkington, 1968). This generation of daughter cells then responds to prolactin by secreting milk proteins. The time course of subsequent subcellular effects is shown in Fig. 8.1 (p. 226). The first observable effect is a rise in labelled nuclear RNA (Turkington, 1970), followed closely by a rise in DNA-dependent RNA polymerase (Turkington and Ward, 1969). The rate of ^{32}P-incorporation into histones is also increased and there is increased phosphorylation of certain non-histone nuclear proteins (Turkington and Riddle, 1969). This leads to a subsequent increase in the synthesis of specific milk proteins. The appearance of specific casein phosphoproteins is correlated directly with the addition of prolactin to the medium. Experiments with cultured mammary cells suggest that prolactin can only produce these changes in the new generation of differentiated cells, and that the synthetic ability of these cells decreases as they age, casein synthesis being initiated from a 'zero-baseline' in each subsequent generation of cells. Following casein phosphoprotein-induction there is an increase in the enzymic activity of the lactose-synthetase system (Turkington *et al.*, 1968; Turkington and Hill, 1969). These changes in milk protein synthesis can also be inhibited by actinomycin D or cyclohexamide, indicating their dependence on the specific induction of RNA.

So far, then, it would appear that in the mammary gland, the time course of events following prolactin stimulation is:

a. Binding of prolactin to a high-affinity binding site on the cell membrane.
b. Translation in same manner of this effect to the nuclear DNA/RNA.
c. Altered genome activity leading to transcription of specific ribosomal, transfer and messenger RNA.
d. Synthesis of milk proteins.

It remains to see what role the prolactin-induced protein kinases play in this pattern of events. It appears from the work of Turkington and his colleagues that these protein kinases are involved in the phosphorylation of certain specific proteins in ribosomes, cell membranes, and nuclear fragments of the mammary epithelial cell. Endogenous protein kinase activity has been observed in highly purified preparations of cell membranes or ribosomes, and this activity was correlated with phosphorylation of endogenous protein in these preparations. These protein kinases appear to produce a higher rate of phosphorylation of the ribosomal proteins than of either cell membrane proteins or soluble nuclear histones (Turkington *et al.*, 1973). Acrylamide gel electrophoresis demonstrated that cells incubated with insulin and prolactin showed a higher rate of phosphorylation of cell membrane proteins than did cells treated with insulin and hydrocortisone. This technique applied to the ribosomal proteins detected the appearance of eight different phosphorylated proteins, the phosphorylation rate of four of these being increased by prolactin. Phosphorylation with ^{32}P appears to be associated with the serine and threonine residues of these proteins (Turkington *et al.*, 1973). The prolactin-stimulated effects agree closely with the changes observed in mammary gland explants from pregnant mice (Turkington *et al.*, 1973).

Induction of the cAMP-dependent protein kinase and the cAMP-binding subunit *in vitro* reached half maximal levels at 30 minutes after the addition of prolactin, and peak levels were reached by one to two hours (Turkington *et al.*, 1973). Phosphorylation of the specific proteins then followed at a slower rate, plasma-membrane proteins showing maximal phosphorylation at eight hours, ribosomal proteins at 16 hours, and nuclear proteins at 24 hours. Phosphorylation in the nucleus is associated with the F_2a_2 and F_2b histone fractions and with a wide range of acidic chromatin species (Turkington and Riddle, 1970; Kadohama and Turkington, 1973).

This sequence of events suggests that the initial changes in mRNA and DNA–RNA polymerase are associated with increased production of the specific protein kinases (Turkington, 1972) and that the protein kinases may be involved in stimulating the synthesis of specific milk proteins.

8.4 The 'second messenger' mediating prolactin action in the mammary gland

Based on the Turkington model outlined above (Turkington *et al.*, 1973), it seems that cyclic nucleotides are involved in the activation of the specific protein kinases, but that they do not act as the 'second messenger' between the prolactin-receptor complex and the nucleus. Cyclic nucleotides (cAMP and cGMP) are involved intimately in the initiation of lactation and the production of milk proteins (Sapag-Hagar and Greenbaum, 1974*a*,*b*), and the level of cGMP increased relative to the level of cAMP in the rat mammary gland immediately after parturition. Dibutyryl cAMP administered to explants of rat mammary tissue attenuated or inhibited the effects of prolactin, insulin and corticosterone, and it also markedly suppressed the production of DNA and RNA in the early stages of lactation (Sapag-Hagar *et al.*, 1974).

Rillema (1974) has shown that cGMP mimics the action of prolactin in stimulating RNA synthesis in the explanted rat mammary gland, and that cAMP and phosphodiesterase-inhibitors oppose this action. This suggests that the early stages of prolactin action in the mammary epithelial cell might be mediated by an increase in cGMP levels and a decrease in the levels of cAMP.

Rillema (1975) showed that prostaglandin F_2a mimics the action of prolactin on the rat mammary gland, while prostaglandin E and prostaglandin A_2 abolish this effect. The effects of prolactin and prostaglandin F_2a, given together, were not additive, which suggests that prostaglandin F_2a may act as an intermediary for prolactin and that it was producing its maximal effect at the dose used. Indomethacin, which is known to specifically inhibit prostaglandin synthesis, abolished the effect of prolactin (Rillema, 1975). Kuchl (1974) has postulated that actions of prostaglandins may be mediated by cyclic nucleotides, and that cGMP may be related specifically to the actions of prostaglandins of the F series. Thus, it is possible that activation of the prolactin receptor site stimulates an increase in membrane-bound prostaglandin F_2a synthesis, and that this in its turn elevates the intracellular cGMP/cAMP ratio which then constitutes the second messenger, and acts on the genome.

References

Abraham, M. (1971). The ultrastructure of the cell types and of the neurosecretory innervation in the pituitary of *Mugil cephalus* L. from freshwater, the sea and a hypersaline lagoon. *Gen. comp. Endocr.*, **17**, 334–350.

Abraham, S., Cady, P. & Caikoff, L. (1960). Glucose and acetate metabolism and lipogenesis in mammary glands of hypophysectomised rats in which lactation was hormonally induced. *Endocrinology*, **66**, 280–288.

Abu-Fadil, S., DeVane, G., Siler, T. & Yen, S. S. C. (1976). Effects of oral contraceptive steroids on pituitary prolactin secretion. *Contraception*, **13**, 79.

Adams, A. E. (1940). Sexual condition in *Triturus viridescens*. III. The reproductive cycle of the adult aquatic form of both sexes. *Am. J. Anat.*, **66**, 235–275.

Adler, N. T. (1969). The effect of the males copulatory behaviour on successful pregnancy of the female rat. *J. comp. Physiol. Psychol.*, **69**, 613–622.

Adler, N. T., Resko, J. A. & Gay, R. W. (1970). The effect of copulatory behaviour on the hormonal change in the female rat prior to implantation. *Physiol. Behav.*, **5**, 1003–1007.

Adler, R. A., Noel, G. L., Wartofsky, L. & Frantz, A. G. (1975). Failure of oral water loading and intravenous saline to suppress plasma prolactin in man. *J. clin. Endocr. Metab.*, **41**, 383.

Ahrèn, K., Fuxe, H. & Hamberger, L. (1971). Turnover changes in the tuberoinfundibular dopamine neurons during the ovarian cycle of the rat. *Endocrinology*, **88**, 1415–1424.

Ajika, K., Krulich, L., Fawcett, C. P. & McCann, S. M. (1972). Effects of estrogen on plasma and pituitary gonadotrophins and prolactin and on hypothalamic releasing and inhibiting factors. *Neuroendocrinology*, **9**, 304–315.

Aler, G. M. (1971). The study of prolactin in the pituitary gland of the Atlantic eel (*Anguilla anguilla*) and by the Atlantic salmon (*Salmo salar*) by immunofluorescence technique. *Acta zool.*, **52**, 145–156.

Aler, G. M., Bage, G. & Fernholm, B. (1971). On the existence of prolactin in cyclostomes. *Gen. comp. Endocr.*, **16**, 498–503.

de Allende, I. L. C. (1939). Accion de la prolactina sobre el oviducts de los batrachios. *Rev. Soc. argent. Biol.*, **15**, 190–193.

Alloiteau, J. J. & Vignall, A. (1958*a*). Pseudogestation après injection de progesterone chez la Ratte. *C.r. Acad. Sci., Paris*, **246**, 2804–2807.

Alloiteau, J. J. & Vignall, A. (1958*b*). Pseudogestation aprés traitement luteotrophe de courte durée chez la Ratte. *C.r. Acad. Sci., Paris*, **247**, 2465–2467.

Amenomori, Y., Chen, C. L. & Meites, J. (1970). Serum prolactin levels in rats during different reproductive states. *Endocrinology*, **86**, 506–510.

Amenomori, Y. & Meites, J. (1969). Radioimmunoassay for rat prolactin. *Proc. Soc. exp. Biol. Med.*, **143**, 793.

Amenomori, Y. & Meites, J. (1970). Effect of a hypothalamic extract on serum prolactin levels during the estrous cycle and lactation. *Proc. Soc. exp. Biol. Med.*, **134**, 492–495.

Amoroso, Z. C. (1955). Endocrinology of pregnancy. *Br. med. Bull.*, **11**, 117–125.

Anden, N. E., Butcher, S. G., Corrodi, H., Fuxe, K. & Ungerstadt, U. (1970). Receptor activity and turnover of dopamine and noradrenaline after neuroleptics. *Europ. J. Pharmac.*, **11**, 303–314.

Anden, N. E., Dahlstrom, A., Fuxe, K., Larsson, K., Olson, L. & Ungerstadt, U. (1966). Ascending monoamine neurons to the telencephalon and diencephalon. *Acta physiol. scand.*, **67**, 313.

Anderson, C. O., Zarrow, M. X., Fuller, G. B. & Denenberg, V. H. (1971). Pituitary involvement in maternal nest-building in the rabbit. *Horm. Behav.*, **2**, 1–7.

Antliff, H. R., Prasad, M. R. N. & Meyer, R. K. (1960). Action of prolactin on seminal vesicles of the guinea pig. *Proc. Soc. exp. Biol. Med.*, **103**, 77–80.

Apostolakis, M., Kapentakis, S., Lazos, G. & Madena-Pyrgaki, A. (1972). Plasma prolactin activity in patients with galactorrhoea after treatment with psycotrophic drugs. In *Lactogenic Hormones*, ed. Wolstenholme, G. E. & Knight, J. pp. 349–354. London: Ciba Foundation.

Aragona, C. & Friesen, H. G. (1975). Specific prolactin binding sites in the prostate and testis of rats. *Endocrinology*, **97**, 677.

Aragona, C., Bohnet, N. G. & Friesen, H. G. (1976). Prolactin binding sites in the male rat liver following castration. *Endocrinology*, **99**, 1017–1023.

Arai, Y., Susuki, Y. & Masuda, S. (1972). Effects of ergocornine on reserpine-induced lactogenic response of male rat mammary glands. *Endocrl. jap.*, **19**, 111–114.

Archer, D. F. (1977). Current concepts of prolactin physiology in normal and abnormal conditions. *Fert. Steril.*, **28**, 125–134.

Archer, D. F. & Josimovich, J. B. (1976). Ovarian response to exogenous gonadotrophins in women with elevated serum prolactin. *Obstet. Gynaec.*, **48**, 155.

Archer, D. F., Sprong, J. W., Nankin, H. R. & Josimovich, J. B. (1976). Pituitary gonadotrophin response in women with idiopathic hyperprolactinemia. *Fert. Steril.*, **27**, 1158.

Archer, D. F., Nanking, H. R., Gabos, P. F., Maroon, J., Nosetz, S., Washwa, S. R. & Josimovich, J. B. (1974). Serum prolactin levels in patients with inappropriate lactation. *Am. J. Obstet. Gyn.*, **119**, 466.

Arimura, A., Saito, M. & Wakabayashi, W. R. (1969). Abstr. 31. Program. p. 46. 51st Ann. Mg Endocr. Soc., New York.

Armstrong, D. T. (1968). Gonadotrophins, ovarian metabolism and steroid biosynthesis. *Rec. Prog. Horm. Res.*, **24**, 255–307.

Armstrong, D. T. & King, E. R. (1971). Uterine progesterone metabolism and progestational response; effects of estrogens and prolactin. *Endocrinology*, **89**, 191–197.

Armstrong, D. T., Knudsen, K. A. & Miller, L. S. (1970). Effects of prolactin upon cholesterol metabolism and progesterone biosynthesis in corpora lutea of rats hypophysectomised during pseudo-pregnancy. *Endocrinology*, **86**, 634–641.

Armstrong, B., Stevens, N. & Dol, S. R. (1974). Retrospective study of the association between the use of rauwolfian derivatives and breast cancer in English women. *Lancet*, **2**, 672.

Ash, R. W., Pearce, J. W. & Silver, A. (1969). An investigation of the nerve supply to the salt gland of the duck. *Q. Jl. exp. Physiol.*, **54**, 281–295.

Asano, M. (1965). Basic experimental studies of the pituitary prolactin – prostate interrelationships. *J. Urol.*, **93**, 87–93.

Assairi, L. Delous, C., Gaye, P., Houdebine, L-M., Ollivier-Bousquet, M. & Denamur, R. (1974). Inhibition by progesterone of the lactogenic effect of prolactin in the pseudopregnant rabbit. *Biochem. J.*, **144**, 245–252.

Assenmacher, I. A., Tixier-Vidal, A. & Boissin, J. (1962). Contenu en hormones gonadotrophes et en prolactine de l'hypophyse du Canard soumis a un traitement lumineux ou reserpinique. *C.r. Soc. Biol.*, **156**, 1555–1560.

Assies, J., Schellekens, A. P. M. & Touber, J. L. (1974). Prolactin secretion in patients with chromophobe adenoma of the pituitary gland. *Neth. J. Med.*, **17**, 163–173.
Astwood, E. B. (1941). The regulation of the corpus luteum function by hypophysial luteotrophin. *Endocrinology*, **29**, 309–319.
Aubert, M. C., Grumbach, M. M. & Kaplan, S. L. (1975). The ontogenesis of human foetal hormones. III. Prolactin. *J. clin. Invest.*, **56**, 155.
Auty, R., Branch, R. A., Cole, E. N., Levine, D. & Ramsay, L. (1976). Prolactin, diuretics and urinary electrolytes in normal subjects. *J. Endocr.*, **70**, 173–181.
Ayalan, D., Peyser, R., Toaff, R., Cordova, T., Harrell, A., Franchimont, P. & Linder, H. R. (1974). Effect of L-dopa on galactopoesis and gonadotrophin levels in the inappropriate lactation syndrome. *Obstet. Gynaec.*, **44**, 159.

Bailey, R. E. (1950). Inhibition with prolactin of light induced gonadal increase in White-crowned sparrows. *Condor*, **52**, 247–251.
Bailey, R. E. (1952). The incubation patch of passerine birds. *Condor*, **54**, 121–136.
Baker, B. I. & Ingleton, P. M. (1973). Factors affecting the release and synthesis of teleost pituitary polypeptide hormones. *J. Endocr.*, **55**, xi.
Baldwin, R. L. & Milligan, L. P. (1966). Enzymatic changes associated with the initiation and maintenance of lactation in the rat. *J. Biol. Chem.*, **241**, 2058.
Ball, J. N. (1965). Effects of autotransplantation of different parts of the pituitary gland on freshwater survival in the teleost *Poecilia latipinna. J. Endocr.*, **33**, v-vi.
Ball, J. N. (1969*a*). Prolactin (fish prolactin or paralactin) and growth hormone. In *Fish Physiology*, ed. Hoar, W. S. & Randall, D. J. pp. 207–240. New York: Academic Press.
Ball, J. N. (1969*b*). Prolactin and osmoregulation in teleost fishes: a review. *Gen. comp. Endocr.* Suppl., **2**, 10–25.
Ball, J. N. & Baker, B. I. (1969). The pituitary gland: anatomy and histophysiology. In *Fish Physiology*, ed. Hoar, W. S. & Randall, D. J. vol. II, pp. 1–10. London: Academic Press.
Ball, J. N. & Ensor, D. M. (1965). Effect of prolactin on plasma sodium in the teleost *Poecilia latipinna. J. Endocr.*, **32**, 267–270.
Ball, J. N. & Ensor, D. M. (1967). Specific action of prolactin on plasma sodium levels in hypophysectomised *Poecilia latipinna.* (Teleosteii). *Gen. comp. Endocr.*, **8**, 432–440.
Ball, J. N. & Ensor, D. M. (1969). Aspects of the action of prolactin on sodium metabolism in cyprinodont fishes. *C.N.R.S. Colloq. Intern.* Paris, **177**, 215–223.
Ball, J. N. & Hawkins, E. F. (1976). Adrenocortical (interrenal) responses to hypophysectomy and adenohypophysial hormones in the teleost *Poecilia latipinna. Gen. comp. Endocr.*, **28**, 59–71.
Ball, J. N. & Ingleton, P. M. (1973). Adaptive variations in prolactin secretion in relation to external salinity in the teleost *Poecilia latipinna. Gen. comp. Endocr.*, **20**, 312–325.
Ball, J. N. & Olivereau, M. (1964). Rôle de la prolactine dans la survie en eau douce de *Poecilia latipinna* hypophysectomisée et arguments en favour de sa synthése par les cellules erythrosinopiles de l'hypophyse des Teleosteans. *C.r. Acad. Sci., Paris*, **259**, 1443.
Ball, J. N., Olivereau, M., Slicher, A. M. & Kallman, K. D. (1965). Functional capacity of ectopic pituitary transplants in the teleost *Poecilia formosa* with comparative discussion on the transplanted pituitary. *Phil. Trans. R. Soc. Ser. B* **249**, 69–99.
Ball, J. N., Baker, B. I., Olivereau, M. & Petre, R. E. (1972). Investigations on hypothalamic control of adenohypophysial function in teleost fishes. *Gen. comp. Endocr.* Suppl., **3**, 11–21.
Ballantyne, B. & Bunch, G. A. (1966). The neurophysiology of the quiescent mammary tissue in *Lepus alba. J. comp. Neurol.*, **127**, 471–487.
Bardin, C. W., Liebelt, A. G. & Liebelt, R. A. (1962). The direct effect of pituitary isografts on mammary gland development in the mouse. *Proc. Soc. exp. Biol. Med.*, **110**, 716–718.

Barnawell, E. B. (1965). A comparative study of the responses of the mammary tissues from several mammalian species to hormones *in vitro*. *J. exp. Zool.*, **160**, 189–206.

Barnett, S. A. & Coleman, E. M. (1959). The effect of low environmental temperature on the reproductive cycle of female mice. *J. Endocr.*, **19**, 232.

Barraclough, C. A. & Sawyer, C. H. (1959). Induction of pseudo-pregnancy in the rat by reserpine and chlorpromazine. *Endocrinology*, **65**, 563–571.

Bartke, A. (1965). Influence of luteotrophin on fertility of dwarf mice. *J. Reprod. Fert.*, **10**, 93–103.

Bartke, A. (1966*a*). Reproduction of female dwarf mice treated with prolactin. *J. Reprod. Fert.*, **11**, 203–206.

Bartke, A. (1966*b*). Influence of prolactin on male fertility in dwarf mice. *J. Endocr.*, **35**, 419–420.

Bartke, A. (1969). Prolactin changes cholesterol stores in the mouse testes. *Nature, Lond.*, **224**, 700–701.

Bartke, A. (1971). Effects of prolactin and luteinising hormone on the cholesterol stores in the mouse testis. *J. Endocr.*, **49**, 317–324.

Bartke, A. (1974). Effects of inhibitors of pituitary prolactin on testicular cholesterol stores, seminal vesicle weight, fertility and lactation in mice. *Biol. Reprod.*, **11**, 319–325.

Bartke, A. (1976). Pituitary-testes relationships: role of prolactin in the regulation of testicular function. In *Sperm Action*, ed. Hubinot, P. O. pp. 136–152. Basel: Karger.

Bartke, A. & Lloyd, C. W. (1970). Influence of prolactin and pituitary isografts on spermatogenesis in dwarf mice and hypophysectomised rats. *J. Endocr.*, **46**, 313–320.

Bartke, A., Croft, B. T. & Dalterio, S. (1976). Prolactin restores plasma testosterone levels and stimulates testicular growth in hamsters exposed to short daylength. *Endocrinology*, **97**, 1601–1604.

Bartosik, D. B. & Romanoff, E. B. (1969). The luteotrophic process; effects of prolactin and LH on sterol and progesterone metabolism in bovine luteal ovaries perfused *in vitro*. In *The Gonads*, ed. McKerns, K. W. pp. 211–244. New York: Appleton.

Bartosik, D. B., Romanoff, E. B., Watson, D. J. & Scricco, E. (1967). Luteotrophic effects of prolactin in the bovine ovary. *Endocrinology*, **81**, 186–194.

Basu, S., Bern, H. A. & Chen, H. (1965). Effects of prolactin, growth hormone and gonadal steroids on the oviduct of Salientia. *Anat. Rec.*, **151**, 441–442.

Bates, R. W., Lahr, E. L. & Riddle, O. (1935). The gross action of prolactin and FSH on the mature ovary and sex accessories of the fowl. *Am. J. Physiol.*, **111**, 361–368.

Bates, R. W., Riddle, O. & Lahr, E. (1937). Mechanism of the antigonadal action of prolactin in adult pigeons. *Am. J. Physiol.*, **111**, 361–368.

Bates, R. W., Milkovic, S. & Garrison, M. M. (1964). Effects of prolactin, growth hormone and ACTH alone and in combination, upon organ weight and adrenal function in normal rats. *Endocrinology*, **74**, 714–723.

Bates, R. W., Miller, R. A. & Garrison, M. M. (1962). Evidence in the hypophysectomised pigeon of synergism among prolactin, growth hormone thyroxine and prednisone upon weight of the body, digestive tract, kidney and body stores. *Endocrinology*, **71**, 345–360.

Batten, T. F. C. & Ball, J. N. (1976). Ultrastructural studies on the prolactin cells and their innervation in the teleost *Poecilia latipinna*. *J. Endocr.*, **69**, 35P–36P.

Batten, T. F. C., Ball, J. N. & Benjamin, M. (1975). Ultrastructure of the adenohypophysia in the teleost *Poecilia latipinna*. *Cell Tiss. Res.*, **161**, 239.

Batten, T. F. C., Ball, J. N. & Grier, T. M. (1976). Circadian changes in prolactin cell activity in the pituitary of the teleost *Poecilia latipinna* in fresh water. *Cell Tiss. Res.*, **165**, 267–280.

Bauman, G. & Loriaux, D. L. (1976). The effect of endogenous prolactin on renal salt and water excretion and adrenal function in man. *J. clin. Endocr. Metab.*, **43**, 643.

Bauman, G., Marynick, G. F., Winters, S. J. & Loriaux, D. L. (1977). The effect of osmotic stimuli on prolactin secretion and renal water excretion in normal man and in chronic hyperprolactinemia. *J. clin. Endocr. Metab.*, **44**, 199–202.

Baylé, J. D. & Assenmacher, I. A. (1965). Absence de stimulation du jabot du pigeon après autogreffe hypophysaires. *C.r. Acad. Sci., (Paris) D*, **261**, 5667–5670.

Baylis, P. F. G. & Vant Hoff, W. (1969). Amenorrhoea and galactorrhoea associated with hypothyroidism. *Lancet*, **2**, 1399.

Beach, F. A. & Wilson, J. H. (1963). Effects of prolactin, progesterone and oestrogen on reactions of non-pregnant rats to foster young. *Psychol. Rep.*, **13**, 231–239.

Beach, J. E., Tyrey, L. & Everett, J. W. (1975) Serum prolactin and LH in early phases of delayed versus direct pseudo-pregnancy in the rat. *Endocrinology*, **96**, 1241–1246.

Beams, H. W. & Meyer, R. K. (1931). The formation of pigeon 'milk'. *Physiol. Zool.*, **4**, 486–500.

Beatty, D. D. (1966). A study of the succession of visual pigments in the Pacific salmon. (*Oncorhynchus*). *Can. J. Zool.*, **44**, 429–455.

Beck, J. S. (1970). Time of appearance of human placental lactogen in the embryo. *New Engl. J. Med.*, **283**, 189–190.

Beck, J. S. & Currie, A. R. (1967). Immunofluorescence localisation of growth hormone in the human pituitary gland and a related antigen in the syncitcotrophoblast. *Vitam. Horm.*, **25**, 89.

Beck, J. S. & Daughaday, W. H. (1967). Human placental lactogen: studies of its acute metabolic effects and disposition in normal man. *J. clin. Invest.*, **46**, 103.

Beck, J. S., Gonda, A., Hamid, M. A., Morgan, R. O., Rubenstein, D. & McGarry, E. E. (1964). In *Proteins and polypeptides.* ed. Astwood, E. B. & Beck, J. C. pp. 144–170. New York: Grune & Stratton.

Becker, D. J., Vinik, A. I., Pimstone, B. L. & Paul, M. (1975). Prolactin response to TRH in protein caloric malnutrition. *J. clin. Endocr. Metab.*, **41**, 782–783.

Behrman, H. R., Orczyke, G. P., MacDonald, G. J. & Greep, R. O. (1970). Prolactin induction of enzyme controlling luteal cholesterol ester turnover. *Endocrinology*, **87**, 1251–1256.

Ben-David, M., Danon, A. & Sulman, F. G. (1970). Acute changes in blood and pituitary prolactin after a single injection of perpehnazine. *Neuroendocrinology*, **6**, 336–342.

Ben-David, M., Danon, A., Benveniste, R., Weller, C. P. & Sulman, F. G. (1971). Results of radioimmunoassays of rat pituitary and serum prolactin after adrenalectomy and perphenazine treatment in rats. *J. Endocr.*, **50**, 599–606.

Ben-David, M., Becker, R., Rolland, D. & Chrombach, A. (1974). Isolation of human prolactin from amniotic fluid: a pilot study in preparative isoelectric focusing and electrophoresis on acrylamide gel. *Endocr. Res. Comm.*, **1**, 211–228.

Bengmark, S. & Hesselojo, R. (1963). The combined effect of prolactin and androstene dione on the growth of the rat seminal vesicle *in vitro*. *Urol. Intern.*, **16**, 387–390.

Bengmark, S. & Hesselojo, R. (1964). Endocrine dependance of rat seminal vesicle tissue in tissue culture. *Urol. Intern.*, **17**, 84–92.

Benjamin, F., Caspar, D. J. & Kolodny, H. H. (1969). Immunoreactive human growth hormone in conditions associated with galactorrhoea. *Obstet. Gynaec.*, **34**, 34–39.

Benson, G. K., Cowie, A. T., Cox, C. P. & Goldzweig, S. A. (1957). Effects of oestrogen and progesterone on mammary development in the guinea pig. *J. Endocr.*, **15**, 126–144.

Bentley, P. J. (1971). *Endocrines and osmoregulation. A comparative account of the regulation of water and salt in the vertebrates.* Heidelberg-New York-Berlin: Springer-Verlag.

Bergenstahl, D. M. & Lipsett, M. B. (1960). Metabolic effects of human growth and growth hormone of other species in man. *J. clin. Endocr. Metab.*, **23**, 1427.

Berle, P., Finsterwalder, E. & Apostolakis, M. (1974). Comparative studies on the effect of human growth hormone, human prolactin and human placental lactogen on lipid metabolism. *Horm. Metab. Res.*, **6**, 347–350.

Berman, R., Bern, H. A., Nicoll, C. S. & Strohman, R. C. (1964). Growth promoting effects of mammalian prolactin and growth hormone in tadpoles of *Rana catesbiana*. *J. exp. Zool.*, **156**, 353–360.

Bern, H. A. (1975). Prolactin and osmoregulation. *Am. Zool.*, **15**, 937–949.

Bern, H. A. & Nicoll, C. S. (1968). The comparative endocrinology of prolactin. *Recent Prog. Horm. Res.*, **24**, 681–720.

Bern, H. A. & Nicoll, C. S. (1969). The taxonomic specificity of prolactins. *C.N.R.S. Colloq. Intern. Paris*, **177**, 193–203.

Bern, H. A. & Nicoll, C. S. (1972). On the actions of prolactin among vertebrates: is there a common denominator. In *Lactogenic Hormones*, ed. Wolstenholme, G. E. & Knight, J. pp. 299–317. London: Ciba Foundation.

Bern, H. A. & Nicoll, C. S. & Strohman, R. C. (1967). Prolactin and tadpole growth. *Proc. Soc. exp. Biol. Med.*, **126**, 518–520.

van Berswordt-Wallrabe, P., Moon, R. C. & Turner, O. W. (1960). Effect of lactogenic hormone, growth hormone and thyroxine in the lactating albino rat. *Proc. Soc. exp. Biol. Med.*, **104**, 530–531.

Besser, G. M. & Thorner, M. O. (1975). Prolactin and gonadal function. *Path. Biol.*, **23**, 779.

Besser, G. M., Parke, L., Edwards, C. R. W., Forsyth, I. A. & McNeilly, A. S. (1972). Galactorrhoea: successful treatment with reduction of plasma prolactin levels by bromergocryptine. *Brit. med. J.*, **iii**, 669.

Besses, G. S., Burrow, G. N., Spalding, S. W. & Donabedian, D. W. (1975). Dopamine infusion acutely inhibits the TSH and prolactin response to TRH. *J. clin. Endocr. Metab.*, **41**, 985.

Bewley, T. A., Brovetto-Cruz, J. & Li, C. H. (1969). Human pituitary growth hormone. Physiological investigations of the native and reduced-alkylated protein. *Biochemistry*, **8**, 4701–4708.

Bewley, T. A. & Li, C. H. (1970). Primary structures of human pituitary growth hormone and sheep pituitary lactogenic hormone compared. *Science*, **168**, 1361–1362.

Beyer, C., Cruz, M. L. & Martinez-Manautau, J. (1970). Effect of chloramidone acetate on mammary development and lactation in the rabbit. *Endocrinology*, **86**, 1172–1174.

Beyer, C., Anguiano, L. G. & Mena, F. (1961). Oxytocin release in response to stimulation of cingulate gyrus. *Am. J. Physiol.*, **200**, 625–627.

Biglieri, E. G., Watlington, C. O. & Forsham, P. H. (1961). Sodium retention with human growth hormone and its subfactors. *J. clin. Endocr.*, **21**, 361–370.

Billeter, E. & Fluckiger, E. (1971). Evidence for a luteolytic function of prolactin in the intact cyclic rat using 2 Br-α-ergocryptine (CB 154). *Experentia*, **27**, 464–465.

Birge, C. A., Jacobs, L., Hammer, C. & Daughaday, W. H. (1970). Catecholamine inhibition of prolactin secretion by isolated rat adenohypophysis. *Endocrinology*, **86**, 120–130.

Birkenhager, J. C., Upton, G. V., Seldenrath, H. J., Krieger, D. T. & Tashjian, A. H. (1976). Medullary thyroid carcinoma; ectopic production of peptides with ACTH-like, CRF-like and prolactin-production stimulating properties. *Acta Endocr.* (*Copnh*), **83**, 280–292.

Birkinshaw, M. & Falconer, I. R. (1972). The localisation of prolactin labelled with radioiodine in rabbit mammary tissue. *J. Endocr.*, **55**, 323–334.

Bishop, W., Fawcett, C. P., Krulich, L. & McCann, S. M. (1972). Acute and chronic effects of hypothalamic lesions on the release of FSH, LH and prolactin in intact and castrated rats. *Endocrinology*, **91**, 643.

Blake, C. A. (1974). Stimulation of pituitary prolactin and TSH release in lactating and proestrus rats. *Endocrinology*, **94**, 503–508.

Blake, C. A. (1976). Effects of intravenous infusion of catecholamines on rat plasma luteinising hormone and prolactin concentrations. *Endocrinology*, **98**, 99–103.

Blanc-Livni, N. & Abraham, M. (1970). The influence of environmental salinity on the prolactin and gonadotrophins – secreting regions in the pituitary of *Mugil* (Teleostei). *Gen. comp. Endocr.*, **14**, 184–197.

Blandau, R. J. (1945). On the factors involved in sperm transport through the cervix uteri of the albino rat. *Am. J. Anat.*, **77**, 253–272.

Blask, D. E., Vaughan, M. K., Reiter, R. J., Johnson, L. Y. & Vaughan, G. M. (1976). Prolactin release and release inhibiting factor activities in the bovine, rat and human pineal gland, *in vitro* and *in vivo* studies. *Endocrinology*, **99**, 152–162.

Blatt, L. M., Slickers, K. A. & Kim, K. H. (1969). Effect of prolactin on thyroxine induced metamorphosis. *Endocrinology*, **85**, 1213–1215.

Bleick, C. R. (1975). Hormonal control of the nuchal hump in the cichlid fish *Cichlasoma*. *Gen. comp. Endocr.*, **26**, 192–198.

Blüm, V. (1973). Experimente mit Teleosteen-prolaktin. *Zool. J. Physiol.*, **77**, 335–347.

Blüm, V. (1974). Prolactin and behaviour in fish. *Acta Endocr.* (*Copnh.*), **151**, 172.

Blüm, V. & Fiedler, K. (1964). Hormonal control of reproductive behaviour in some cichlid fish. *Gen. comp. Endocr.*, **5**, 186–196.

Blüm, V. & Weber, K. M. (1968). The influence of prolactin on the activity of the steroid 3B-ol-dehydrogenase in the ovary of the cichlid fish *Aequidens pulcher*. *Experentia*, **24**, 1259–1260.

Bohnet, H. G. (1976). Induction of lactogenic receptors. I. In the liver of hypophysectomised female rats. *Endocr. Res. Comm.* **3**, 187–198.

Boisseau, J. P. (1964). Effets de la castration et de l'hypophysectomie sur l'incubation de l'Hippocampe male (*Hippocampus hippocampe*). *C.r. Acad. Sci., Paris*, **259**, 4839–4840.

Boisseau, J. P. (1965). Action de quelques hormones sur l'incubation del Hippocampus male normaux ou castres, ou hypophysectomisés. *C.r. Acad. Sci., Paris*, **260**, 313–314.

Boisseau, J. P. (1967). *Les regulationes hormonales de l'incubation chez un vertebre male recherches sur la reproduction del'hippocampe*. Thesis, Faculty of Science, University of Bordeaux.

Boisseau, J. P. (1969). Prolactin et incubation chez l'hippocampe. *C.N.R.S. Colloq. Intern. Paris*, **177**, 205–215.

Bolander, F. F. & Fellows, R. E. (1976). Purification and characterisation of bovine placental lactogen. *J. Biol. Chem.*, **251**, 2703–2708.

Bolton, C. E. (1971). Effect of prolactin and corticosterone on lipid synthesis by rabbit mammary gland explants. *J. Endocr.*, **51**, xxxi–xxxii.

Bolton, N. J., Scanes, C. G. & Chadwick, A. (1976). A radioimmunoassay for prolactin in the circulation of birds. *J. Endocr.*, **67**, 51P.

Boot, L. M. (1969). *Induction by prolactin of mammary tumours in mice*. Amsterdam: North Holland.

Bornacin, M. Cuthbert, A. E. & Maetz, J. (1972). The effects of calcium on branchial sodium fluxes in the sea water adapted eel, *Anguilla anguilla* L. *J. Physiol. Lond.*, **222**, 487–496.

Boschwitz, D. (1969). Influence of prolactin on the ultimo branchial bodies of *Bufo viridis*. *Israel. J. Zool.*, **18**, 277–289.

Boschwitz, D. & Bern, H. A. (1971). Prolactin, calcitonin and blood calcium in the toads *Bufo boreas* and *Bufo marinus*. *Gen. comp. Endocr.*, **17**, 586–588.

Botte, V., d'Istria, M., Delrio, G. & Cheiffi, G. (1972). Hormonal regulation in thumb pads of males in *Rana esculenta*. *Gen. comp. Endocr.*, **18**, 577.

Bourne, R. A. & Bryant, J. A. (1974). Effects of prolactin on the synthesis of casein in rabbit mammary tissue. *Int. J. Biochem.*, **5**, 45–48.

Bowers, C. Y., Friesen, H., Cheng, J. K. & Folkers, K. (1972). In *Proc. 4th Int. Congr. Endocr.* Washington D.C., ed. Scow, R. Amsterdam: Excerpta Medica.

Bowers, C. Y., Friesen, H. G. & Folkers, K. (1973). Further evidence that TRH is also a physiological regulator of PRL in man. *Biochem. biophys. Res. Comm.*, **51**, 512–521.

Bowers, C. Y., Friesen, H. G., Hwang, P., Guyda, H. J. & Folkers, K. (1971). Prolactin and thyrotropin release in man by synthetic pyroglutamyl-histidyl-prolinamide. *Biochem. biophys. Res. Comm.*, **45**, 1035–1041.

Boyar, R. N., Kapan, S., Finklestein, J. W., Perlow, M., Sassin, J. F., Fukoshima, D. K. Weitman, E. D. & Hollman, L. (1974). Hypothalamic pituitary function in diverse hyperprolactinaemic states. *J. clin. Invest.*, **53**, 1588.

Boyd, A. E., Spencer, E., Jackson, I. M. D. & Reichlin, S. (1976). Prolactin releasing factor (PRF) in procine hypothalamic extract distinct from TRH. *Endocrinology*, **99**, 861.

Boyns, A. R., Cole, E. N., Golder, M. P., Danutra, V., Harper, M. E., Brownsey, B., Cowley, T., Jones, G. E. & Griffiths, K. (1972). Prolactin studies with the prostate. In *Prolactin and carcinogenesis*, Proc. 4th Tenovus Workshop, ed. Boyns, A. R. & Griffiths, K. pp. 207–216. Cardiff: Alpha Omega Alpha.

Bradshaw, S. D. (1975). Osmoregulation and pituitary-adrenal function in desert reptiles. *Gen. comp. Endocr.*, **25**, 230–248.

Brant, J. W. A. & Nalbandov, A. V. (1956). Role of sex hormones in albumen secretion by the oviduct of chickens. *Poult. Sci.*, **35**, 692–700.

Brauman, J., Brauman, H. & Pasteels, J. L. (1964). Immunoassay of growth hormone in cultures of human hypophysis by the method of complement fixation: comparison of the growth hormone secretion and the prolactin activity. *Nature, Lond.*, **202**, 1116–1118.

Braun, E. H. & Dantzler, W. D. (1972). Function of mammalian type and reptilian type nephrons in the kidney of desert quail. *Am. J. Physiol.*, **222**, 617–629.

Brazeau, R., Vale, W., Burgos, R., Ling, N., Butcher, N., Rivier, J. & Gulleman, R. (1973). Hypothalamic polypeptide that inhibits the secretion of immunoreactive pituitary growth hormone. *Science*, **179**, 77–79.

Breitenbach, R. P. & Meyer, R. K. (1959). Pituitary prolactin levels in laying, incubating and brooding pheasants (*Phasianus colchicus*). *Proc. Soc. exp. Biol.*, **101**, 16–19.

Breuer, C. B. (1969). Stimulation of DNA synthesis in cartilage of hypophysectomised rats by native and modified placental lactogen and anabolic hormones. *Endocrinolgy*, **85**, 989–999.

Brewer, K. J. (1976). *Osmoregulation in the Chelonia*. Ph.D. Thesis, University of Liverpool.

Bridges, C. D. B. (1972). The rhodopsin-porphropsin visual system. In *Handbook of Sensory Physiology*, VII/1. Photochemistry of Vision. ed. Dartnell, H. J. A. pp. 417–480. Heidelburg: Springer-Verlag.

Bridges, R. S. & Goldman, B. D. (1975). Ovarian control of prolactin secretion during late pregnancy in the rat. *Endocrinology*, **97**, 196–198.

Bridges, R. S., Goldman, B. D. & Bryant, L. P. (1974). Serum prolactin concentrations and the initiation of maternal behaviour in the rat. *Horm. Behav.*, **5**, 219–226.

Brockway, B. F. (1968). Budgerigars are not determinate layers. *Wilson Bull.*, **80**, 106–107.

Brown, P. S. & Brown, S. C. (1973). Prolactin and thyroid hormone interactions in salt and water balance in the newt *Notophthalamus viridescens*. *Gen. comp. Endocr.*, **20**, 456–466.

Brown, P. S. & Frye, B. E. (1968). Effects of prolactin and growth hormone on growth and metamorphosis of tadpoles of the frog *Rana pipiens*. *Gen. comp. Endocr.*, **13**, 126–138.

Brown, P. S. & Frye, B. E. (1968*b*). Effects of hypophysectomy, prolactin and growth hormone on growth of post-metamorphic frogs. *Gen. comp. Endocr.*, **13**, 139–145.

Browning, H. C. & White, W. D. (1965). Local stimulation of mammary glands by pituitary or ovarian grafts in the mouse. *Texas Rep. Biol. Med.*, **23**, 26–37.

Bryant, G. D. & Greenwood, F. C. (1968). Radioimmunoassay for ovine, caprine and bovine prolactin in plasma and tissue extracts, *Biochem. J.*, **109**, 831.

Bryant, G. D., Linzell, J. L. & Greenwood, F. C. (1970). Plasma prolactin in goats measured by radioimmunoassay, the effects of teat stimulation, mating behaviour, stress, fasting and of oxytocin, insulin and glucose injections. *Hormones*, **1**, 26.

Bryant, G. D., Silver, J., Greenwood, F. C., Pasteels, J. L., Robyn, C. & Hubinot, P. O. (1971). Radioimmunoassay of a human pituitary prolactin in plasma. *Hormones*, **2**, 139–152.

Buckman, M. T. & Peake, G. T. (1973*a*). Estrogen potentiation of phenothiazine-induced prolactin secretion in man. *J. clin. Endocr. Metab*, **37**, 977.

Buckman, M. T. & Peake, G. T. (1973*b*). Osmolar control of prolactin secretion in man. *Science*, **181**, 755.

Bullough, W. A. & Wallis, M. (1974). An *in vitro* bioassay for prolactin based on stimulation of casein synthesis by a dispersed cell preparation from mouse mammary gland. *J. Endocr.*, **62**, 463–472.

Burden, C. E. (1956). The failure of hypophysectomised *Fundulus heteroclitus* to survive in freshwater. *Biol. Bull.*, **110**, 8–29.

Burman, K. D., Dimond, R. C., Noel, G. L., Earll, J. M., Frantz, A. G. & Wartofsky, L. (1975). Klinefelters syndrome: examination of thyroid function and the TSH and PRL response to thyrotropin-releasing hormone prior to and after testosterone treatment. *J. Clin. Endocr. Metab.*, **41**, 1161–1166.

Burns, J. T. & Meier, A. H. (1971). Daily variations in the pigeon crop-sac responses to prolactin. *Experentia*, **27**, 572–575.

Burns, J. W. & Richards, J. S. (1974). Effects of mammalian and host gonadotrophins on the ovaries of female Texas horned lizards, *Phyronosoma cornulium*. *Comp. Biochem. Physiol.*, **47A**, 655–661.

Burstyn, P. G. (1975). Water and sodium accumulation in rabbits after administration of prolactin. *J. Endocr.*, **68**, 15P.

Burstyn, P. G., Horrobin, D. F. & Manku, M. S. (1972). Saluretic effect of aldosterone in the presence of increased salt intake and restoration of normal action by prolactin or by oxytocin. *J. Endocr.*, **55**, 369–376.

Butcher, R. L., Fuge, N. W. & Collins, W. E. (1972). Semi-circadian rhythm in plasma levels of prolactin during early gestation in the rat. *Endocrinology*, **90**, 1125–1127.

Butcher, G. A., Robertson, J. & Sutherland, T. J. (1971). The role of cyclic AMP in certain biological control systems. Ciba Foundn. Symp., **25**. London: Churchill Livingstone.

Butle, H. L. (1974). Seasonal variation of prolactin in plasma of male goats. *J. Reprod. Fert.*, **37**, 95–99.

Butle, H. L. & Forsythe, I. A. (1976). Placental lactogen in the cow. *J. Endocr.*, **68**, 141–146.

Butle, H. L., Forsythe, I. A. & Knaggs, G. S. (1972). Plasma prolactin measured by radioimmunoassay and bioassay in pregnant and lactating goats and the occurrence of a placental lactogen. *J. Endocr.*, **53**, 483–491.

Cade, T. J. & Greenwald, C. (1966). Nasal salt secretion in falconiform birds. *Condor*, **68**, 338–350.

Callard, I. P. & Chan, S. W. C. (1974). Reptilian ovarian steroidogenesis and the influence of mammalian gonadotrophins (follicle stimulating hormone and luteinising hormone) *in vitro*. *J. Endocr.*, **55**, 143–147.

Callard, I. P. & Ziegler, H. (1970). Inhibitory effects of prolactin upon gonadotrophin stimulated ovarian growth in the iguanid lizard, *Dipsosaurus dorsalis*. *J. Endocr.*, **47**, 131–132.

Callard, I. P., McChesney, I. & Scanes, C. G. (1975). Influence of mammalian and avian gonadotrophins and prolactins on steroidogenesis in turtle (*Chrysemys picta*) follicular and luteal tissues. *J. Endocr.*, **65**, 22P.

Camanni, F., Massara, F., Belforte, L. & Molinatti, G. M. (1975*a*). Changes in plasma growth hormone levels in normal and acromegalic subjects following administration of 2 Br-α-ergocryptine. *J. clin. Endocr. Metab.*, **40**, 363.

Camanni, F., Massara, F., Belaforte, L., Rostello, A. & Molinatti, G. M. (1977). Effect of dopamine on plasma growth hormone levels in normal and acromegalic subjects. *J. clin. Endocr. Metab.*, **44**, 465–473.

Camanni, F., Massara, F., Fassio, V., Molinatti, G. M. & Muller, E. (1975*b*). Effect of five dopaminergic drugs on plasma growth hormone levels in acromegalic subjects. *Neuroendocrinology*, **19**, 227.

Campantico, E., Giunta, C., Guardabassi, A. & Vietti, M. (1972). The stabilising action of prolactin on the lysosomes in tails from *Xenopus laevis* Daudin tadpoles. *Gen. comp. Endocr.*, **18**, 396–399.

Canales, C. S., Soria, J., Zarate, A., Mason, M. & Molina, M. (1976). The influence of pyridoxine on prolactin secretion and milk-secretion in women. *Br. J. Obstet. Gynaec.*, **83**, 378–379.

Canfield, C. J. & Bates, R. W. (1965). Nonpuerperal galactorrhoea. *New Engl. J. Med.*, **273**, 897–902.

Carlsson, S., Kollander, S. & Muller, E. R. (1972). The distribution of ^{125}I marked bovine prolactin and human chorionic gonadotrophin in rats with experimental ovarian tumours. *Acta obstet. gynaecol. scand.*, **51**, 175–182.

Carlsson, H. E., Jacobs, L. S. & Daughaday, W. H. (1973). Growth hormone, thyrotrophin and prolactin responses to thyrotropin releasing hormone following diethylstilboestrol pretreatment. *J. clin. Endocr. Metab.*, **37**, 488.

Carr, L. A., Conway, P. M. & Vooght, J. L. (1975). Inhibition of brain catecholamine synthesis and release of prolactin and luteinising hormone in the ovariectomised rat. *J. Pharmac. exp. Ther.*, **192**, 15–21.

Cassady, J. M., Li, G. S., Spitzner, E. B. & Floss, H. G. (1974). Ergot alkaloids, ergolines and related compounds as inhibitors of prolactin release. *J. med. Chem.*, **17**, 300–307.

Cathcart, E. P., Gairns, F. W. & Garven, H. S. D. (1948). The innervation of the human quiescent nipple, with notes on pigmentation, erection and hyperneury. *Trans. R. Soc. Edinb.*, 699–717.

Catt, K. & Moffat, B. (1967). Isolation of internally labelled rat prolactin by preparative disc electrophoresis. *Endocrinology*, **80**, 324–328.

Cerute, R. A. & Lyons, W. R. (1966). Mammogenic activities of the mid-gestational mouse placenta. *Endocrinology*, **67**, 884–887.

Chadwick, A. (1966). Prolactin-like activity in the pituitary of the frog. *J. Endocr.*, **34**, 247–255.

Chadwick, A. (1968). Prolactin-like activity in the pituitary gland of fishes and amphibians. *J. Endocr.*, **35**, 75–81.

Chadwick, A. (1969). Effects of prolactin in homiothermic vertebrates. *Gen. Comp. Endocr.*, Suppl., **2**, 63–68.

Chadwick, A. & Jordan, B. J. (1971). The lipids of the crop epithelium of pigeons after injection with prolactin from the pituitary of different vertebrate classes. *J. Endocr.*, **49**, 51–58.

Chadwick, A., Folley, S. J. & Gemzell, C. A. (1961). Lactogenic activity of human pituitary growth hormone. *Lancet*, **ii**, 241–143.

Chadwick, C. S. (1940). Identity of prolactin with water drive factor in *Triturus viridescens*. *Proc. Soc. exp. Biol. Med.*, **45**, 335–337.

Chadwick, C. S. (1941). Further observations on the water drive in *Triturus viridescens*, II. Induction of the water drive with lactogenic hormone. *J. exp. Zool.*, **86**, 175–187.

Chadwick, C. S. & Jackson, H. R. (1948). Acceleration of skin growth and moulting in the red eft, *Triturus viridescens*. *J. exp. Zool.*, **147**, 127–131.

Challis, J. R. G. & Linzell, J. L. (1971). The concentration of unconjugated oestrogens in the plasma of pregnant goats. *J. Reprod. Fert.*, **37**, 95.

Chambolle, P. (1966). Recherches sur l'allongement de la durée de survie après hypophysectomie chez *Gambusia. C.r. Acad. Sci., Paris*, **262**, 1750–1753.

Chambolle, P. (1969). Influence de l'injection d'ACTH et de prolactine sur la gestation de femelles de Gambusia privées d'hypophyse. *C.r. Acad. Sci., Paris*, **268**, 1215–1217.

Chamness, G. C., Costlow, M. E. & McGuire, W. L. (1975). Estrogen receptor in rat liver and its dependence on prolactin. *Steroids*, **26**, 363–371.

Chan, D. K. O., Chester Jones, I. & Moseley, W. (1968). Pituitary and adrenocortical factors in the control of water and electrolyte composition of the freshwater European eel, *Anguilla anguilla*. L. *J. Endocr.*, **42**, 91–98.

Chan, D. K. O., Callard, I. P. & Chester Jones, I. (1970). Observations on the water and electrolyte composition of the iguanid lizard *Dipsosaurus dorsalis dorsalis* (Baird and Girard) with special reference to the control by the pituitary gland and adrenal cortex. *Gen. comp. Endocr.*, **15**, 374–387.

Chan, D. K. O. & Chester Jones, I. (1968). Regulation and distribution of plasma and calcium and inorganic phosphate in the European eel (*Anguilla anguilla*). *J. Endocr.*, **42**, 109–117.

Chavin, W. (1956). Pituitary-adrenal control of melanisation in xanthic goldfish, *Carassius auratus* L. *J. exp. Zool.*, **133**, 1–45.

Chen, C. L. & Meites, J. (1970). Effects of oestrogen and progesterone on serum and pituitary prolactin levels in ovariectomised rats. *Endocrinology*, **86**, 503–505.

Chen, H. T. & Meites, J. (1975). Effects of biogenic amines and TRH on release of prolactin in the rat. *Endocrinology*, **96**, 10–14.

Chen, J. S., Robertson, H. A. & Friesen, H. G. (1976). The purification and characterisation of ovine placental lactogen. *Endocrinology*, **98**, 65–76.

Chen, C. L., Amenomori, Y., Lu, K. H., Voogt, J. L. & Meites, J. (1970). Serum prolactin levels in rats with pituitary transplants or hypothalamic lesions. *Neuroendocrinology*, **6**, 220–227.

Cherms, F. L., Herrick, R. B., McShan, W. H. & Hymer, W. C. (1962). Prolactin content of the anterior pituitary gland of turkey hens in different reproductive stages. *Endocrinology*, **71**, 288–292.

Chester, R. V. & Zucker, I. (1970). Influence of male copulatory behaviour on sperm transport, pregnancy and pseudo-pregnancy in female rats. *Physiol. Behav.*, **5**, 35–43.

Chester-Jones, I. (1963). *The adrenal cortex*. Cambridge: Cambridge University Press.

Chi, C. C. (1970). Afferent connections to the ventromedial nucleus of the hypothalamus in the rat. *Brain Res.*, **17**, 439.

Chidambaram, S., Meyer, R. K. & Hasler, A. D. (1972). Effects of hypophysectomy, pituitary autografts, prolactin, temperature and salinity of the medium on survival and natremia in the bullhead, *Ictalurus melas*, *Comp. Biochem. Physiol.*, **43A**, 443–457.

Child, D. F., Gordon, H., Mashuter, K. & Joplin, G. F. (1975). Pregnancy, prolactin and pituitary tumours. *Brit. med. J.* **4**, 87–89.

Chiodini, P. G., Liuzzi, A., Muller, E. E., Botalla, G., Oppizzi, G., Verde, G. & Silvestrini, F. (1976). Inhibitory effect of an ergoline derivative, methergoline, on growth hormone and prolactin levels in acromegalic patients. *J. clin. Endocr. Metab.*, **43**, 356–363.

Chiu, K. W. & Phillips, J. G. (1971). The effect of prolactin on the sloughing cycle in the Tokay, (*Gecko gecko*, Lacertilia). *J. Endocr.*, **39**, 463–472.

Chiu, K. W. & Phillips, J. G. (1972). The role of prolactin in the sloughing cycle of the lizard, *Gecko gecko* L. *J. Endocr.*, **49**, 625–634.

Chochinov, R. H., Katupanya, A., Mary, I. K., Underwood, L. E. & Daughaday, W. H. (1976). Amniotic fluid reactivity detected by somatomedin C radioactivity assay;

correlation with growth hormone, prolactin and fetal renal maturation. *J. clin. Endocr. Metab.*, **42**, 983–986.

Choudrhay, J. B. & Greenwald, G. S. (1969). Luteotrophic complex of the mouse. *Anat. Rec.*, **163**, 373–388.

Christian, J. J. (1964). Effect of chronic ACTH treatment on maturation of female mice. *Endinocrinology*, **74**, 669.

Clarke, N. B. & Kaltenbach, J. C. (1961). Direct action of thyroxine on skin of the adult newt. *Gen. comp. Endocr.*, **49**, 619–624.

Clarke, R. H. & Baker, B. L. (1964). Circadian periodicity in the concentration of prolactin in the rat hypophysis. *Science*, **143**, 375.

Clemens, J. A. & Meites, J. (1971). Neuroendocrine status of old constant estrous rats. *Neuroendocrinology*, **7**, 249.

Clemens, J. A. & Sawyer, B. D. (1974). Evidence that methadone stimulates prolactin release by dopamine receptor blockade. *Endocr. Res. Comm.*, **1**, 373–378.

Clemens, J. A., Shaar, C. J., Kleber, J. W. & Tandy, W. A. (1971*a*). Reciprocal control by the preoptic area of LH and prolactin. *Expl Brain Res.*, **12**, 250.

Clemens, J. A., Shaar, C. J., Tandy, W. A. & Roush, M. E. (1971*b*). Effects of hypothalamic stimulation on prolactin secretion in steroid treated rats. *Endocrinology*, **89**, 1317.

Clemens, J. A., Reys, F. I., Winter, J. S. & Faiman, C. (1977). Studies on human sexual development IV. Fetal pituitary and serum and amniotic fluid concentrations of prolactin. *J. clin. Endocr. Metab.*, **44**, 408–413.

Clifton, K. H. & Furth, J. (1960). Ducto-alveolar growth in mammary glands of adreno-gonadectomised male rats bearing mammotrophic pituitary tumours. *Endocrinology*, **66**, 893–897.

Cohen, D. C., Greenberg, J. A., Licht, P., Bern, H. A. & Zipser, R. D. (1972). Growth and inhibition of metamorphosis in the newt *Taricha torosa* by mammalian hypophysial and placental hormones. *Gen. comp. Endocr.*, **18**, 384–390.

Cole, E. N., Evered, D. & Horrobin, D. F. (1975). Is prolactin a fluid and electrolyte regulating hormone in man. *J. Physiol.*, Lond., **252**, 54–55P.

Connelly, T. G., Tassava, R. A. & Thornton, C. S. (1968). Limb regeneration and survival of prolactin treated hypophysectomised adult newts. *J. Morph.*, **126**, 365–372.

Convey, E. M., Tucker, H. A., Smith, U. G. & Zolman, J. (1972). Prolactin, thyroxine and corticoid after TRH. *J. Anim. Sci.*, **35**, 258, Abstr.

Copinschi, G., L'Hermite, M., Le Clercq, R., Godstein, J., van Halst, L., Vivasavo, E. & Robyn, C. (1975). Effects of glucocorticoids on pituitary hormonal responses to hypoglycaemia. Inhibition of prolactin release. *J. clin. Endocr. Metab.*, **40**, 442–449.

Coppedge, R. L. & Segaloff, A. (1951). Urinary prolactin excretion in man. *J. clin. Endocr. Metab.*, **11**, 465–476.

Coppola, J. A., Leonardi, R. G., Lippman, W., Perrine, J. W. & Ringler, I. (1965). Induction of pseudopregnancy in rats by depletion of endogenous catecholamines. *Endocrinology*, **77**, 485–490.

Corenblum, B., Webster, B. R., Mortimer, C. B., Ezrin, C. & Shewchuck, A. (1975). Possible anti-tumour effect of 2 Br-α-ergocryptine (CB 154, Sandoz) in two patients with large prolactin-secreting pituitary adenomas. *Clin. Res.*, **23**, 614A.

Corenblum, B., Pairaudeau, N. & Schechuck, A. B. (1976). Prolactin hypersecretion and short luteal phases. *Obstet. Gynaec.*, **47**, 486.

Corvilain, J. & Abramow, M. (1962). Growth hormone and renal physiopathology. *J. Urol. Nephrol., Paris*, **69**, 1–12.

Cotes, P. H., Crichton, J. A., Folley, S. J. & Young, F. G. (1949). Galactopoietic activity of purified anterior pituitary growth hormone. *Nature, Lond.*, **164**, 992–993.

Cotta, Y., Terqui, M. & Pelletier, J. (1974). Testosterone et LH plasmatique chez l'agneau de la naissance à la puberté. *C.r. Acad. Sci., Paris*, D, **280**, 1473–1476.

Cowie, A. T. (1966). Anterior pituitary function in lactation. In *The Pituitary Gland*, ed. Harris, G. W. & Donovan, B. T. Vol. 12. London: Butterworths.

Cowie, A. T. (1969). Variations in the yield and composition of the milk during lactation in the rabbit and the galactopoietic effect of prolactin. *J. Endocr.*, **44**, 437–450.

Cowie, A. T. & Tindall, J. S. (1961). The maintenance of lactation in the rat after hypophysial anterior lobectomy during pregnancy. *J. Endocr.*, **22**, 403–408.

Cowie, A. T. & Tindall, J. S. (1971). *The physiology of lactation.* London & Southampton: Edward Arnold Ltd.

Cowie, A. T. & Watson, S. C. (1966). The adrenal cortex and lactogenesis in the rabbit. *J. Endocr.*, **35**, 213–214.

Cowie, A. T., Hartmann, P. E. & Turvey, A. (1969). The maintenance of lactation in the rabbit after hypophysectomy. *J. Endocr.*, **43**, 651–652.

Cowie, A. T. & Tindall, J. S. & Yokoyama, A. (1966). The induction of mammary growth in the hypophysectomised goat. *J. Endocr.*, **34**, 185–195.

Crim, J. W. (1972). Studies on the possible regulation of plasma sodium by prolactin in the amphibia. *Comp. Biochem. Physiol.*, **43A**, 349–357.

Crim, J. W. (1975*a*). Prolactin induced modification of visual pigments in the eastern red spotted newt *Notphthalamus viridescens. Gen. comp. Endocr.*, **26**, 233–242.

Crim, J. W. (1975*b*). Prolactin-thyroxine antagonism and the metamorphosis of visual pigments in *Rana catesbiana* tadpoles. *J. exp. Zool.*, **192**, 355–362.

Cristy, M. (1974). Effect of prolactin and thyroxine on the visual pigments of trout *Salmo gairdneri. Gen. comp. Endocr.*, **23**, 58–62.

Cumming, I. A. (1975). The ovine and bovine oestrus cycle. An appreciation of J. R. Godings contribution to the understanding of the oestrus cycle. *J. Reprod. fert.*, **43**, 583–596.

Cunningham, G. R. & Huckins, C. (1977). An FSH and prolactin-secreting pituitary tumour: Pituitary dynamics and testicular histology. *J. clin. Endocr. Metab.*, **44**, 248.

Cuthbert, A. W. (1972). Neurohypophysial hormones and sodium transport. *Proc. R. Soc. B*, **262**, 103–111.

Cuthbert, A. W. & Maetz, J. (1972). Amiloride and sodium fluxes across fish gills in freshwater and seawater. *Comp. Biochem. Physiol.*, **43A**, 227–232.

Cuthbert, A. W. & Shum, W. K. (1974). Amiloride and the sodium channel. *Naunyn-Schmiedebergs Arch. exp. Path. Pharmak.*, **281**, 261–269.

Dalton, T. R. & Snart, R. (1969). Effect of prolactin on the active transport of sodium by the isolated toad bladder. *J. Endocr.*, **43**, vi-vii.

Dannies, P. S. & Tashjian, A. H. (1973). Effects of thyrotropin releasing hormone and hydrocortisone on synthesis and degradation of prolactin in a rat pituitary cell strain. *J. biol. Chem.*, **248**, 6174–6179.

Dannies, P. S., Gautvik, K. M. & Tashjian, A. H. (1976). A possible role of cyclic AMP in mediating the effects of thyrotropin releasing hormone on prolactin release and on prolactin and growth hormone synthesis in pituitary cells in culture. *Endocrinology*, **98**, 1147.

Dao, T. L. (1964). Carcinogenesis of mammary gland in rat. *Prog. exp. Tum. Res.*, **5**, 157–216.

Dao, T. L. & Gawlik, D. (1963). Direct mammotrophic effect of a pituitary homograft in rats. *Endocrinology*, **72**, 884–892.

Daughaday, W. H., Lowenstein, T. E., Jacobs, L. S., Malarkey, W. B. & Mariz, I. K. (1971). In *Gonadotrophins*, ed. Saxena, S. p. 460. New York: Wiley.

Daughaday, W. H. & Jacobs, L. S. (1972). Normal and pathological secretion of prolactin in man. *Proc. 4th Int. Congr. Endocrinol.* Amsterdam: Excerpta Medica.

Davies, S. L. & Borger, M. I. (1972). The effect of prolactin on sexual function in cattle. *J. Anim. Sci.*, **35**, 239.

Davies, S. L., Reichart, L. E. & Niswender, G. D. (1971). Serum levels of prolactin in sheep as measured by radioimmunoassay. *Biol. Reprod.* **4**, 145.

Dawson, A. B. (1954). Differential staining of two types of acidophil in the anterior pituitary of the rat.

Dawson, A. B. & Friedgood, H. B. (1938). Differentiation of two classes of acidophiles in the anterior pituitary of the female rabbit and cat. *Stain Technol.*, **13**, 17–21.

Debnam, E. S. & Snart, R. S. (1974). Water transport response of toad to prolactin. *Comp. Biochem. Physiol.*, **52A**, 75–76.

Defeo, V. J. (1963). Temporal aspects of uterine sensitivity in the pseudopregnant or pregnant rat. *Endocrinology*, **72**, 305–316.

Deis, R. P. (1968). The effect of an exteroreceptive stimulus on milk ejection in lactating rats. *J. Physiol.*, **197**, 37–46.

Delitalia, G., Masala, A., Alagna, S., Devilla, L. & Lotti, G. (1976). Growth hormone and prolactin release in acromegalic patients following metergoline administration. *J. clin. Endocr. Metab.*, **43**, 1382–1386.

Delitalia, G., Lodico, G., Masala, A., Alagna, S. & Devilla, L. (1977). Action of metergoline suppression of prolactin release induced by mechanical breast emptying. *J. clin. Endocr. Metab.*, **44**, 763–765.

Delouis, C. & Denamur, R. (1972). Induction of lactose synthesis by prolactin in rabbit mammary gland extract. *J. Endocr.*, **52**, 311–319.

Denamur, R. (1969). Comparative aspects of hormonal control of lactogenesis. In *Progress in Endocrinology*, ed. Gual, C. & Ebling, F. G. J. pp. 959–972. Amsterdam: Excerpta Medica.

Denamur, R., Martinet, J. & Short, R. V. (1973). Pituitary control of the ovine corpus luteum. *J. Reprod. Fert.*, **32**, 207–220.

Denenberg, V. H., Taylor, R. E. & Zarrow, M. X. (1969). Maternal behaviour in the rat: An investigation and quantification of nest building. *Behaviour*, **34**, 1–16.

Dent, J. N. (1967). Maintenance of thyroid function in newts with transplanted pituitary glands. *Gen. comp. Endocr.*, **6**, 401–408.

Dent, J. N. (1975). Integumentary effects of prolactin in the lower vertebrates. *Am. Zool.* **15**, 923–937.

Dent, J. N., Eng, L. A. & Forbes, M. S. (1973). Relations of prolactin and thyroid hormones to moulting skin texture and cutaneous secretion in the red spotted newt. *J. exp. Zool.*, **184**, 369–382.

Derby, A. & Etkin, W. (1968). Thyroxine induced tail resorption *in vitro* as affected by anterior pituitary hormones. *J. exp. Zool.*, **169**, 1–8.

Dessaur, H. C. (1955). Effect of season on appetite and food composition of the lizard *Anolis carolinensis*. *J. exp. Zool.*, **128**, 1–12.

Dhariwal, A. P. S., Grosvenor, C. E., Antines-Rodriguez, J. & McCann, S. M. (1968). Studies on the purification of ovine prolactin-inhibiting factor. *Endocrinology*, **82**, 1236–1241.

Dharmamba, M. (1970). Studies on the effect of hypophysectomy and prolactin on plasma osmolarity and plasma sodium in *Tilapia massambica*. *Gen. Comp. Endocr.*, **14**, 256–259.

Dharmamba, M. & Maetz, J. (1972). Effects of hypophysectomy and prolactin on the sodium balance of *Tilapia mossambica* in fresh water. *Gen. comp. Endocr.*, **19**, 175–183.

Dharmamba, M., Handin, R. I., Nandi, J. & Bern, H. A. (1967). Effect of prolactin on fresh water survival and on plasma osmotic pressure of hypophysectomised *Tilapia mossambica*. *Gen. comp. Endocr.*, **9**, 295–303.

Dharmamba, M., Mayer-Gostan, N., Maetz, J. & Bern, H. A. (1973). Effect of prolactin on sodium movement in *Tilapia mossambica* adapted to seawater. *Gen. comp. Endocr.*, **21**, 179–187.

Dickerman, S., Clark, J., Dickerman, E. & Meites, J. (1972). Effects of haloperidol on serum and pituitary prolactin and on hypothalamic PIF in rats. *Neuroendocrinology*, **9**, 332–340.

Dickerman, S., Kledzik, G., Gelate, M., Chen, H. J. & Meites, J. (1974). Effects of haloperidol on serum and pituitary prolactin, LH and FSH and hypothalamic PIF and LRF. *Neuroendocrinology*, **11**, 332–340.

Dilley, W. G. (1971). Morphogenic and mitogenic effects of prolactin on rat mammary gland *in vitro*. *Endocrinology*, **88**, 514–517.

Dixon, J. S. & Li, C. H. (1966). Retention of the biological activity of human pituitary growth hormone after reduction and carboamidomethylation. *Science*, **154**, 785–786.

Dodd, J. M. (1960). Gonadal and gonadotrophic hormones in lower vertebrates. In *Marshalls physiology of reproduction*, ed. Parkes, A. A. Vol. I. pp. 417–582. Boston: Little, Brown & Co.

Dodd, M. H. I. & Dodd, J. M. (1976). The biology of metamorphosis In *The Physiology of Amphibia*, ed. Lofts, B. London and New York: Academic Press.

Doerr-Schott, J. (1966). Etude aux microscopes optique et electronique des differents types de cellules de la pars distalis et de la pars intermedia de *Triturus marmoralis. Latr. Ann. Endocrinol. Paris*, **27**, 101–109.

Doerr-Schott, J. (1968). Cytologie et cytophysiologie de l'adenohypophyse des amphibiens. *Ann. Biol.*, **7**, 189–225.

Dohler, K. D. & Wuttke, W. (1974). Total blockade of phasic pituitary prolactin release in rats: effect on serum LH and progesterone during the oestrus cycle and pregnancy. *Endocrinology*, **94**, 1595–1600.

Dolnik, V. R. (1961). Mekhanizm enereticheskoi podgotovki ptitisk pereletic in Jaktory, se opredelyayeschie. In *Ekologyia i migratsü ptits Pribattiki*, pp. 281–288. Akad. Nauk. lat. SSR.

Domanski, E., Skreczkowski, E., Stupnicka, E., Fitko, R. & Dobrowolski, W. (1967). Effect of gonadotrophins on the secretion of progesterone and oestrogens by the sheep ovary perfused *in situ*. *J. Reprod. Fert.*, **14**, 365–372.

Doneen, B. A. (1974). Bioassay of prolactins and structurally related hormones in organ cultured teleost urinary bladder. *Am. Zool.*, **14**, 1245 Abstr.

Doneen, B. A. (1976). Water and ion movements in the urinary bladder of the gobiid teleost *Gillichthys mirabilis* in response to prolactins and cortisol. *Gen. comp. Endocrin.*, **28**, 33–41.

Doneen, B. A. & Bern, H. A. (1974). *In vitro* effects of prolactin and cortisol on water permeability of the urinary bladder of the teleost *Gillichthys mirabilis*. *J. exp. Zool.*, **187**, 173–179.

Doneen, B. A. & Nagahama, Y. (1973). Roles of prolactin and cortisol in osmoregulatory functions of the urinary bladder of the euryhaline teleost *Gillicthys mirabilis*. *Am. Zool.*, **13**, 1278 Abstr.

Donofrio, R. J. & Reiter, R. J. (1972). Depressed pituitary prolactin levels in blinded anosmic female rats: role of the pineal gland. *J. Reprod. Fert.*, **31**, 159.

Donoso, A. O., Banzan, A. M. & Barcaglioni, J. C. (1974). Further evidence on the direct action of L-dopa on prolactin release. *Neuroendocrinology*, **15**, 236–239.

Donoso, A. O., Bishop, W. & McCann, S. M. (1973). The effects of drugs which modify catecholamine synthesis on serum prolactin in rats with median eminence lesions. *Proc. Soc. Exp. Biol. Med.*, **143**, 360–363.

Donovan, B. T. (1963). Pituitary stalk section and corpus luteum function in the ferret. *J. Physiol.*, **165**, 23P-24P.

Douglas, D. S. & Neely, S. M. (1969). The effect of dehydration on salt gland performance. *Am. Zool.*, **9**, 1095.

Drouin, J., De Lean, A., Rainville, D., LaChance, R. & LaBrie, F. (1976). Characteristics of the interaction between thyrotropin releasing-hormone and somatostatin for thyrotropin and prolactin release. *Endocrinology*, **98**, 514.

Duff, D. W. & Fleming, W. R. (1972*a*). Sodium metabolism of the freshwater cyprinodont, *Fundulus catenatus. J. comp. Physiol.*, **80**, 179–191.

Duff, D. W. & Fleming, W. R. (1972*b*). Some aspects of sodium balance in the freshwater cyprinodont, *Fundulus olivaceus. J. comp. Physiol.*, **80**, 191–201.

Dular, R., LaBella, F., Vivian, S. & Eddie, L. (1974). Purification of prolactin releasing and inhibiting factors from beef. *Endocrinology*, **94**, 563.

Dumont, J. N. (1965). Prolactin induced cytologic changes in the mucosa of the pigeon crop during crop 'milk' formation. *Z. Zellforsch. mikrosk. Anat.*, **68**, 755–782.

Dunn, J. D., Arimura, A. & Scheving, L. E. (1972). Effects of stress on circadian periodicity in serum Lh and prolactin concentration. *Endocrinology*, **90**, 29.

Dussault, J. H., Turcotte, R., & Guyda, H. (1976). The effect of acetylsalicylic acid on TSH and PRL secretion after TRH stimulation in the human. *J. clin. Endocr. Metab.*, **42**, 232–235.

Dusseau, J. W. & Meier, A. H. (1971). Diurnal and seasonal variations of plasma adrenal steroid hormone in the white throated sparrow *Zonotrichia albicolis. Gen. comp. Endocr.*, **16**, 399–408.

Eayrs, J. T. & Baddley, R. M. (1956). Neural pathways in lactation. *J. Anat.*, **90**, 161–171.

Edwards, C. R. W., Forsyth, I. A. & Besser, G. (1971). Amenorrhoea, galactorrhoea and primary hypothyroiism with high circulating levels of prolactin. *Brit. med. J.*, **3**, 462.

Edwards, R. P., Diver, M. J. & Davies, J. C. (1976). Plasma oestriol and human placental lactogen measurements in patients with high risk pregnancies. *Br. J. Obstet. Gynaec.*, **83**, 229–237.

Ehara, Y., Yen, S. C. C. & Siler, T. M. (1975). Serum prolactin levels during puberty. *Am. J. Obstet. Gynec.*, **121**, 995–997.

Ehara, Y., Silver, T., Vandenberg, G., Sinha, Y. N. & S. S. C. (1973). Circulating prolactin levels during the menstrual cycle: episodic release and diurnal variation. *Am. J. Obstet. Gynec.*, **117**, 962.

Ehni, G. & Eckles, N. E. (1959). Interruption of the pituitary stalk in the patient with mammary cancer. *J. Neurosurg.*, **16**, 628–652.

Eisner, E. (1960). The relationship of hormones to the reproductive behaviour of birds, referring especially to parental behaviour. A review. *Anim. Behav.*, **8**, 153–181.

El-Minairi, M. F. & Foda, G. M. (1971). Post-partum lactation and amenorrhoea. Endometrial pattern and reproductive ability. *Am. J. Obstet. Gynec.*, **111**, 17–21.

Emmart, E. W. (1969). The localisation of endogenous 'prolactin' in the pituitary gland of the goldfish *Carassius auratus. Gen. comp. Endocr.*, **11**, 550–594.

Ensor, D. M. (1975). Prolactin and adaptation. *Symp. zool. Soc. Lond.*, **35**, 129–148.

Ensor, D. M. & Ball, J. N. (1968*a*). A bioassay for fish prolactin (paralactin). *Gen. comp. Endocr.*, **11**, 104–110.

Ensor, D. M. & Ball, J. N. (1968*b*). Prolactin and freshwater sodium fluxes in *Poecilia latipinna* (Teleostei). *J. Endocr.* **41**, xvi.

Ensor, D. M. & Ball, J. N. (1972). Prolactin and osmoregulation in fishes. *Fedn. Proc.*, **31**, 1615–1622.

Ensor, D. M. & Phillips, J. G. (1970). The effect of salt loading on the pituitary prolactin levels of the duck *Anas platyrhynchos* and juvenile herring or lesser black backed gulls (*Larus argentatus* or *Larus fuscus*). *J. Endocr.*, **48**, 167–172.

Ensor, D. M. & Phillips, J. G. (1971). The effect of environmental stimuli on the circadian rhythm of prolactin production in the duck (*Anas platyrhynchos*). *J. Endocr.*, **48**, lxxi–lxxii.

Ensor, D. M., Phillips, J. G. & O'Halloran, M. J. (1976). The effect of extreme cold stress on nasal gland function in the domestic duck *Anas platyrhynchos*. *Gen. comp. Endocr.*, **31**, 317–329.

Ensor, D. M. & Phillips, J. G. (1972*a*). The effect of age and environment on extrarenal salt excretion in juvenile gulls (*Larus argentatus* or *Larus fuscus*). *J. Zool., Lond.* **168**, 119–126.

Ensor, D. M. & Phillips, J. G. (1972*b*). The effect of dehydration on salt and water balance in gulls (*Larus argentatus* or *Larus fuscus*). *J. Zool., Lond.*, **168**, 127–137.

Ensor, D. M., Edmondson, M. R. & Phillips, J. G. (1972*a*). Prolactin and dehydration in rats. *J. Endocr.*, **53**, lix–lx.

Ensor, D. M., Simons, I. M. & Phillips, J. G. (1972*b*). The effect of hypophysectomy and prolactin replacement on salt and water metabolism in *Anas platyrhynchos*. *J. Endocr.*, **57**, xi.

Ensor, D. M., Thomas, D. H. & Phillips, J. G. (1971). The possible role of the thyroid in extrarenal secretion following a hypertonic salt load in the duck *Anas platyrhynchos*. *J. Endocr.*, **48**, lxxi–lxxi.

Erdheim, J. & Stumme, E. (1909). Beitr. path. Anat., **46**, 1.

Erlig, D. (1972). Salt transport across isolated frog skin. *Proc. R. Soc. B*, **262**, 153–163.

Espinosa-Campos, J., Butler, W. R. & Knobil, E. (1975). Inhibition of corpus luteum function in the rhesus monkey by 2 brom-α-ergocryptine (CB 154). p. 63. *57th Ann. Mtg, Endocr. Soc.* New York.

Etkin, W. (1966). How a tadpole becomes a frog. *Scient. Am.*, **214**, 76–88.

Etkin, W. (1970). The endocrine mechanism of amphibian metamorphosis, an evolutionary achievement. *Mem. Soc. Endocr.*, **18**, 137–155.

Etkin, W. & Gona, A. G. (1967). Antagonism between prolactin and thyroid hormone in amphibian development. *J. exp. Zool.*, **165**, 249–258.

Etkin, W. & Lehrer, R. (1960). Excess growth in tadpoles after transplantation of the adenohypophysis. *Endocrinology*, **67**, 457–466.

Ettigi, P., Lal, S. & Friesen, H. (1973). Prolactin, phenothiazines, admission to mental hospital and breast cancer. *Lancet*, **2**, 266.

Euker, J. E., Meites, J. & Reigle, G. D. (1973). Serum LH and prolactin following restraint stress in the rat. *Physiologist, Wash.*, **16**, 307.

Euker, J. E., Meites, J. & Reigle, G. D. (1975). Effect of acute stress on serum LH and prolactin in intact, castrate and dexamethasone treated rats. *Endocrinology*, **96**, 85–92.

Evans, A. J. (1962). The *in vitro* effect of B-glucuronidase in the testis of the rat. *J. Endocr.*, **24**, 233–244.

Evans, A. J. (1966). Further investigations of an assay *in vitro* for prolactin. *J. Endocr.* **34**, 319–328,

Evans, H. M., Simpson, M. E., Lyons, W. R. & Turpeinen, K. (1941). Anterior pituitary hormones which favour the production of traumatic uterine placentomata. *Endocrinology*, **28**, 933–945.

Everett, J. W. (1952). Presumptive hypothalamic control of spontaneous ovulation. *Ciba Fdn. Colloq. Endocr.*, **4**, 167–177.

Everett, J. W. (1954). Luteotrophic function of autografts of the rat hypophyses. *Endocrinology*, **54**, 685–690.

Everett, J. W. (1963). Pseudo-pregnancy in the rat from brief treatment with progesterone: effect of isolation. *Nature, Lond.*, **198**, 695–696.

Everett, J. W. (1964). Central neural control of reproduction functions of the hypothalamus. *Physiol. Rev.*, **44**, 373–431.

Everett, J. W. (1967). Provoked ovulation or long delayed pseudo-pregnancy from coital stimuli in barbiturate blocked rats. *Endocrinology*, **80**, 145–154.

Everett, J. W. & Nikitowitch-Winer, M. (1963). In *Advances in Neuroendocrinology*, ed. Nalbandov, A. V. p. 289. Urbana, Illinois: Univ. Illinois Press.

Eyeson, K. N. (1970). Cell types in the distal lobe of the pituitary of the West African rainbow lizard, *Agama agama* L. *Gen. comp. Endocr.*, **14**, 357–367.

Faglia, G., Ferrari, C., Beck-Rossa, P., Spada, A., Travagliani, P. & Ambrose, B. (1973). Reduced plasma thyrotropin response to thyrotropin releasing hormone after dexomethasone administration in normal subjects. *Horm. Metab. Res.*, **5**, 289.

Fajer, A. B. (1971). Loci of action of prolactin and luteinising hormone in the hamster ovary during lactation; the interstitial tissue. In *Symp. Proc. 7th Pan Am. Congr. Endocr.* Int. Congr. Ser. **99**, 128–133. Excerpta Medica Fdn.

Fajer, A. B. & Barraclough, C. A. (1967). Ovarian secretion of progesterone and 20-alpha-hydroxypregn-4-en-3-one during pseudo-pregnancy and pregnancy in rats. *Endocrinology*, **81**, 617–622.

Falconer, I. R. (1972). Uptake and binding of ^{125}I-labelled prolactin by mammary tissue. *Biochem. J.*, **126**, 8P–9P.

Falconer, I. R. & Fiddler, T. J. (1970). Effects of intraductal administration of prolactin, actinomycin D and cyclohexamide on lipoprotein lipase activity in the mammary glands of pseudopregnant rabbits. *Biochim. Biophys. Acta*, **218**, 508–514.

Falconer, I. R. & Rowe, J. M. (1975). Possible mechanism for action of prolactin on mammary cell sodium transport. *Nature*, **256**, 327–328.

Fang, V. S. & Kim, M. H. (1975). Study on maternal, fetal and amniotic prolactin at term. *J. clin. Endocr. Metab.*, **41**, 1030–1034.

Farmer, S. W. (1970). In vivo *and* in vitro *studies on luteal function in rats*. Ph.D. Thesis, Berkeley, University of California.

Farmer, S. W., Bewley, J. A., Russel, S. M. & Nicoll, C. S. (1976). Comparison of secreted and extracted forms of rat pituitary prolactin. *Biochim. biophys. Acta*, **437**, 562–570.

Farmer, S. W., Papkoff, H., Bewley, T. A., Hagashida, T., Nishioker, R. S., Bern, H. A. & Li, C. H. (1977). Isolation and properties of teleost prolactin. *Gen. comp. Endocr.*, **31**, 60–71.

Farner, D. S. (1955). The annual stimulus for migration: experimental and physiologic aspects. In *Recent Studies in Avian Biology*, ed. Wolfson, A., pp. 198–238. University of Illinois.

Farner, D. S. (1960). Metabolic adaptations in migration. *Proc. Int. Ornithol. Congr.*, Helsinki, 197–208.

Farooq, A., Denenberg, V. H., Ross, S., Savin, P. B. & Zarrow, M. X. (1969). Maternal behaviour in the rabbit: Endocrine factors involved in hair loosening. *Am. J. Physiol.*, **204**, 271.

Feder, H. H., Resko, J. A. & Gay, R. W. (1968). Progesterone levels in arterial plasma of preovulatory and ovariectomised rats. *J. Endocr.*, **41**, 563–569.

Felker, J. P., Zaragossa, N. & Benuzzi-Badoni, J. (1972). The double effect of human chorionic sematomammotrophin (HCG) and pregnancy on lipogenesis and on lipolysis in the rat epididymal fat pad and fat pad cells. *Horm. Metab. Res.*, **4**, 293–296.

Ferguson, K. A. & Wallace, A. L. S. (1961). Prolactin activity of human growth hormone. *Nature, Lond.*, **190**, 632–633.

Fiddler, T. J., Birkinshaw, M. & Falconer, I. R. (1971). Effects of intraductal prolactin on some aspects of the ultrastructure and bichemistry of mammary tissue in the pseudopregnant rabbit. *J. Endocr.*, **49**, 459–469.

Fiedler, K. (1962). Die wirkung von Prolactin auf das Verhalten des Lippfisches, *Crenilabrus ocellatus* (Forskal). *Zool. Jb. Physiol. Bd.*, **69**, 609–620.

Findlay, A. L. R. (1968). The effect of teat anaesthesia on the milk ejection reflex in the rabbit. *J. Endocr.*, **40**, 127–128.

Fitzgerald, D. K., Brodbeck, U. & Kigoshawa, I. (1970). Alpha lactalbumin and the lactose synthetase reaction. *J. biol. Chem.*, **245**, 2103–2108.

Fleming, W. R. & Ball, J. N. (1972). The effect of prolactin and ACTH on the sodium metabolism of *Fundulus kansae* held in deionised water, sodium enriched seawater and concentrated seawater. *Z. Vergl. Physiol.*, **76**, 125–134.

Flint, D. J. & Ensor, D. M. (1976). Ergocryptine and prolactin replacement therapy on lactation and pregnancy in the rat. *J. Endocr.*, **71**, 41P.

Floquet, A., Grignon, G. & Legait, E. (1963). Etude morphologique de l'évolution du follicle post-ovulaire chez la poule. *Gen. comp. Endocr.*, **3**, 699.

Florini, J. R., Tonelli, G., Breur, C. B., Coppola, J., Ringler, I. & Bell, P. H. (1966). Characterisation and biological effects of purified placental protein (human). *Endocrinology*, **79**, 692–708.

Fontaine, M. (1956). The hormonal control of water and salt-electrolyte metabolism in fish. *Mem. Soc. Endocr.*, **5**, 69–82.

Forbes, M. S., Dent, J. N. & Singhas, C. A. (1975). The developmental cytology of the nuptial pad in the red spotted newt. *Devl Biol.*, **46**, 56–78.

Ford, J. J. & Yoshinga, K. (1975). The role of prolactin in the luteotrophic process of lactating rats. *Endocrinology*, **96**, 335–339.

Forster, R. C. (1975). Changes in the urinary bladder and kidney function in the starry flounder (*Platicthys stellatus*) in response to prolactin and freshwater transfer. *Gen. comp. Endocr.*, **27**, 116–120.

Forsyth, I. A. (1967). In *Hormones in Blood*, ed. Gray, C. H. & Bacharach, A. L. Vol. I, 2nd edn. pp. 233–272. London: Academic Press.

Forsyth, I. A. (1973). Secretion of a prolactin like hormone by the placenta in ruminants. In *Le Corps Jaune*. pp. 239–255. Colloque de la Société Nationale pour l'étude de la Sterilité et de Fecondité, Paris.

Forsyth, I. A. (1974). In *Lactogenic Hormones, Fetal Nutrition and Lactation*, ed. Josimovich, J. B. p. 49. New York: John Wiley.

Forsyth, I. A. & Blake, L. A. (1976). Placental lactogen (chorionic mammotrophin) in the field vole, *Microstus agreslis* and the bank vole, *Cleithrionmys glareolus*. *J. Endocr.*, **70**, 19–23.

Forsyth, I. A., Strong, C. R. & Dibs, R. (1972). Interaction of insulin, corticosterone and prolactin in promoting milk fat synthesis by mammary explants from pregnant rabbits. *Biochem. J.*, **127**, 929–935.

Fortier, C., Delgado, A. & Ducommon, P. (1970). Functional interrelationships between the adenohypophysis, thyroid, adrenal cortex and gonads. *Can. med. Assoc. J.*, **103**, 864–874.

Foukas, M. D. (1973). An antilactogenic effect of pyridoxine. *J. Obstet. Gynaec. Br. Commonw.*, **80**, 718.

Fox, W. & Dessaur, H. C. (1958). Photoperiodic stimulation of appetite and growth in the male lizard, *Anolis carolinensis*. *J. exp. Zool.*, **134**, 557–575.

Franks, S. & Brooks, C. G. D. (1976). Basal and stimulated prolactin levels in childhood. *Horm. Res.*, **7**, 65–76.

Franks, S., Jacobs, H. S. & Nabarro, J. D. H. (1975*a*). Studies of prolactin secretion in pituitary disease. *J. Endocr.*, **67**, 55p.

Franks, S., Jacobs, H. S. & Nabarro, J. D. H. (1976). Prolactin concentrations in patients with acromegaly; clinical significance and response to surgery. *Clin. Endocr.*, **5**, 63.

Franks, S., Murray, M. A. F., Jequier, A. M., Steele, S. J., Nabarro, J. D. N. & Jacobs, H. S. (1975*b*). Incidence and significance of hyperprolactinemia in women with amenorrhea. *Clin. Endocr.*, **4**, 597.

Frantz, A. G. (1976). Prolactin, growth hormone and human placental lactogen. In *Peptide hormones*, ed. Parsons, J. A., pp. 199–229. London: Macmillan.

Frantz, A. G. & Kleinberg, D. L. (1970). Prolactin: evidence that it is separate from growth hormone in human blood. *Science*, **170**, 745–747.

Frantz, A. G., Rabkin, M. T. & Friesen, H. (1965). Human placental lactogen in choriocarcinoma of the male. Measurement by radioimmunoassay. *J. clin. Endocr. Metab.*, **25**, 1136.

Frantz, A. G., Kleinberg, D. L. & Noel, G. L. (1972*a*). Studies on prolactin in man. *Rec. Prog. Horm. Res.*, **28**, 527–573.

Frantz, A. G., Kleinberg, D. L. & Noel, G. L. (1972*b*). Physiological and pathological secretion of human prolactin studied by *in vitro* bioassay. In *Lactogenic Hormones*, ed. Wolstenholme, G. E. & Knight, J. pp. 137–150. London: Ciba Foundation.

Frantz, A. G., Habif, D. V., Hyman, G. A., Such H. K., Sassin, J. F., Zimmerman, E. A., Noel, G. L. & Kleinberg, D. L. (1973). Physiological and pharmacological factors affecting prolactin secretion, including its suppression by L-dopa in breast cancer. In *Human Prolactin*, ed. Pasteels, J. L. & Robyn, C. p. 273. Amsterdam: Exerpta Medica.

Frantz, W. L. & Rillema, J. A. (1968). Prolactin stimulated uptake of amino acids – ^{14}C and ^{3}HOH in pigeon crop mucosa. *Am. J. Physiol.*, **215**, 762–767.

Frantz, W. L. & Turkington, R. W. (1972). In *Prolactin and carcinogenesis*, Proc. 4th Tenovus Worshop, ed. Boyns. A. R. & Griffiths, K. pp. 47–63. Cardiff: Alpha Omega Alpha.

Fredrickson, H. (1939). Endocrine factors involved in the development and function of the mammary glands of female rabbits. *Acta Obstet. Gynec. scand.*, **19**, Suppl. 1.

Freeman, J. J., Hilf, R. & Iovino, A. J. (1964). Effects of steroids upon the weight, protein and nucleic acid concentration of the preputial glands of the female rat. *Endocrinology*, **74**, 990–993.

Freeman, M. E. & Niell, J. D. (1972). The pattern of prolactin secretion during pseudo-pregnancy in the rat: a daily nocturnal surge. *Endocrinology*, **90**, 1292.

Freeman, M. E., Smith, M. S. & Neill, J. D. (1974). Prolactin secretion during pseudo-pregnancy in the rat: ovarian control. *Endocrinology*,

Freeman, M. E., Dupke, K. C. & Croteau, C. M. (1976). Extinction of the estrogen induced daily signal for LH release in the rat: a role for the proestrus surge of progesterone. *Endocrinology*, **99**, 223–229.

Freyer, M. E. & Evans, H. M. (1923). Participation of the mammary gland in the changes of pseudo-pregnancy in the rat. *Anat. Rec.*, **25**, 108.

Friedgood, H. B. & Dawson, A. B. (1940). Physiological significance and morphology of the carmine cell in the cat's anterior pituitary. *Endocrinology*, **26**, 1022–1031.

Friesen, H. G. (1965). Purification of a placental hormone with immunological and chemical similarity to human growth hormone. *Endocrinology*, **76**, 369–381.

Friesen, H. G. (1966). Lactation induced by human placental lactogen and cortisone acetate in rabbits. *Endocrinology*, **79**, 212–215.

Friesen, H. G. (1968). Biosynthesis of placental proteins and placental lactogen. *Endocrinology*, **83**, 744–753.

Friesen, H. G. (1974). Placental lactogens. In *American Handbook of Physiology*. Sect. 7. Vol. 2. Part II. American Society of Physiologists.

Friesen, H. G. & Guyda, H. (1971). Biosynthesis of monkey growth hormones and prolactin *in vitro*. *Endocrinology*, **88**, 1353–1362.

Friesen, H. G. & Hwang, P. (1973). Human prolactin. *A. Rev. Med.*, **24**, 251–270.

Friesen, H. G., Suwa, S. & Parc, P. (1969). Synthesis and secretion of placental lactogen and other proteins by the placenta. *Rec. Prog. Horm. Res.*, **25**, 161–205.

Friesen, H. G., Guyda, H. & Hardy, J. (1970). Biosynthesis of human growth hormone and prolactin. *J. clin. Endocr. Metab.*, **31**, 611–624.

Friesen, H. G., Guyda, H. & Hwang, P. (1971). Synthesis of primate prolactin. *Nature*, **232**, 19–20.

Friesen, H. G., Hwang, P. & Guyda, H. (1972*a*). In *Prolactin and Carcinogenesis*. Proc. 4th Tenovus Workshop, ed. Boyns, A. R. & Griffiths, K. pp. 64–80. Cardiff: Alpha Omega Alpha.

Friesen, H. G., Belanger, C., Guyda, H. & Hwang, P. (1972*b*). The synthesis and secretion of placental lactogen and pituitary prolactin. In *Lactogenic Hormones*, ed. Wolstenholme, G. E. & Knight, J. pp. 83–103. London: Ciba Foundation.

Friesen, H. G., Guyda, H., Hwang, P., Tyson, J. E. & Barbeau, A. (1972*c*). Functional evaluation of prolactin secretion. A guide to therapy. *J. clin. Invest.*, **51**, 706–709.

Friesen, H. G., Webster, B. R., Hwang, P., Guyda, H., Munro, R. E. & Read, L. (1972*d*). Prolactin secretion and synthesis in a patient with Forbes-Albright syndrome *J. clin. Endocr. Metab.*, **34**, 192–199.

Frohman, L. A. & Szabo, M. (1975). 57th Ann. Mtg Endocr. Soc., 93A. New York. Abstr.

Furnham, F. A. & Ussing, H. H. (1950). Characteristic response of isolated frog skin potential to neurohypophysial principles and its relation to transport of sodium and water. *Fedn Proc.*, **9**, 46.

Furth, J., Clifton, K. H. & Gadsden, E. L. (1956). Dependant and autonomous mammotropic pituitary tumours in rats: their somatotropic features *Cancer Res.*, **16**, 608.

Fuxe, K., Hokfelt, T. & Nilson, O. (1969). Castration, sex hormones and tuberoinfundibular dopamine neurons. *Neuroendocrinology*, **5**, 107–120.

Gachev, E. (1963). Rol'prolaktina v podderzhaini urovynya laktozȳ v moloka. *Zh. obshch. Biol.*, **24**, 382–383.

Gachev, E. (1968). Duration of prolactin – induced lactose synthesis. *C.R. Acad. bulg. Sci.*, **21**, 577–579.

Gadklari, S. V., Chapekar, T. N. & Ranadive, K. J. (1968). Response of mouse mammary gland to hormonal treatment. *Indian J. Exp. Biol.*, **6**, 75–79.

Gala, R. R. & Reece, R. P. (1965). Influence of hypothalamic fragments and extracts on lactogen production *in vitro*. *Proc. Soc. exp. Biol. Med.*, **117**, 833.

Gala, R. R., Subraimanian, M. G., Peters, J. A. & Jaques, S. (1976). Influence of cholinergic receptor blockade on drug induced prolactin release in the monkey. *Horm. Res.*, **7**, 118–128.

Galgano, M. (1942). Richerche sperimentali intoro al ciclo sessionale stagionale di *Triton cristatus* laur. VI. Intoro al determinismo dei caratheri sesuali stagionali nelle femmine. Icaratheri ambesessuali. *Monit. Zool. ital.*, **53**, 1–12.

Garrison, M. M. & Scow, R. O. (1973). Effect of prolactin on lipoprotein lipase in crop sac and adipose tissue of pigeons. *Am. J. Physiol.*, **228**, 1542–1544.

Gartside, M. W. & Tindall, J. S. (1975). The prognostic value of human placental lactogen (HPL) levels in threatened abortion. *Br. J. Obstet. Gynaec.*, **82**, 303–309.

Gautvik, K. M., Tashjian, A. H., Kourides, I. A., Weintraub, B. D., Gracker, C. T., Maloof, F., Susuki, K. & Zuckerman, J. E. (1974). Thyrotropin releasing factor is not the sole mediator of prolactin release during suckling. *New Engl. J. Med.*, **290**, 1162–1165.

Gay, V. L., Midgeley, A. R. & Niswender, G. D. (1970). Patterns of gonadotrophin secretion associated with ovulation. *Fedn Proc.*, **29**, 1880.

Genazzani, A. R., Benuzzi-Badeni, M. & Felber, J. P. (1969). Human chorionic somatomammotrophine (HCSM): Lipolytic action of a pure preparation on isolated fat cells. *Metabolism*, **18**, 121–127.

Giunta, C., Campantico, R. & Guardabassi, C. (1973). Biochemical response elicited in normal or prolactin treated *Xenopus laevis* (Daudin) specimens by environmental changes. *Comp. Biochem. Physiol.*, **47A**, 323–331.

Giunta, C., Campantico, E., Vietti, M. & Guastalla, A. (1972). Prolactin stabilising action on the lysosomes in the tail and gut from *Rana temporaria* tadpoles at prometamorphosis. *Gen. comp. Endocr.*, **18**, 568–571.

Glass, M. R., Shaw, R. W., Butt, W. R., Logan Edwards, R. & London, D. R. (1975). An abnormality of oestrogen feedback in amenorrhoea – galactorrhoea. *Br. med. J.*, 3, 274.

Glass, M. R., Williams, J. W., Butt, W. R., Logan Edwards, R. & London, D. R. (1976). Basal serum prolactin values and responses to the administration of thyrotrophin releasing hormone (TRH) in women with amenorrhoea. *Br. J. Obstet. Gynaec.*, 83, 495–501.

Glass, R. L., Troolim, H. A. & Jenness, R. (1967). Comparative biochemical studies of milks IV. Constituent fatty acids of milk fats. *Comp. Biochem. Physiol.*, **22**, 415–425.

Golder, M. P., Boyns, A. R., Harper, M. E. & Griffiths, K. (1972). An effect of prolactin on prostatic adenylate cyclase activity. *Biochem. J.*, **128**, 725–727.

Golubuff, C. G. & Ezrin, C. (1969). Effect of pregnancy on the somatotroph and prolactin cell of the human adenohypophysis. *J. clin. Endocr. Metab.*, **29**, 1533–1538.

Gona, A. G. (1967). Prolactin as agoitrogenic agent in amphibia. *Endocrinology*, **81**, 748–754.

Gona, A. G. (1968). Radioiodine studies of prolactin action in tadpoles. *Gen. comp. Endocr.*, **11**, 278-283.

Gona, A. G. (1971). Prolactin induced stimulation of ^{3}H-proline incorporated into the basal lamella of tadpole skin: light and electron microscopic study. *Tiss. Cell*, **3**, 557–566.

Gona, A. G. & Etkin, W. (1970). Inhibition of maetamorphosis in *Ambystoma tigrinum* by prolactin. *Gen. comp. Endocr.*, **14**, 589–591.

Gona, A. G., Pearlman, T. & Etkin, W. (1970). Prolactin thyroid interaction in the newt, *Diemictylus viridescens. J. Endocr.*, **48**, 585–590.

Gona, A. G., Pearlman, T. & Gona, O. (1973). Effects of prolactin and thyroxine in hypophysectomised and thyroidectomised red efts of the newt *Notophthalmus* (*Diemictylus*) *viridescens. Gen. comp. Endocr.*, **21**, 377–380.

Goodridge, A. G. (1964). *The effect of hormones on lipid metabolism and related metabolic activity in some passerine birds.* Ph.D. Thesis, University of Michigan, Ann Arbor, Michigan.

Goodridge, A. G. & Ball, E. C. (1967). The effect of prolactin on lipogenesis in the pigeon. *In vitro* studies. *Biochemistry*, **6**, 2335–2343.

Gorski, M. & Lawton, I. E. (1972). Timing of the prepubertal adrenal-ovarian interaction in the rat. *Biol. Reprod.*, **7**, 112.

Gourdji, D. (1970). Prolactine et relations photosexuelles chez les oiseaux. *C.N.R.S. Colloq. Intern. Paris*, **172**, 233–258.

Gourdji, D. & Tixier-Vidal, A. (1966). Variations du contenu hypophysaire en prolactine chez le canard Pekin male au cours du cycle sexuelle et de la photostimulation experimentale. *C.r. Acad. Sci., Paris*, D **262**, 1746–1749.

Govan, A. D. T. & Mukherjee, C. L. (1950). Maternal toxaemia and foetal ovarian function. *J. Obstet. Gynaec. Br. Commonw.*, **57**, 525.

Grandison, L. & Meites, J. (1972). Luteolytic action of prolactin during oestrous cycle of the mouse. *Proc. Soc. exp. Biol. Med.*, **140**, 32–325.

Grant, W. C. (1961). Special aspects of the metamorphic process, second metamorphosis. *Am. Zool.*, **1**, 163–171.

Grant, W. C. & Banks, P. M. (1968). *Bull. Mt. Desert Island Biol. Labs*, **8**, 31.

Grant, W. C. & Cooper, G. (1965). Behaviour and integumentary changes associated with induced metamorphosis in Diemictylus. *Biol. Bull.*, **129**, 510–522.

Grant, W. C. & Grant, J. A. (1958). Water drive studies on hypophysectomised efts of *Diemictylus viridescens.* I. The role of the lactogenic hormone. *Biol. Bull.*, **114**, 1–9.

Grant, D. B., Kaplan, S. L. & Grumbach, M. M. (1970). Studies on a monkey placental protein with immunochemical similarity to human growth hormone and human chorionic sommatomammotrophin. *Acta. Endocrinol.* (*Kbh.*), **63**, 736–746.

Grant, W. C. & Pickford, G. E. (1959). Presence of the red eft water drive factor (prolactin) in the pituitaries of teleosts. *Biol. Bull.*, **116**, 429–435.

Grayhack, J. T. (1963). Pituitary factors influencing the growth of the prostate. *Natl Cancer Inst. Monographs*, **12**, 189–199.

Grayhack, J. T. & Lebowitz, J. M. (1967). Effect of prolactin on citric acid of the lateral lobe of the prostate of Sprague-Dawley rats. *Inves. Urol.*, **5**, 87–94.

Grayhack, J. T., Bunce, P. I., Kearns, J. W. & Scott, W. W. (1955). Influence of the prostatic response to androgen in the rat. *Bull. John Hopkins Hosp.*, **96**, 154–164.

Greenwald, G. S. (1969). Evidence for a luteotrophic complex in the hamster and other species. In *Progress in endocrinology*, ed. Gual, C. pp. 921-926. Amsterdam: Excerpta Medica Foundation.

Greenwald, C. S. & Rothchild, I. (1968). Prolactin and corpus luteum function in rats. *J. Anim. Sci.*, **27**, Suppl. 139.

Greenwood, F. C., Bryant, G. D., Siler, T., Morgenstem, B., Robyn, C., Hubinot, P. O. & Pasteels, J. L. (1971). Human prolactin radioimmunoassay preliminary report. In *Radioimmunoassay Methods*, ed. Kirkham, K. E. & Hunter, W. M. pp. 218–222. Edinburgh & London: Churchill Livingstone.

Greep, R. O. & Hisaw, F. L. (1938). Pseudo-pregnancies from electrical stimulation of the cervix in dioestrus. *Proc. Soc. exp. Biol. Med.*, **39**, 359–360.

Greibrokk, T., Currie, B. L. & Johanson, N. K. (1974). Purification of a prolactin-inhibiting hormone and the revealing of hormone D-GH 1H which inhibits the release of growth hormone. *Biochem. Biophys. Res. Comm.*, **59**, 704–709.

Greibrokk, T., Hausen, J. & Knudsen, R. (1975). On the isolation of a prolactin inhibiting factor (hormone). *Biochem. Biophys. Res. Comm.*, **67**, 338–334.

Griffiths, R. W. (1974). Pituitary control of adaptation to freshwater in the teleost genus *Fundulus*. *Biol. Bull.*, **146**, 357–376.

Grignon, G. (1963). Cytophysiologie del'adenohypophyse des reptiles. *C.N.R.S. Colloq. Intern. Paris*, **128**, 287–300.

Grosvenor, C. E. (1965). Effect of nursing and stress upon prolactin inhibiting activity of the rat hypothalamus. *Endocrinology*, **77**, 1037.

Grosvenor, C. E. & Mena, F. (1973). Evidence for a time delay between prolactin release and the resulting rise in milk secretion rate in the rat. *J. Endocr.*, **58**, 31–39.

Grosvenor, C. E. & Turner, C. W. (1958). Pituitary lactogenic hormone concentration and milk secretion in lactating rats. *Endocrinology*, **63**, 535.

Grosvenor, C. E. & Turner, C. W. (1960). Pituitary lactogenic hormone concentration during pregnancy in the rat. *Endocrinology*, **66**, 96–99.

Grosvenor, C. E., McCann, S. M. & Naller, R. (1965). Inhibition of nursing induced and stress induced fall in pituitary prolactin concentration in lactating rats by injection of acid extracts of bovine hypothalamus. *Endocrinology*, **76**, 883–889.

Grosvenor, C. E. & Whitworth, H. S. (1976). Incorporation of rat prolactin into rat milk *in vivo* and *in vitro*. *J. Endocr.*, **70**, 1–11.

Grota, L. J. & Eik-Nes, K. B. (1967). Plasma progesterone concentrations during pregnancy and lactation in the rat. *J. Reprod. Fert.*, **13**, 83–91.

Grumbach, M. M., Kaplan, S. L., Sciarra, J. T. & Burr, I. M. (1968). Chorionic growth hormone prolactin (CGP) secretion, disposition, biological activity in man and postulated function as the growth hormone of the second half of pregnancy. *Ann. N.Y. Acad. Sci.*, **148**, 501–531.

Gua, C. & Chard, T. (1975). The distribution of placental lactogen in the human term placenta. *Br. J. Obstet. Gynaec.*, **82**, 790–793.

Gul, B. & Dils, R. (1969). Enzymic changes in rabbit and rat mammary gland during the lactation cycle. *Biochem. J.*, **112**, 293–301.

Guraya, S. S. (1975). Histochemical observations on the lipid changes in rat corpora lutea during various reproductive states after treatment with hormones. *J. Reprod. Fert.*, **43**, 67–75.

Gusden, J. P., Leake, N. H. Van Dyke, A. H. & Atkins, W. (1970). Immunochemical comparison of human placental lactogen and placental proteins from other species. *Am. J. Obstet. Gynec.*, **107**, 441–444.

Guyda, H. & Friesen, H. G. (1971). The separation of monkey prolactin from monkey growth hormone by affinity chromatography. *Biochem. Biophys. Res. Comm.*, **42**, 1068–1075.

Guyda, H., Hwang, P. & Friesen, H. G. (1971). Immunologic evidence for monkey and human prolactin (MPr & HPr). *J. clin. Endocr. Metab.*, **32**, 120–123.

Hadfield, G. & Young, S. (1958). The controlling influence of the pituitary on the growth of the normal breast. *Br. J. Surg.*, **46**, 265–273.

Hafiez, A. A., Philpott, J. E. & Bartbe, A. (1971). The role of prolactin in the regulation of testicular function; the effect of prolactin and luteinising hormone on 3 hydroxysteroid dehydrogenase activity in the testes of mice and rats. *J. Endocr.*, **50**, 619–623.

Hafiez, A. A., Lloyd, C. W. & Bartbe, A. (1972). The role of prolactin in the regulation of testis function, synergistic effects of prolactin and luteinising hormone on the plasma levels of testosterone and androstenedione in hypophysectomised rats. *J. Endocr.*, **52**, 327–332.

Hagen, T. C., Gaunsing, A. R. & Sill, A. J. (1976). Preliminary evidence for a human prolactin releasing factor. *Neuroendocrinology*, **21**, 255–261.

Hagiro, N., Watanabe, M. & Goldzieher, J. W. (1969). Inhibition by adrenocorticotrophin of gonadotrophin-induced ovulation in mature female rats. *Endocrinology*, **84**, 308.

Hall, T. R., Chadwick, A. & Callard, I. P. (1975). Control of prolactin secretion in the terrapin (*Chrsemys picta*). *J. Endocr.*, **67**, 52–54P.

Hallowes, R. C. & Wang, D. Y. (1971). The effect of prolactin and growth hormone on the ultrastructure of mouse mammary gland in organ culture. *J. Endocr.*, **49**, v.

Hallowes, R. C. & Wang, D. Y. & Lewis, D. J. (1973). The lactogenic effects of prolactin and growth hormone on mammary gland explants from virgin and pregnant Sprague Dawley rats. *J. Endocr.*, **57**, 253–264.

Hamosh, M. & Seow, R. O. (1971). Lipoprotein lipase activity in guinea pig and rat milk. *Biochim. Biophys. Acta*, **231**, 283–289.

Hamosh, H., Clary, T. R., Chornik, S. S. & Seow, R. O. (1970). Lipoprotein lipase activity of adipose tissue and mammary tissue and plasma triglyceride in pregnant and lactating rats. *Biochim. biophys. Acta*, **210**, 473–482.

Hanwell, A., Linzell, J. L. & Peaker, M. (1971). Cardiovascular responses to salt-loading in conscious domestic geese. *J. Physiol.*, **213**, 389–398.

Hanwell, A., Linzell, J. L. & Peaker, M. (1972). Mature and localisation of the receptors for salt gland secretion in the goose. *J. Physiol.*, **226**, 389–398.

Hardwerger, S., Pang, E. C. & Aloj, G. M. (1971). Correlations in the structure and function of human placental lactogen and human growth hormone I. Modification of the disulphide bonds. *Endocrinology*, **91**, 721–727.

Harms, P. G., Ojeda, S. R. & McCarn, S. M. (1974). Prostaglandin induced release of pituitary gonadotrophins; central nervous system and pituitary sites of action. *Endocrinology*, **94**, 1459.

Hart, I. C. (1973*a*). Effect of 2-bromo-α-ergocryptine on milk yield and the level of prolactin and growth hormone in the blood of the goat at milking. *J. Endocr.*, **57**, 179.

Hart, I. C. (1973*b*). *Hormonal studies in the control of lactation in the goat with special reference to prolactin*. Ph.D. University of Reading.

Hart, I. C. (1974). The relationship between lactation and the release of prolactin and growth hormone in the goat. *J. Reprod. Fert.*, **39**, 485–499.

Hartman, P. E., Cowie, A. T. & Hosking, Z. D. (1970). Changes in enzymatic activity chemical composition and histology of the mammary gland and blood metabolities of lactating rabbits after hypophysectomy and replacement therapy with sheep

prolactin, human growth hormone or bovine growth hormone. *J. Endocr.*, 48, 433–448.
Hartree, A. S., Kovacic, N. & Thomas, M. (1965). Growth promoting and luteotrophic activities of human growth hormone. *J. Endocr.*, 33, 249–258.
Hashimoto, I., Hendricks, D. M., Anderson, L. L. & Melampy, R. R. (1968). Progesterone and pregn.-4-en-20-ol-3-one in ovarian venous blood during various reproductive states in the rat. *Endocrinology*, 82, 333–341.
Haun, C. K. & Sawyer, C. H. (1960). Initiation of lactation in rabbits following placement of hypothalamic lesions. *Endocrinology*, 67, 270.
Hayman, R. H. (1972). *Bos indians* and *Bos taurus* crossbred dairy cattle in Australia. Effect of calf removal and prolactin – treatment on lactation in crossbred *Bos indians* and *Bos taurus* females. *Aust. J. Agric. Res.*, 23, 519.
Hayward, J. L., Atkins, H. J. B., Falconer, M. A., McLean, K. S., Salmen, L. F., Schuur, P. H. & Shaheen, C. H. (1970). Clinical trials comparing transfrontal hypophysectomy with adrenalectomy and with transethmoidal hypophysectomy. In the *Clinical Management of Advanced Breast Cancer*, ed. Joslin, E. A. F. & Gleave, E. N. pp. 50–53. Cardiff: Alpha Omega Alpha.
Hefco, E. & Krulich, L. & Donhoff, I. E. (1972). Plasma levels of FSH and prolactin after complete hypothalamic deafferentation in the rat. *Fedn Proc.* 31, 211.
Heimer, L. & Manta, W. J. H. (1969). The hypothalamic distribution of the stria terminalis in the rat. *Brain Res.*, 13, 284.
Heinonen, O. P., Shapiro, S., Tuominen, L. & Twienner, M. I. (1974). Reserpine use in relation to breast cancer. *Lancet*, 2, 675.
Heitzman, R. J. (1969). The induction by exogenous hormones of enzymes metabolising glucose-6-phosphate in the mammary gland of the pseudo-pregnant rabbit. *J. Dairy Res.*, 36, 47–52.
Helbock, H., Sauls, H. S., Reynolds, J. W. & Brown, D. R. (1971). Effects of human growth hormone preparations on sodium transport in isolated frog skin. *J. clin. Endocr.*, 33, 903–907.
Helms, C. W. & Drury, W. H. (1960). Winter and migratory weight and fat: field studies on some North American buntings. *Bird-Banding*, 31, 1–40.
Hensleigh, P. A. & Johnson, D. C. (1971). Heat stress effects during pregnancy. Retardation of fetal rat growth. *Fert. Steril.*, 22, 522.
Herbert, D. C. & Hayeshida, T. (1970). Prolactin localisation in the primate by immunofluorescence. *Science*, 169, 378–379.
Herlant, M. (1960). Etude critique de deux techniques nouvelles destinées a mettre en evidence les differentes categories cellulaires presentes dans la glande pituitaire. *Bull. Microsc. appl.*, 10, 37–44.
Herlant, M. & Pasteels, J. L. (1967). Histophysiology of human anterior pituitary. In, *Methods and Achievements in Experimental Pathology*, ed. Bagusz, E. & Jasmin, G. Vol. 3, pp. 250–305. Basel: Karger.
Herlant, M. & Radacot, S. (1957). Lobe anterieur de l'hypophyse de la chatte au cours de la gestation et de la lactation. *Archs Biol., Paris*, 58, 217–248.
Herlitz, H., Hamberger, L. & Rosberg, S. (1974). Cyclic AMP in isolated corpora lutea of the rat: Influence of gonadotrophins and prostaglandins. *Acta Endocr., Copnh.*, 77, 737–752.
Herlyn, U., Geller, M. F., Van Bersordt-Wallrabe, I. & von Berswordt-Wallrabe, R. (1965). Pituitary lactogenic hormone release during the onset of pseudo-pregnancy in intact rats. *Acta Endocr., Copnh.*, 48, 220–224.
L'Hermite, M. (1973). The present status of prolactin assays in clinical practice. *Clins Endocr. Metab.*, 2, 3, 423–449.
L'Hermite, M. & Lefranc, L. (1972). Recherches sur les voies monoaminoergiques de l'encephale d'Anguille vulgaris. *Arch. Anat. microsc. Morph. exp.*, 61, 139–152.
L'Hermite, M. & Robyn, C. (1972). Prolactine hypophysaire humaine; detection

radioimmunologique et taux au cours de la grossesse. *Annls Endocr.*, **33**, 357–360.

L'Hermite, M., Delvoye, P., Nobin, J., Vebemans, M. & Robyn, C. (1972). Human prolactin secretion as studied by radioimmunoassay; some aspects of its regulation. In *Prolactin and carcinogenesis*, ed. Boyns, A. R. & Griffiths, K. pp. 81–97. Cardiff: Alpha Omega Alpha.

L'Hermite, M., Heusen, J.C. & Robyn, C. (1973). Increased serum prolactin levels by ethynytestradiol in postmenopausal women with mammary cancer. (Cited in L'Hermite, 1974.)

Herrera, F. C. & Curran, P. F. (1963). The effect of Ca and antidiuretic hormone on Na transport across frog skin. 1. Examinations of inter-relationships between Ca and hormone. *J. gen. Physiol.*, **46**, 999–1010.

Heusen, J. C. (1973). The role of prolactin inhibition in the treatment of advanced breast cancer. In *Hypothalamic control of breast cancer*, ed. Stoll, B. A. Philadelphia and Toronto: Pittmen & Lippincott.

Heusen, J. C., Van Waelbroeck, C. & LeGross, N. (1970). Growth inhibition of rat mammary carcinoma and endocrine changes produced by 2-Br-alpha-ergocryptine a suppressor of lactation and nidation. *Eur. J. Cancer*, **6**, 353.

Heusen, J. C., Waelbroeck, C., LeGros, M., Galloz, G., Robyn, C. & L'Hermite, M. (1972). Inhibition of DMBA-induced mammary carcinogenesis in the rat by 2-Br-α-ergocryptine (CB 154) an inhibitor of prolactin and by nafoxidine (U-11, 100A) an estrogen antagonist. *Gynaec. Invest.*, **2**, 130–137.

Higgins, K. M. & Ball, J. M. (1972). Failure of ovine prolactin to release TSH in *Poecilia latipinna*. *Gen. comp. Endocr.*, **18**, 597.

Hill-Samli, M. & MacLeod, R. M. (1974). Interaction of thyrotropin-releasing hormone and dopamine on the release of prolactin from the rat anterior pituitary *in vitro*. *Endocrinology*, **95**, 1189–1192.

Hill-Samli, M. & MacLeod, R. M. (1975). TRH blockade of the ergocryptine and apomorphine inhibition of prolactin release *in vitro*. *Proc. Soc. exp. Biol. Med.*, **149**, 511–514.

Hilliard, J., Spies, H. G. & Sawyer, C. H. (1969). Hormonal factors regulating ovarian cholesterol mobilisation and progestin secretion in intact and hypophysectomised rabbits. In *The Gonads*, ed. McKerns, K. W. pp. 55–92. New York: Appleton.

Hinkle, P. M. & Tashjian, A. H. (1974). Interaction of thyrotropin releasing hormone with pituitary cells in culture. In *Hormones in Cancer*, ed. McKerns, K. W. pp. 203–227. New York: Academic Press.

Hinde, R. A. & Steele, E. (1964). Effect of exogenous hormones on the tactile sensitivity of the canary brood patch. *J. Endocr.*, **30**, 355–359.

Hinde, R. A., Bell, R. Q. & Steele, E. (1963). Changes in sensitivity of the canary brood patch during the natural breeding season. *Anim. Behav.*, **11**, 553–560.

Hirano, T. (1974). Effect of prolactin on water and electrolyte movements in the isolated urinary bladder of the flounder, *Kareins bicoloratus*. *Gen. comp. Endocr.*, **22**, 382.

Hirano, T. (1975). Effects of prolactin on osmotic and diffusion permeability of the urinary bladder of the flounder *Platichthys flesus*. *Gen. comp. Endocr.*, **27**, 88–94.

Hirano, T., Johnson, D. W. & Bern, H. A. (1971). Control of water movement in flounder urinary bladder by prolactin. *Nature*, **230**, 469–471.

Hirano, T. & Utida, S. (1968). Effects of ACTH and cortisol on water movements in isolated intestine of the eel *Anguilla japonica*. *Gen. comp. Endocr.*, **11**, 373–380.

Hirvonen, E., Ranta, T. & Seppala, M. (1976). Prolactin and thyrotropin-releasing hormone in patients with secondary amenorrhoea, the effect of bromocryptine. *J. clin. Endocr. Metab.*, **42**, 1024–1030.

Hixon, J. E. & Clegg, M. T. (1969). Influence of the pituitary on ovarian progesterone output in the ewe: effects of hypophysectomy and gonadotrophic hormones. *Endocrinology*, **84**, 828–834.

Hixon, J. E., Raud, H. R. & Armstrong, D. T. (1973). Prolactin induced luteolysis in rats following ovarian x-irradiation. *Endocrinology*, **92**, 1667–1672.

Hodgen, G. D., Dufau, M. L., Catt, K. J. & Tullner, W. W. (1972). Estrogens, progesterone and chorionic gonadotrophin in pregnant rhesus monkeys. *Endocrinology*, **91**, 896.

Hoffman, B., Schams, D. & Bapp. R. (1974). Luteotrophic factors in the cow; evidence for LH rather than prolactin. *J. Reprod. Furt.*, **4D**, 77–85.

Höhn, E. O. & Cheng, S. C. (1965). Prolactin and the incidence of brood patch formation and incubation behaviour of the two sexes in certain birds with special reference to phalaropes. *Nature*, **208**, 197.

Hökfelt, T. (1967). The possible ultrastructural identification of tuberoinfundibular dopamine containing nerve endings in the median emminence of the rat. *Brain Res.*, **5**, 121–123.

Hökfelt, T. & Fuxe, K. (1972). Effect of prolactin and argot alkaloids on the tubero-infundibular dopamine (DA) neurons. *Neuroendocrinology*, **9**, 100–122.

Holcombe, L. C. (1970). Prolonged incubation behaviour of the red winged blackbird incubating several egg sizes. *Behaviour*, **36**, 74–83.

Holmes, R. L. & Ball, J. N. (1974). *The pituitary gland: a comparative account*. London: Cambridge University Press.

Holmes, W. N., Phillips, J. G. & Butler, D. G. (1961). The effect of adrenocortical steroids on renal and extrarenal response of the domestic duck (*Anas platyrhynchos*) after hypertonic saline loading. *Endocrinology*, **69**, 483–495.

Holmes, W. N., Fletcher, G. L. & Stewart, D. J. (1968). The patterns of renal electrolyte excretion in the duck (*Anas platyrhynchos*) maintained on freshwater and on hypertonic saline. *J. exp. Biol.*, **48**, 487–508.

Holt, J. A., Richards, J. S. & Midgeley, A. R. (1976). Effect of prolactin on LH receptor in rat luteal cells. *Endocrinology*, **94**, 904–909.

Hong, C. S., Park, C. S. & Park, Y. S. (1968). Seasonal changes of antidiuretic hormone action on sodium transport across frog skin. *Am. J. Physiol.*, **215**, 439–443.

Hopkins, C. R. (1969). The fine structural localisation of acid phosphatase in the prolactin cell of the teleost pituitary the stimulation and inhibition of secretory activity. *Tiss. Cell*, **1**, 653–671.

Horrobin, D. F. (1974). *Prolactin 1974*. London: Medical and Technical Publishers.

Horrobin, D. F. (1975). Prolactin and preclampsia. *Br. med. J.*, **1**, 629.

Horrobin, D. F., Lloyd, I. J., Lipton, A., Burstyn, P. G., Durbin, N. & Muiruiri, K. (1971). Actions of prolactin on human renal function. *Lancet*, **2**, 352.

Horrobin, D. F., Burstyn, P. G. & Manku, M. S. (1973*a*). Saluretic action of aldosterone in the presence of excess cortisol; restoration of salt retaining action by prolactin. *J. Endocr.*, **56**, 343–344.

Horrobin, D. F., Manku, M. S. & Burstyn, P. G. (1973*b*). Effect of intravenous prolactin infusion on arterial blood pressure in rabbits. *Cardiovasc. Res.*, **7**, 585.

Horrobin, D. F., Manku, M. S. & Robertshaw, D. (1973*c*). Water-losing action of antidiuretic hormone in the presence of excess cortisol; restoration of normal action by prolactin or by oxytocin. *J. Endocr.*, **58**, 135–136.

Houssay, B. A. (1947). Ovulacion y postura del sapo *Bufo arenarum* Hensel. V. Transporto de les ovules par el oviducte y el utero. *Rev. Soc. argent. Biol.*, **23**, 275–287.

Howard, K. & Ensor, D. M. (1976). Effect of prolactin on sodium transport across frog skin *in vitro*. *J. Endocr.*, **67**, 56P–57P.

Howard, K. & Ensor, D. M. (1977*a*). The effect of prolactin on sodium transport in the frog *Rana temporaria*. *J. Endocr.*, (in press).

Howard, K. & Ensor, D. M. (1977*b*). Sexual dimorphism in the response of isolated frog skin to prolactin. *J. Endocr.*, (in press).

Howard, K. & Ensor, D. M. (1977*c*). Interaction of lysine vasopressin and prolactin on isolated frog skin. *J. Endocr.*, (in press).

Huggins, C. E., Parsons, F. M. & Jensen, E. V. (1955). Promotion of growth of preputial glands by steroids and the pituitary growth hormone. *Endocrinology*, **57**, 25–32.

Hutchinson, R. E., Hinde, R. A. & Bendon, B. E. (1967). Oviduct development and its relation to other aspects of reproduction in domesticated canaries. *J. Zool., Lond.*, **155**, 87–102.

Hwang, P., Guyda, H. & Friesen, H. G. (1971). A radioimmunoassay for human prolactin. *Proc. natn. Acad. Sci. USA*, **68**, 1902–1906.

Hwang, P., Guyda, H. & Friesen, H. G. (1972). Purification of human prolactin. *Biol. Chem.*, **247**, 1955–1958.

Ichikawa, T., Kobayashi, H., Zimmerman, P. & Muller, U. (1973). Pituitary response to environmental osmotic changes in the larval guppy, *Lebistes reticulatus* (Peters). *Z. Zellforsch. mikrosk. Anat.*, **141**, 161–180.

Ichinose, R. R. & Mandi, S. (1966). Influence of hormones on lobulo-alveolar differentiation of mouse mammary glands *in vitro*. *J. Endocr.*, **35**, 331–340.

Ingalls, W. G., Convey, E. M. & Hafs, H. D. (1973). Bovine serum LH, GH and prolactin during late pregnancy and early lactation. *Proc. Soc. exp. Biol. Med.*, **143**, 161–164.

Ingleton, P. M., Baker, B. I. & Ball, J. N. (1973). Secretion of growth hormone and prolactin by teleost pituitaries *in vitro* 1. Effect of sodium and osmotic pressure during short term incubations *J. comp. Physiol.*, **87**, 317–328.

Ingleton, P. M., Ensor, D. M. & Hancock, M. P. (1971). Identification of prolactin in foetal rat pituitary. *J. Endocr.*, **51**, 799–800.

Ingleton, P. M., Batten, T. & Ball, J. N. (1977). Observations on the catecholaminergic innervation of the prolactin cells in the teleost *Poecilia latipinna*. *J. Endocrinol.*, **73**, 9P.

Innes, I. R. & Niebersen, M. (1970). In *The Pharmacological Basis of Therapeutics*, ed. Goodman, L. S. & Gilman, A. pp. 478–523. New York: Macmillan.

Inoguchi, S. & Sato, Y. (1962). Acid and aldehyde fuchsinophillic granules in so called basophils of the anterior pituitaries of capons and laying hens. *Archs Histol. Jap.*, **22**, 273–280.

Ivey, J. L. & Tashjian, A. H. (1975). Regulation of prolactin and growth hormone production by clonal strains of functional pituitary cells in culture. *Am. Zool.*, **15**, 249–256.

Jacobs, L. S. & Daughaday, W. H. (1973). Pathophysiology and control of prolactin secretion in patients with pituitary and hypothalamic disease. In *International Symposium on Human Prolactin*, ed. Pasteels, J. L. Robyn, C. pp. 84–97. Amsterdam: Excerpta Medica.

Jacobs, L. S., Mariz, I. K. & Daughaday, W. H. (1972). A mixed heterologus radio immunoassay for human prolactin. *J. clin. Endocr. Metab.*, **34**, 484–490.

Jacobs, L. S., Snyder, P. J., Utiger, R. D. & Daughaday, W. H. (1973). Prolactin response to thyrotropin-releasing hormone in normal subjects. *J. clin. Endocr. Metab.*, **36**, 1069.

Jacobs, L. S., Snyder, P. J., Wilber, J. F., Utiger, R. D. & Daughaday, W. H. (1971). Increased serum prolactin after administration of synthetic thyrotropin releasing hormone (TRH) in man. *J. clin. Endocr. Metab.*, **33**, 996–998.

Jacobs, H. S., Franks, S., Murray, M.A.F., Hull, M.G.R., Steele, S. J. & Nabarro, J. D. M. (1976). Clinical and endocrine features of hyperprolactinaemic amenorrhoea. *Clin. Endocr.* **5**,

Jaffe, R. C. & Geschwind, I. I. (1974*a*). Studies on prolactin inhibition of thyroxine-induced metamorphosis in *Rana catesbiana* tadpoles. *Gen. comp. Endocr.*, **22**, 289–295.

Jaffe, R. C. & Geschwind, I. I. (1974*b*). Influence of prolactin on thyroxine-induced changes in hepatic and tail enzymes and nitrogen metabolism in *Rana catesbiana* tadpoles. *Proc. Soc. exp. Biol. Med.*, **146**, 961–966.

Jaffe, R. B., L'Hermite, M. & Midgeley, A. R. (1972). Discussion on studies on prolactin in man. *Rec. Prog. Horm. Res.*, **28**, 574.

Jaffe, R. B., Yuen, B. H., Keye, W. R. & Midgeley, A. R. (1973). Physiologic and pathologic profiles of circulating human prolactin. *Am. J. Obstet. Gynec.*, **117**, 757.

Jarden, V. C., Kaernev, S. & Robinson, C. (1975). Inhibition of oestrogen-stimulated prolactin release by anti-oestrogens. *J. Endocr.*, **65**, 151–152.

Jensen, E. V. & De Sombre, E. R. (1972). Mechanism of action of female sex hormones. *A. Rev. Biochem.*, **41**, 203–230.

Jepson, J. H. & Lowenstein, L. (1964). Effect of prolactin on erythropoiesis in the mouse. *Blood*, **24**, 726–738.

Jewelewicz, R., Dynenfurth, I., Warren, M., Frantz, A. G. & Varde Withe, R. L. Effect of thyrotropin releasing hormone (TRH) upon the menstrual cycle in women. *J. clin. Endocr. Metab.*, **39**, 387.

Jobin, M., Ferland, L., Cote, J. & Labre, F. (1975). Effect of exposure to cold on hypothalamic TRH activity and plasma levels of TSH and prolactin in the rat. *Neuroendocrinology*, **18**, 204–212.

Johke, T. (1969). Prolactin release in response to milking stimulus in the cow and goat estimated by radioimmunoassay. *Endocr. jap.*, **16**, 179.

Johke, T. (1970). Factors affecting the plasma prolactin level in the cow and goat estimated by radioimmunoassay. *Endocr. jap.*, **17**, 393.

John, J. E. & Pfeiffer, E. W. (1963). Testosterone-induced incubation patches of phalarope birds. *Science*, **140**, 1225–1226.

John, T. M., Meier, A. H. & Buyant, E. E. (1972). Thyroid hormones and the circadian rhythms of fat and cropsac responses to prolactin in the pigeon. *Physiol. Zool.*, **45**, 34–42.

Johnson, D. W. (1973). Endocrine control of hydromineral balance in teleosts. *Am. Zool.*, **13**, 799–818.

Johnson, D. W. (1974). Temporal augmentation of LH by prolactin in stimulation of androgen production by the testes of hypophysectomised male rats. *Proc. Soc. exp. Biol. Med.*, **145**, 610–613.

Johnson, D. W., Hirano, T., Bern, H. A. & Conte, F. P. (1972). Hormonal control of water and sodium movements in the urinary bladder of the starry flounder, *Platichthys stellatus*. *Gen. comp. Endocr.*, **19**, 115–128.

Johnson, D. W., Hirano, T., Sage, M., Foster, R. C. & Bern, H. A. Time course of response of starry flounder (*Platichthys stellatus*) urinary bladder to prolactin and to salinity transfer. *Gen. comp. Endocr.*, **24**, 373–380.

Jones, R. E. (1969*a*). Effect of prolactin and progesterone on gonads of breeding Californian quail. *Proc. Soc. exp. Biol. Med.*, **131**, 172–174.

Jones, R. E. (1969*b*). Hormonal control of incubation patch development in the California quail, *Lephertyx californicus*. *Gen. comp. Endocr.*, **13**, 1–13.

Jones, R. E. (1971). The incubation patch of birds. *Biol. Rev.*, **46**, 315–339, 403.

Jones, E. A. & Cowie, A. T. (1972). The effect of hypophysectomy and subsequent replacement therapy with sheep prolactin or bovine growth hormone on the lactose synthetase activity of rabbit mammary gland. *Biochem. J.* **130**, 997–1002.

Jørgensen, C. B. & Larsen, L. O. (1964). Further observations on moulting and its hormonal control in *Bufo bufo* L. *Gen. comp. Endocr.*, **9**, 49–63.

Joseph, M. M. & Meier, A. H. (1971). Daily variations in the fattening response to prolactin in *Fundulus grandis* held on different photoperiods. *J. exp. Zool.*, **178**, 59–62.

Joseph, M. M. & Mukabo, H. B. (1975). Extraovarian effect of prolactin on the traumatised uterus in the rat. *J. Reprod. Fert.*, **45**, 413–414.

Joseph, M. M. & Siwela, A. A. (1976). Effect of prostaglandin F_2a on the response of traumatised uterus to prolactin in the ovariectomised rat. *J. Endocr.*, **68**, 167–168.

Josimovich, J. B. (1966). Potentiation of somatotrophic and diabetogenic effects of growth hormone by human placental lactogen (HPL). *Endocrinology*, **78**, 707–714.

Josimovich, J. B. (1973). Placental protein hormones during pregnancy. *Clin. Obstet. Gynaec.*, **16**, 46–65.

Josimovich, J. B. & MacLaren, L. A. (1962). Presence in the human placenta and term serum of a highly lactogenic substance immunologically related to human growth hormone. *Endocrinology*, **71**, 209–220.

Josimovich, J. B. & Merisko, K. (1975). *Gynaec. Invest.*, **6**, 6. Abstr.

Josimovich, J. B. & Minzy, D. H. (1968). Biological and immunological studies on human placental lactogen. *Ann. N.Y. Acad. Sci.*, **148**, 488–500.

Josimovich, J. B., Koser, B. & Boccola, L. (1970). Placental lactogen in maternal serum as an index of fetal health. *Obstet. Gynaec.*,**36**, 244–250.

Josimovich, J. B., Levitt, M. J. & Stevens, V. C. (1973). Comparison of baboon and human placental lactogens. *Endocrinology*, **93**, 242–244.

Josimovich, J. B., Stock, R. J. & Tobon, J. (1974*a*). Effects of primate placental lactogen upon lactation. In *Lactogenic Hormones, Fetal Nutrition and Lactation*, ed. Josimovich, J. B., Reynolds, M. & Cobo, E. J. p. 335. New York: Wiley and Sons.

Josimovich, J. B., Weniso, G. & Hutchinson, D. (1974*b*). Sources and diposition of pituitary prolactin in maternal circulation, amniotic fluid, fetus and placenta in the pregnant rhesus monkey. *Endocrinology*, **94**, 1364–1371.

Juergens, W. G., Stockdale, F. E., Topper, Y. J. & Elias, J. J. (1965). Hormone dependent differentiation of mammary gland *in vitro*. *Proc. natn. Acad. Sci. USA*, **54**, 629–634.

Juhn, M. & Harris, P. C. (1956). Responses in molt and lay of fowl to progestens and gonadotrophins. *Proc. Soc. exp. Biol. Med.*, **92**, 709–711.

Kadohama, M. & Turkington, R. W. (1973). Changes in acidic chromatin proteins during mammary cell differentiation. *Can. J. Biochem.*, **51**, 1167–1176.

Kagawa, C. M. (1964). Action of anti-aldosterone compounds in the laboratory. In *Hormonal Steroids*, ed. Martini, L. & Pecile, A. Vol. I. pp. 445–456. New York & London: Academic Press.

Kahn, G. (1971). Variations des concentrations plasmatiques de l'hormone luteinisante et de le prolactine au cours du cycle oestrien de la brebis. *C.r. Acad. Sci., Paris*, **272**, 2934.

Kahn, G. & Denamur, R. (1974). Possible role of prolactin during the oestrus cycle and gestation in the ewe. *J. Reprod. Fertil.*, **39**, 473–483.

Kalra, S. P., Ajika, K., Krulich, L., Fawcett, C. P., Quijada, M. & McCann, S. M. (1970). Effects of hypothalamic and preoptic electrochemical stimulation on prolactin and gonadotrophin release in proestrus rats. *Endocrinology*, **88**, 1150–1158.

Kamberi, I. A., Mical, R. S. & Porter, J. C. (1970). Prolactin-inhibiting activity in hypophysial stalk blood and elevation by dopamine. *Experentia*, **26**, 1150.

Kamberi, I. A., Mical, R. S. & Porter, J. C. (1971*a*). Effect of anterior pituitary perfusion and intraventricular injection of catecholamines on prolactin release. *Endocrinology*, **88**, 1012.

Kamber, I. A., Mical, R. A. & Porter, J. C. (1971*b*). Effects of melatronin and serotonin on the release of FSH and prolactin. *Endocrinology*, **88**, 1288.

Kanematsu, S., Hilliard, J. & Sawyer, C. H. (1963*a*). Effect of hypothalamic lesions on pituitary prolactin content in the rabbit. *Endocrinology*, **73**, 345.

Kanematsu, S., Hilliard, J. & Sawyer, C. H. (1963*b*). Effect of reserpine on pituitary prolactin content and its hypothalamic site of action in the rabbit. *Acta Endocr., Copnh.*, **44**, 467–474.

Kann, G. & Denamur, R. (1974). Possible role of prolactin during the oestrus cycle and gestation in the ewe. *J. Reprod. Fert.*, **39**, 473–483.
Kano, K. & Oba, T. (1976). Polyamine transport and metabolism in mouse mammary gland. General properties and hormonal regulation. *J. biol. Chem.*, **251**, 2795–2800.
Kanuijer, M. (1972). Hormonal effect on Na-K-ATPase activity in the gill of the Japanese eel, *Anguilla japonica* with special reference to seawater adaptation. *Endocr. jap.*, **19**, 489–493.
Kaplan, S. L. & Grumbach, M. M. (1964). Studies of a human and simian placental hormone with growth hormone-like and prolactin-like activities. *J. clin. Endocr. Metab.*, **24**, 80–100.
Kaplan, S. L. & Grumbach, M. M. (1965). Serum chorionic 'growth hormone-prolactin' and serum pituitary growth hormone in mother and fetus at term. *J. clin. Endocr. Metab.*, **25**, 1370–1374.
Kaplan, S. L., Gurpide, E. & Sciarra, J. J. (1968). Metabolic clearance rate and production of growth hormone-prolactin in late pregnancy. *J. clin. Endocr. Metab.*, **28**, 1450–1460.
Karg, H. & Schaus, D. (1970). Discussion on prolactin levels in bovine blood under different physiological conditions. In *Lactation*, ed. Falconer, I. R. p. 141. London: Butterworths.
Karg, H. & Schaus, D. (1974). Prolactin release in cattle. *J. Reprod. Fert.*, **39**, 463–472.
Karg, H., Schaus, D. & Reinhardt, V. (1972). Effect of 2 Br-α-ergocryptine on plasma prolactin level and milk yield in cows. *Experentia*, **28**, 574–576.
Karmali, R. *et al.* (1976). Plasma prolactin levels during a simulated dive. *Br. med. J.*, **2**(6029), 237.
Kaspi, T., Ayalon, D. & Nebel, L. (1976). *In vitro* activity of syncytiotrophoblast from human term placentas: hormonal determinations. *Israel J. med. Sci.*, **12**, 82.
Kato, Y., Mabai, Y., Imura, K., Chihara, K. & Ohgo, S. (1974). Effect of 5-hydroxy-trytophan (5HTP) on plasma prolactin levels in man. *J. clin. Endocr. Metab.*, **38**, 695.
Kawakani, M. & Terasawa, E. (1967). Differential control of sex hormone and oxytocin upon the evoked potentials in the hypothalamus and mid-brain reticular formation. *Jap. J. Physiol.*, **17**, 65–93.
Kedam, O. & Leaf, A. (1966). The relation between salt and ionic transport coefficients. *J. gen. Physiol.*, **49**, 655–662.
Keenan, E. J. & Thomas, J. A. (1975). Effects of testosterone and prolactin or growth hormone on the accessory sex organs of castrated mice. *J. Endocr.*, **64**, 111–115.
Keenan, E. J., Thomas, G. A., Lloyd, J. W. & Mawhinney, M. S. (1973). Effect of bovine prolactin on androgen accumulation by the mouse prostrate gland. *Pharmacologist*, **15**, 256.
Kelly, P. A., Posner, B. I. & Friesen, H. G. (1975). Effects of hypophysectomy, ovariectomy and cyclohexamide on specific binding sites for lactogenic hormones in rat liver. *Endocrinology*, **97**, 1408–1415.
Kelly, P. A., Posner, B. I. & Tsushima, Y. (1974). Studies of insulin growth hormone and prolactin binding; ontogenesis, effect of sex and pregnancy. *Endocrinology*, **96**, 532–539.
Kelly, P. A., Shiu, R. P., Friesen, H. G. & Robertson, H. A. (1973*a*). Placental lactogen levels in several species throughout pregnancy. *Endocrinology*, **92**, Suppl. A-233. Abstr.
Kelly, P. A., Shiu, R. P. C., Robertson, M. C. & Friesen, H. G. (1973*b*). *Fedn Proc.*, **32**, 213.
Kennedy, T. G. & Armstrong, D. T. (1972). Extra-ovarian action of prolactin in the regulation of uterine human fluid accumulation in rats. *Endocrinology*, **90**, 1503–1509.
Kennedy, T. G. & Armstrong, D. T. (1973). Lack of specificity for the extra-ovarian prolactin effect of vaginal mucification in rats. *Endocrinology*, **92**, 847–852.
Kerdelhue, B., Catin, S., Kerden, C. V. & Jutiz, M. (1975). Delayed effects of *in vivo* LHRH immunoneutralisation on gonadotrophins and prolactin secretion in the female rat. *Endocrinology*, **98**, 1539.

Kerotelter, J. H., Kirschner, L. B. & Rafuse, D. (1970). On the mechanism of sodium ion transport by the irrigated gills of rainbow trout. (*Salmo gairdnerii*). *J. gen. Physiol.*, **56**, 342–359.

Kerr, T. (1965). Histology of the distal lobe of the pituitary of *Xenopus laevis* Daudin. *Gen. comp. Endocr.*, **5**, 232–240.

Kerr, T. (1966). The development of the pituitary in *Xenopus laevis* Daudin. *Gen. comp. Endocr.*, **6**, 303–311.

Keye, W. R., Ho Yuen, B., Knopf, R. F. & Jaffe, R. B. (1976). Amenorrhoea, hyperprolactinemia and pituitary enlargement secondary to primary hypothyroidism. Successful treatment with thyroid replacement. *Obstet. Gynaec.*, **48**(6), 697–702.

Kibuyama, S., Nabaro-Otani, R. & Yasumasu, I. (1976). Hormonal control of the secretory activity of the cloacal glands in the newt, *Triturus pyrrhogaster. Endocr. jap.*,

Kilpatrick, R., Armstrong, D. T. & Greep, R. O. (1963). Maintenance of the corpus luteum. *Lancet*, **ii**, 462.

Kilpatrick, R., Armstrong, D. T. & Greep, R. O. (1964). Maintenance of the corpus luteum by gonadotrophins in the hypophysectomised rabbit. *Endocrinology*, **74**, 453–461.

King, J. R. & Farner, D. S. (1956). Bioenergetic basis of light-induced fat deposition in the white crowned sparrow. *Proc. Soc. exp. Biol. Med.*, **93**, 354–359.

Kitay, J. L. & Altschule, M. D. (1959). *The Pineal Gland.* Cambridge: Harvard University Press.

Klee, W. A. & Klee, C. B. (1970). The role of alpha-lactalbumin in lactose synthetase. *Biochem. biophys. Res. Comm.*, **39**, 833–841.

Kleinberg, D. L. & Frantz, A. G. (1971). Human prolactin measurement in plasma by *in vitro* assay. *J. clin. Invest.*, **50**, 1557–1568.

Knight, P. J. & Chadwick, A. (1976). The effect of *in vivo* administration of chicken hypothalamic extract on the crop sac and pituitary gland of the pigeon. *Gen. comp. Endocr.*, **27**, 488–494.

Knudsen, J. F. & Mahesh, V. B. (1975). Initiation of precocious sexual maturation in the immature rat, treated with dehydracpiendrosterone. *Endocrinology*, **96**, 458–468

Koch, Y., Chow, Y. F. & Meites, J. (1971). Metabolic clearance and secretion rates of prolactin in the rat. *Endocrinology*, **89**, 1303.

Koefod-Johnson, V. & Ussing, H. H. (1951). The nature of the frog skin potential. *Acta physiol. scand.*, **42**, 299–308.

Kohmoto, K. (1975). Synthesis of two lactogenic proteins by the mouse placenta *in vitro*. *Endocr. jap.*, **22**, 275–278.

Kohmoto, K. & Bern, H. A. (1970). Demonstration of mammotropic activity of the mouse placenta in organ culture and by transplantation. *J. Endocr.*, **48**, 99–107.

Köhnlein, H. E., Bianchi, L. and Bierman, F. J. (1966). Einfluss von Lactationshormone auf die ischamisiente Rattenmere. *Prog. Drug Res.*, **16**, 480–487.

Koller, G. (1952). Der Mestbau der weissen maus und seine hormonale Auslösung. *Verh. dt. zool. Ges.*, 160–168.

Koller, G. (1956). Hormonale und psychische Steurung bein Mestbau weiser Maüser. *Zool. Anz.* suppl. 19. (Verh. dt. zool. Ges., 1955), 123–132.

Koller, P. O., Grunley, P. M. & O'Malley, B. W. (1969). Estrogen-induced cytodifferentiation of the ovalbumin-secreting glands of the chick oviduct. *J. Cell Biol.*, **40**, 8–27.

Komisaruk, B. R. (1967). Effects of local brain in plants of progesterone on reproductive behaviour in ring doves. *J. comp. Physiol. Psychol.*, **64**, 219–224.

Koprowski, J. A. & Tucker, H. A. (1973). Serum prolactin during various physiological states and its relationship to milk production in the bovine. *Endocrinology*, **92**, 1480–1487.

Koprowski, J. A. & Tucker, H. A. & Cowey, E. M. (1972). Prolactin and growth hormone circadian periodicity in lactating cows. *Proc. Soc. exp. Biol. Med.*, **140**, 1072.

Kraicer, J., Ducammon, P., Tobin, M., Remp, C., Von Rees, G. V. & Fertier, C. (1963). Pituitary and plasma TSH response to stress in the intact and adrenalectomised rat. *Fedn Proc.*, **22**, 507 Abstr.

Kragt, C. & Meites, J. (1966). Separation of rat anterior pituitary hormones by polyacrylamide gel electrophoresis. *Proc. Soc. exp. Biol. med.* **121**, 805–808.

Kragt, C. & Meites, J. (1965). Stimulation of pigeon pituitary prolactin release by pigeon hypothalamic extract *in vitro*. *Endocrinology*, **76**, 1169–1176.

Kramer, K. L., Schreibman, M. P. & Pang, P. V. T. (1973). The effect of 2 Br-α-ergocryptine methane sulfonate on freshwater survival and pituitary cytology in the teleost, *Fundulus heteroclitus*. *Am. Zool.*, **13**, 1279.

Krehbiel, R. H. (1941). The effects of lactation on the implantation of ova of a concurrent pregnancy in the rat. *Anat. Rec.* **81**, 43–63.

Krogh, A. (1939). *Osmotic regulation in aquatic animals.* New York: Dover Publ. Inc.

Krulich, L., Quijada, M. & Illner, P. (1971). Program, 53rd Mtg Endocr. Soc., San Fransisco, 1971, p. 83. (Abstr. 82).

Krulich, L., Mefco, E. & Asehembrenner, J. F. (1975). Mechanism of the effects of hypothalamic deafferentation on prolactin secretion in the rat. *Endocrinology*, **96**, 107–118.

Krulich, L. Mefco, E., Illner, P. & Reed, C. B. (1974). The effects of acute stress on the secretion of LH, FSH, Prolactin and GH in the normal male rat with comments on their statistical evaluation. *Neuroendocrinology*, **16**, 293–312.

Kuchl, P. (1974). Prostaglandins, cyclic nucleotides and cell function. *Prostaglandins*, **5**, 325–340.

Kuhl, C., Gaede, P. & Kleke, J. G. (1975). Human placental lactogen concentration during physiological fluctuations of serum glucose in normal pregnant and gestational diabetogenic women. *Acta Endocr.*, **80**, 365–373.

Kuhn, E. R. & Engelen, H. (1976). Seasonal variation in prolactin and TSH releasing activity in the hypothalamus of *Rana temporaria*. *Gen. comp. Endocr.*, **28**, 277–282.

Kuhn, E., Kvulich, L., Fawcett, C. P. & McCann, S. M. (1974). The ability of hypothalamic extracts to lower blood prolactin levels in lactating rat. *Proc. R. Soc. exp. Biol. Med.*, **146**, 104–108.

Kunz, J. & Keller, P. J. (1976). HCG, HPL, oestradiol, progesterone and AFP in serum in patients with threatened abortion. *Br. J. Obstet. Gynaec.*, **83**, 640–644.

Kuroshima, A., Armura, A., Bowers, C. Y. & Schally, A. V. (1966). Inhibition by pig hypothalamic extracts of depletion of pituitary prolactin in rats following cervical stimulation. *Endocrinology*, **78**, 216–217.

Kwa, H. G., Verhofstad, F. (1967). Prolactin levels in the plasma of female rats. *J. Endocr.*, **39**, 455.

Kwa, H. G., De Jong-Babber, M., Eugleman, E. & Cletan, F. J. (1974). Plasma prolactin in human breast cancer. *Lancet*, **1**, 433.

Labrie, F., Ganthier, M., Pelletier, P., Borgeal, P., Lemay, A. & Gouge, J. J. (1974). Role of the microtubules in basal and stimulated release of growth hormone and prolactin in rat adenohypophysis *in vitro*. *Endocrinology*, **93**, 903.

La Brie, S. J. (1972). Endocrines and water and electrolyte balance in reptiles. *Fedn Proc.*, **31**(6), 1599–1608.

Lahlou, B. & Giordan, A. (1970). Le controle hormonal des echages et de la balance de l'eau chez le teleostean d'eau douce *Carassius auratus*, intacte et hypophysectomisé. *Gen. comp. Endocr.*, **14**, 491–509.

Lahlou, B. & Sawyer, W. H. (1969). Electrolyte balance in hypophysectomised goldfish, *Carassius auratus* L. *Gen. comp. Endocr.*, **12**, 370–377.

Lahr, E. H. & Riddle, O. (1938). Proliferation of cropsac epithelium in incubating and in prolactin injected pigeons studied with the colchicine method. *Am. J. Physiol.*, **123**, 614–619.

Lal, S., Guyda, H. & Bikadoff, S. (1977). Effect of methysergide and primozide on apomorphine induced growth hormone secretion in man. *J. clin. Endocr. Metab.*, **44**, 766–770.

Lal, S., de la Vega, C. E., Sourko, T. & Friesen, H. G. (1973). Effect of apomorphine on growth hormone, prolactin, luteinising hormone and follicle stimulating levels in human serum. *J. clin. Endocr. Metab.*, **37**, 719–724.

Lam, P. & Rothchild, I. (1973). Absence of the luteolytic effect of prolactin in the pregnant rat after hypophysectomy and hysterectomy on day 12. *J. Endocr.*, **56**, 609–610.

Lam, T. J. (1969). The effect of prolactin on osmotic influx of water in isolate gills of the marine three-spine stickleback *Gasterosteus aculeatus* L. form *trachurus*. *Comp. Biochem. Physiol.*, **31**, 909–913.

Lam, T. J. (1972). Prolactin and hydromineral regulation in fishes. *Gen. comp. Endocr.*, Suppl. 3. 328–338.

Lam, T. J. & Hoar, W. G. (1967). Seasonal effects of prolactin on freshwater osmoregulation of the marine form, (*trachurus*) of the stickleback *Gasterosteus aculeatus*. *Can. J. Zool.*, **45**, 509–156.

Lam, T. J. & Leatherland, J. F. (1969). Effects of prolactin on the glomerulus of the marine three spine stickleback *Gasterosteus aculeatus* L. form *trachurus* after transfer from sea water to fresh water, during the late autumn and early winter. *Can. J. Zool.*, **47**, 245–250.

Laws, D. F. & Farner, D. S. (1960). Prolactin and the photoperiodic testicular response in white-crowned sparrows. *Endocrinology*, **67**, 279–281.

Lawson, D. M. & Galen, R. R. (1976). The interaction of dopaminergic and secrotonergic drugs on plasma prolactin in ovariectomised, estrogen-treated rats. *Endocrinology*, **98**, 42.

Leatherland, J. F. (1970). Seasonal variation in the structure and ultrastructure of the pituitary of the marine form (*trachurus*) of the three spined stickleback *Gasterosteus aculeatus* L. maintained in different ambient salinities. *Can. J. Zool.*, **51**, 225–235.

Leatherland, J. F. (1972). Histophysiology and innervation of the pituitary gland of the goldfish *Carassius auratus* a light and electron microscopic study. *Can. J. Zool.* **50**, 835–844.

Leatherland, J. F. & Ensor, D. M. (1973). Activity of the autotransplanted pituitary glands in the goldfish *Carassius auratus* L. maintained in different ambient salinities. *Can. J. Zool.*, **51**, 225–235.

Leatherland, J. F. & Ensor, D. M. (1974). Effect of hypothalamic extracts on prolactin secretion in the goldfish, *Carassius auratus* L. *Comp. Biochem. Physiol.*, **47A**, 419–426.

Leatherland, J. F. & Lam, T. (1969). Effect of prolactin on the density of mucous cells on the gill filaments of the marine form (*trachurus*) of the three spined stickleback, *Gasterosteus aculeatus* L. *Can. J. Zool.*, **47**, 787–792.

Leatherland, J. F. & McKeown, B. (1973). Effect of ambient salinity on prolactin and growth hormone secretion and on hydro-mineral regulation in kokanee salmon smolts (*Oncorhynchus nerka*). *J. comp. Physiol.*, **89**, 215–226.

Leatherland, J. F. & McKeown, B. (1974). Effect of ambient salinity on prolactin and growth hormone secretion and on hydromineral regulation in kokanee salmon smolts (*Oncorhynchus nerka*). *J. comp. Physiol.*, **89**, 215–226.

Leatherland, J. F., Ball, J. N. & Hyder, M. (1974). Structure and fine structure of the

hypophyseal pars distalis in indigenous African species of the genus *Tilapia*. *Cell. Tiss. Res.*, **149**, 245–266.
Leatherland, J. F., Hyder, M. & Ensor, D. M. (1975*a*). Regulation of plasma Na^+ and K^+ concentrations in five African species of *Tilapia* fishes. *Comp. Biochem. Physiol.*, **48A**, 699–710.
Leatherland, J. F., McKeown, B. & John, T. N. (1975*b*). Circadian rhythm of plasma prolactin, growth hormone, glucose and free fatty acid in juvenile kokanee salmon (Oncorhynchus nerka). *Comp. Biochem. Physiol.*, **47A**, 821–828.
Leblond, C. P. & Noble, G. K. (1937). Prolactin-like reaction produced by hypophysis of various vertebrates. *Proc. Soc. exp. Biol. Med.*, **35**, 517–518.
Lee, P. A. (1976). Puberty in girls: correlation of serum levels of gonadotrophins, prolactin, androgens, estrogens and progestins with physical changes. *J. clin. Endocr. Metab.*, **43**, 944–947.
Lee, P. A., Jaffe, R. B. & Midgeley, A. R. (1974). Senum gonadotrophins, testosterone and prolactin concentrations throughout puberty in boys. A longitudinal study. *J. clin. Endocr. Metab.*, **39**, 664–672.
Lee, R. W. & Meier, A. H. (1967). Diurnal variations of the fattening response to prolactin in the golden topminnow, *Fundulus chrysteotus*. *J. exp. Zool.*, **166**, 307–316.
Legait, E. & Legait, H. (1955). Modifications de structure du lobe distal de l'hypophyse au cours de la couvaison chez la poule Rhode Island. *C.r. Ass. Anat.*, **84**, 188–199.
Lehrman, D. S. (1958). Effect of female sex hormones on incubation behaviour in the Ring Dove (*Streptopelia risoria*). *J. comp. Physiol. Psychol.*, **51**, 142–145.
Lehrman, D. S. (1963). On the initiation of incubation behaviour in doves. *Anim. Behav.*, **11**, 433–438.
Lehrman, D. S. & Brody, P. (1957). Oviduct response to estrogen and progesterone in the Ring Dove (*Streptopelia risoria*). *Proc. Soc. exp. Biol. Med.*, **95**, 373–375.
Lemoine, A. M. & Olivereau, M. (1972). Variations de la teneur en acide N-acetyl-neuranimique de l'intestin de l'Anguille apres hypophysectomie et traitement avec la prolactine. *C.r. Soc. Biol.*, **166**, 507–512.
Lemoine, A. M. & Olivereau, M. (1973). Variations de la teneur en acide N-acetyl-neuraminique de la branchie de l'anguille au cours des changements de salinité. *C.r. Soc. Biol.*, **167**, 411–416.
Leonard, & Reece, (1941).
Lestroh, A. J. V. & Li, C. H. (1957). Stimulation of the sex accessories of hypophysectomised male rats by non-gonadotrophic hormones of the pituitary gland. *Acta Endocr.*, **25**, 1–16.
Leung, B. S. & Sasaki, S. H. (1973). Prolactin and progesterone effect on specific estradiol binding in interine and mammary tissues *in vitro*. *Biochem. biophys. Res. Comm.*, **55**, 1180–1187.
Lewis, U. J. & Cleaver, E. (1965). Evidence for two types of conversion reactions for prolactin and growth hormone. *J. Biol. Chem.*, **240**, 247–252.
Lewis, U. J., Singh, R. W. & Seavery, B. K. (1971). Human prolactin: isolation and some properties. *Biochem. biophys. Res. Comm.*, **44**, 1169–1176.
Lewis, U. J., Singh, R. N. P. & Seavery, B. K. (1972). Problems in the purification of human prolactin. In *Prolactin and Carciogenesis*, ed. Boyns, A. R. & Griffiths, K. pp. 124–127. Cardiff: Alpha Omega Alpha.
Lewis, U. J., Singh, R. N. P., Sinha, Y. N. & Van der Laan, W. P. (1971). Electrophoretic evidence for human prolactin *J. clin. Endocr. Metab.*, **33**, 153–156.
Li, C. S. (1970).
Ann. Sclavo. (Siena), **12**, 651.
Li, C. H. (1972). Recent knowledge of the chemistry of lactogenic hormones. In *Lactogenic Hormones*, ed. Wolstenholme, G. E. & Knight, J. pp. 7–22. London: Ciba Foundation.
Li, C. H. (1976).

Li, C. H. & Dixon, J. S. (1971). Human pituitary growth hormone ·32: The primary structure of the hormone; revision. *Arch. Biochem. Biophys.*, **146**, 233–236.

Licht, P. (1967). Interaction of prolactin and gonadotrophins on appetite, growth and tail regeneration, in the lizard. *Anolis carolinensis. Gen. comp. Endocr.*, **9**, 49–53.

Licht, P. (1972). Environmental physiology of reptilian breeding cycles. Role of temperature. *Gen. comp. endocr.*, Suppl. 3, 477–489.

Licht, P. (1974). Luteinising hormone (LH) in the reptilian pituitary gland. *Gen. comp. Endocr.*, **22**, 463–469.

Licht, P. & Hoyer, H. E. (1968). Sematotrophic effects of exogenous prolactin and growth hormone in juvenile lizards (*Lacerta* s. sicula). *Gen. comp. Endocr.*, **11**, 338–346.

Licht, P. & Jones, R. E. (1967). Effects of exogenous prolactin on reproduction and growth in adult males of the lizard. *Anolis carolinensis. Gen. comp. Endocr.*, **8**, 228–244.

Licht, P. & Nicoll, C. S. (1969). Localisation of prolactin in the reptilian pars distalis. *Gen. comp. Endocr.*, **12**, 526–535.

Licht, P. & Papkoff, H. (1974). Phyllogenetic survey of neuraminidase sensitivity of reptilian gonadotrophin. *Gen. comp. Endocr.*, **23**, 415–420.

Licht, P. & Rosenberg, L. L. (1969). Presence and distribution of gonadotrophin and thyrotophin in the pars distalis of the lizard, *Anolis carolinensis. Gen. comp. Endocr.*, **13**, 439–454.

Licht, P., Cohen, D. C. & Bern, H. A. (1972). Somatotrophic effects of mammalian growth hormone and prolactin in larval newts, *Tarisha torosa. Gen. comp. Endocr.*, **18**, 391–415.

Lienhart, R. (1927). Contribution a l'étude de l'incubation. *C.r. Soc. Biol.*, **97**, 1296–1297.

Ling, E. R., Kan, S. K. & Porter, J. W. G. (1961). The composition of milk and the nutritive value of its components. In *Milk; the Mammary Gland and its Secretion*, ed. Kan, S. K. & Cowie, A. T. Vol. 2, Ch. 17. New York & London: Academic Press.

Linkie, D. M. & Niswender, G. D. (1972). Serum levels of prolactin, luteunising hormone and follicle stimulating hormone during pregnancy in the rat. *Endocrinology*, **90**, 532.

Linzell, J. & Peaker, M. (1971). Mechanism of milk secretion. *Physiol. Rev.*, **51**, 564–597.

Linzell, J. L. (1963). Some effects of denervating and transplanting mammary glands. *Q. Jl. exp. Physiol.*, **48**, 34–60.

Linzell, J. L. & Peaker, M. (1973). Changes in mammary gland permeability at the onset of lactation in the goat: an effect on tight junctions. *J. Physiol.*, **230**, 13P.

Linzell, J. L. & Peaker, M. (1974). Changes in colostrium composition and in the permeability of the mammary epithelium at about the time of parturition in the goat. *J. Physiol*, **243**, 129–151.

Linzell, J. L., Peaker, M. & Taylor, J. C. (1975). The effects of prolactin and oxytocin on milk secretion and on the permeability of the mammary epithelium in the rabbit. *J. Physiol.*, **253**, 547–563.

Lisk, R. D., Prettow, R. A. & Friedman, S. M. (1969). Hormonal stimulation necessary for elicitation of maternal nest-building in the mouse (*Mus musculus*). *Anim. Behav.*, **17**, 730–737.

Lloyd, J. A. (1965). Seasonal development of the incubation patch in the starling (*Sturnus vulgaris*). *Condor*, **67**, 67–72.

Lloyd, J. W., Thomas, J. A. & Mawhinney, M. G. (1974). A difference in the *in vitro* accumulation and metabolism of testosterone -1, 2-^{3}H by the rat prostate gland following incubation with or bovine prolactin. *Steroids*, **22**, 473–483.

Lloyd, S. J., Josimovich, J. B. & Archer, D. F. (1975). Amenorrhoea and galactorrhoea: results of therapy with 2-bromo-α-ergocryptine (CB 154). *Am. J. Obstet. Gynecol.*, **122**, 85.

Lo, H. D. F. & Fuchs, S. S. (1962). Failure to induce retrieving by sensitisation or the injection of prolactin. *J. comp. Physiol. Psychol.*, **55**, 1111–1113.

Lockett, M. F. (1964). A comparison of the direct renal actions of pituitary growth and lactogenic hormones. *J. Physiol.*, **181**, 192–199.

Lockett, M. F. & Mail, B. (1965). A comparative study of the renal actions of growth hormone and lactogenic hormones in rats. *J. Physiol.*, **180**, 147–156.

Lockwood, D. H., Stockdale, F. E. & Topper, Y. J. (1967). Hormone-dependent differentiation of mammary gland. Sequence of action of hormones in relation to cell cycle. *Science*, **156**, 945–947.

Lofts, B. & Marshall, A. J. (1956). The effects of prolactin administration on the internal rhythm of reproduction in male birds. *J. Endocr.*, **13**, 101–106.

Lofts, B. & Marshall, A. J. (1958). An investigation of the refractory period of reproduction in male birds by means of exogenous prolactin and follicle stimulating hormone. *J. Endocr.*, **17**, 91–98.

Long, J. A. & Evans, H. R. (1922). The estrous cycle in the rat and its associated phenomena. *Mem. Univ. Calif.*, **6**, 1–45.

Lostroh, A. J. & Li, C. H. (1957). Stimulation of the sex accessory glands of hypophysectomised male rats by non-gonadotrophic hormones of the pituitary gland. *Acta endocrinol. (Kbh.)*, **25**, 1–16.

Lu, K. H. & Meites, J. (1971). Inhibition by L-dopa and monoaminoxidase inhibitors of pituitary prolactin release. Stimulation by methyldopa and d-Amphetamine. *Proc. Soc. exp. Biol. Med.*, **137**, 480–483.

Lu, K. H. & Meites, J. (1972). Effect of L-dopa on serum prolactin and PIF in intact and hypophysectomised pituitary grafted rats. *Endocrinology*, **91**, 868.

Lu, K. H., Koch, Y. & Meites, J. (1971). Direct inhibition by ergocernine of pituitary prolactin release. *Endocrinology*, **89**, 229–233.

Lu, K. H., Shaar, C. J. & Kortright, K. H. (1973). Effects of synthetic TRH on *in vitro* and *in vivo* prolactin release in the rat. *Endocrinology*, **93**, 152–155.

Lu, K. H., Amenomeri, Y., Cher, C. L. & Meites, J. (1970). Effects of control acting drugs on serum and pituitary prolactin levels in rats. *Endocrinolgy*, **87**, 667–670.

Lu, K. H., Shaar, C. J., Kartright, K. H. & Meites, J. (1972). Effects of synthetic TRH on *in vitro* and *in vivo* prolactin release in the rat. *Endocrinology*, **91**, 1540–1545.

Lu, K. H., Cher, H. T., Huang, H. H., Grandisen, L., Marshall, S. & Meites, J. (1976). Relation between prolactin and gonadotrophic secretion in post-partum lactating rats. *J. Endocr.*, **68**, 241–250.

Lucci, M. G., Bengale, H. H. & Assali, H. G. (1975). Suppressive action of prolactin on renal response to volume expansion. *Am. J. Physiol.*, **229**, 81.

Lutterbeck, P. M., Pryor, J. S., Varga, L. & Wenner, R. (1971). Treatment of non-puerperal galactorrhoea with an ergot akaloid. *Br. med. J.*, **3**, 228.

Lyons, W. R. (1939). Preparation and assay of mammotrophic hormone. *Proc. Soc. exp. Biol. Med.*, **35**, 645–648.

Lyons, W. R. (1958). Hormonal synergism in mammary growth. *Proc. R. Soc. B.*, **149**, 303–325.

Lyons, W. R. & Allan, W. M. (1938). Duration of sensitivity of the endometrium during lactation in the rat. *Am. J. Physiol.*, **122**, 624–626.

Lyons, W. R. & Dixon, J. S. (1966). The physiology and chemistry of mammotrophic hormone. In *The Pituitary Gland*, ed. Harmis, S. W. & Donovan, B. T., Vol. I. pp. 527–581. London: Butterworths.

Lyons, W. R., Li, C. H. & Johnson, R. E. (1958). The hormonal control of mammary growth. *Rec. Prog. Horm. Res.*, **14**, 219–248.

McAllister, A. & Wellbourn, R. B. (1962). Stimulation of mammary cancer by prolactin and the clinical response to hypophysectomy. *Brit. med. J.*, **1**, 1669–1670.

McBride, J. & Kern, S. M. (1963). The lipoprotein lipase of mammary gland and the correlation of its activity to lactation. *J. Lipid Res.*, **4**, 17–20.

McCann, S. M., Dhariwal, P. S. & Porter, J. C. (1968). Regulation of the adenohypophysis. *A. Rev. Physiol.*, **30**, 589–640.

McCann, S. M., Krulich, L., Cooper, K. J., Kalra, P. S., Kalra, S. P., Libertum, C., Negro-Viler, A., Ovias, R., Rennekliev, O. & Fawcett, C. P. (1973). Hypothalamic control of gonadotrophin and prolactin secretion, implications for fertility control. *J. Reprod. Fert.* Suppl., **20**, 43.

McKenzie, G. M. (1971). Apomorphine induced aggression in the rat. *Brain Res.*, **34**, 323–330.

McKeown, B. A. (1972). Effect of 2 Br-α-ergocryptine on freshwater survival in the teleosts *Xiphophorous maculatus* and *Poecilia latipinna*. *Experientia*, **28**, 675–676.

McKeown, B. A. & von Overbeeke, A. P. (1971). Immunohistochemical identification of pituitary hormone producing cell in the sockeye salmon (*Oncorhynchus nerka*) (Walbaum). *Z. Zellforsch. mikrosk. Anat.*, **112**, 350–362.

McKeown, B. A., Leatherland, J. F. & John, T. M. (1975). The effect of growth hormone and prolactin on the mobilisation of free fatty acids and glucose in the kokanee salmon, *Oncorhynchus nerka*. *Comp. Biochem. Physiol.*, **50B**. 425–430.

McLean, P. (1958). Carbohydrate metabolism of mammary tissue ·1. Pathways of glucose catabolism in the mammary gland. *Biochim. biophys. Acta*, **30**, 303–310.

McMurty, J. P. & Malven, P. V. (1974). Experimental alterations of prolactin levels in goat milk and blood plasma. *Endocrinology*, **95**, 559–564.

McNathy, K. P., Sowers, R. S. & McNeilly, A. S. (1976). A possible role for prolactin in control of steroid secretion by the human graafian follicle. *Nature*, 250–653.

McNathy, K. P., Hunter, W. M., McNeilly, A. S. & Sawers, R. S. (1975). Changes in the concentration of pituitary and steroid hormones in the follicular fluid of human graafian follicles throughout the menstrual cycle. *J. Endocr.*, **64**, 555.

McNeilly, A. S. (1975). Lactation and the physiology of prolactin secretion. *Postgrad. med. J.*, **51**, 231–235.

McNeilly, A. S. & Chard, T. (1974). Circulating levels of prolactin during the menstrual cycle. *Clin. Endocr.*, **3**, 478.

McNeilly, A. S., Evans, D. G. & Chard, T. (1973). Observations on prolactin levels in the menstrual cycle. In *Human Prolactin.* ed. Pasteels, J. L. & Robyn, C. p. 23. Amsterdam: Excerpta Medica.

McShen, W. H., Davis, J. S., Soukup, S. W. & Meyer, R. K. (1950). The nucleic acid content and succinic dehydrogenase activity of stimulated pigeon crop gland tissue. *Endocrinology*, **47**, 274–280.

MacDonald, S. J. & Greep, R. O. (1968). Maintenance of progestic secretion from rat corpora lutea. *Perspect. biol. Med.*, **11**, 490–497.

MacDonald, S. J., Yoshinga, K. & Greep, R. O. (1970). Maintenance of luteal function in rats by rat prolactin. *Proc. Soc. exp. Biol. Med.*, **136**, 687–688.

MacDonald, S. J., Yoshinga, K. & Greep, R. O. (1973). Luteal regression with rat prolactin. *Proc. Soc. exp. Biol. Med.*, **143**, 1031–1033.

MacFarlane, M. A. A. (1971). Ph.D. Thesis, University of Stirling.

MacFarlane, M. A. A. & Maetz, J. (1974). Effects of hypophysectomy on sodium and water exchange in the eurylaline flounder *Platichthys flesus* L. *Gen. comp. Endocr.*, **22**, 77–89.

Machlin, L. J., Jacobs, L. S., Civulis, M., Kinees, R. & Miller, R. (1974). An assay for growth hormone and prolactin-releasing activities using a bovine cell culture system. *Endocrinology*, **95**, 1350–1358.

MacIndoe, J. H. & Turkington, R. W. (1973). Stimulation of human prolactin secretion by intravenous infusion of L. tryptophan. *J. clin. Invest.*, **52**, 1972.

MacLeod, R. M. (1969). Influence of norepinephrine and catecholamine-depleting agents on the synthesis and release of prolactin and growth hormone. *Endocrinology*, **85**, 916–923.

MacLeod, R. M. (1976). Regulation of prolactin secretion. In *Frontiers in Neuroendo-*

crinology, ed. Martini, L. & Genong, W. F. Vol. 14. pp. 169–194. New York: Raven Press.

MacLeod, R. M. & Fonthan, E. H. (1970). Influence of ionic environment on the *in vitro* synthesis and release of pituitary hormones. *Endocrinology*, **86**, 863–869.

MacLeod, R. M. & Lehmeyer, J. E. (1972). Regulation of the synthesis and release of prolactin. In *Lactogenic Hormones*, ed. Wolstenholme, G. W. and Knight, J. pp. 53–75. Ciba Foundation.

MacLeod, R. M. & Lehmeyer, J. E. (1973). Suppression of pituitary tumour growth and function by ergot alkaloids. *Cancer Res.*, **33**, 849–855.

MacLeod, R. M. & Lehmeyer, J. E. (1974). Studies on the mechanism of the dopamine mediated inhibition of prolactin secretion. *Endocrinology*, **94**, 1077.

MacLeod, R. M., Fonthan, E. H. & Lehmeyer, J. E. (1970). Prolactin and growth hormone production as influenced by catecholamines and agents that affect brain catecholamines. *Neuroendocrinology*, **6**, 283–294.

Maderson, P. F. A. (1967). The histology of the escutcheon scales of Gonatodes (Gekkanidae) with a comment on the squamata sloughing cycle. *Copeia*, 743–752.

Maderson, P. F. A., & Licht, P. (1967). Epidermal morphology and sloughing frequency in normal and prolactin treated *Anolis carolinensis* (Iguamidae, Lacertilia). *J. Morph.*, **123**, 157–171.

Madersen, P. F. A., Chin, K. W. & Phillips, J. G. (1970). Endocrine-epidermal relationships in squamata reptiles. *Mem. Soc. Endocr.*, **18**, 259–285.

Madhwa Raj, H. G., & Moudgal, H. R. (1970). Hormonal control of gestation in the rat. *Endocrinology*, **86**, 874–889.

Maetz, J. (1970). Mechanisms of salt and water transfer across membranes in teleosts in relation to the aquatic environment. *Mem. Soc. Endocr.*, **18**, 3–29.

Maetz, J. (1971). Sodium transport in teleost fish. *Phil. Trans. R. Soc. B.*, **262**, 57–76.

Maetz, J., Mayer, N. & Chartier-Baraduc, M. M. (1967). La balance minerale du sodium chez Anguilla anguilla en eau de mer, en eau douce et au cours de transfert d'un milieu à l'autre: effets de l'hypophysectomie et de la prolactine. *Gen. comp. Endocr.*, **8**, 177–188.

Maetz, J., Sawyer, W. H., Pickford, G. & Mayer, N. (1967). Evolution de la balance minerale du sodium chez *Fundulus heteroclitus* au cours du transfert d'eau de mer en eau douce effets de l'hypophysectomie et de la prolactine. *Gen. comp. Endocr.*, **8**, 163–176.

Magnini, G., Ebiner, J. R., Burckhardt, P. & Felber, J. P. (1976). Study on the relationship between plasma prolactin levels and androgen metabolism in man. *J. clin. Endocr. Metab.*, **43**, 944.

Mahesh, V. B., Dalla Pria, S. & Greenblatt, R. B. (1969). Abnormal lactation with Cushing's syndrome – a case report. *J. clin. Endocr. Metab.*, **29**, 978.

Mainoya, J. R. (1975). Analysis of the role of endogenous prolactin on fluid and sodium chloride absorption by the rat jejurum. *J. Endocr.*, **67**, 343–349.

Mainoya, J. R., Benn, H. A. & Regen, J. W. (1974). Influence of ovine prolactin on transport of fluid and sodium chloride by the mammalian intestine and gall bladder. *J. Endocr.*, **63**, 311–317.

Majumder, G. C. & Turkington, R. W. (1971*a*). Stimulation of mammary cell proliferation *in vitro* by protein factor(s) present in serum. *Endocrinology*, **88**, 1506–1510.

Majumder, G. C. & Turkington, R. W. (1971*b*). Adenosine 3′,5′-monophosphate dependant and independant protein phosphokinase isoenzymes from mammary gland. *J. biol. Chem.*, **246**, 2650–2657.

Malaise, W. F., Malaise Lagae, J. F., Picard, C. & Flament Durand, J. (1969). Effects of pregnancy and chorionic growth hormone upon insulin secretion. *Endocrinology*, **84**, 41–44.

Malarkey, W. R. & Panbratz, K. (1974). Evidence for prolactin releasing activity in human plasma not associated with TSH release. *Clin. Res.*, **22**, 600A (Abstr.)

Malarky, W. B., Jacobs, L. S. & Daughaday, W. H. (1971). Levodopa suppression of prolactin in nonpuerperal galactorrhea. *New Engl. J. Med.*, **285**, 1160–1163.

Mallanipati, R. S. & Johnson, D. C. (1973). Serum and pituitary prolactin, LH and FSH in andorgenised female and normal rats treated with various doses of estradiol benzoate. *Neuroendocrinology*, **11**, 46–56.

Malven, P. V. (1969*a*). Hypophysial regulation of luteolysis in the rat. In *The Gonads*, ed. McKerns, K. W., New York: Appleton.

Malven, P. V. (1969*b*). Luteotrophic and luteolytic responses to prolactin in hypophysectomised rats. *Endocrinology*, **84**, 1224–1229.

Malven, P. V. & Portens, S. E. (1973). Increased serum prolactin induced in hypophysectomised rats bearing ectopic pituitaries. *Proc. Soc. exp. Biol. Med.*, **144**, 956–959.

Malven, P. V. & Sawyer, C. H. (1966). A luteolytic action of prolactin in hypophysectomised rats. *Endocrinology*, **79**, 268–274.

Malven, P. V., Cousar, G. J. & Row, E. H. (1969). Structural luteolysis in hypophysectomised rats. *Am. J. Physiol.*, **216**, 421–424.

Manku, M. S., Nassar, B. A. & Horrobin, D. F. (1973). Effects of prolactin on the responses of the rat aortic wall strips to noradrenaline and angiotensin. *Lancet*, **2**, 991–994.

Manku, M. S., Mtabaji, J. P. & Horrobin, D. F. (1975*b*). Effect of cortisol, prolactin and ADH on the amniotic membrane. *Nature*, **258**, 78–80.

Marchlewska-Koj, A. & Krulich, L. (1975). The role of central monoamines in the stress induced prolactin release in the rat. *Fedn Proc.*, **34**, 252.

Marshall, S., Kledzik, G. S. & Gelato, M. (1976). Effects of oestrogens and testosterone on specific prolactin binding sites in the kidneys and adrenal of rats. *Steroids*, **27**, 187–195.

Martin, Y. B. & Friesen, H. G. (1969). Effect of human placental lactogen on the isolated islets of Langerhans *in vitro*. *Endocrinology*, **84**, 619–621.

Martin, J. B., Lal, S., Tolis, G., & Friesen, H. G. (1974). Inhibition by apomorphine of prolactin secretion in patients with elevated serum prolactin. *J. clin. Endocr.*, **39**, 180–182.

Masur, S. K. (1969). Fine structure of the autotransplanted pituitary in the red eft *Notophthalmus viridescens*. *Gen. comp. Endocr.*, **12**, 12–32.

Mathias, D. L. (1967). Studies of the luteotrophic and mammotrophic factors in the trophoblast and maternal peripheral blood of the rat at mid-pregnancy. *Anat. Rec.*, **159**, 55–67.

Mathias, D. L. (1974). Evidence for a hamster placental lactogen. *Anat. Rec.*, **159**, 55–61.

Mattheij, J. A. M. & Sprangers, J. A. P. (1969). The site of prolactin secretion in the adenohypophysis of the stenohaline teleost *Anoptichthys jordani* and the effect of this hormone on mucous cell. *Z. Zellforsch. mikrosk. Anat.*, **105**, 91–106.

Mattheij, J. A. M. & Stoband, H. W. J. (1971). The effects of osmotic experiments and prolactin on the mucous cells in the skin and the ionocytes in the gills of the teleost *Cichlasoma biocellatum*. *Z. Zellforsch. mikrosk. Anat.*, **121**, 93–101.

Mati, J. K., Mugambi, M. & Murinbi, P. B. (1974). Effect of prolactin on isolated rat myometrium. *J. Endocr.*, **60**, 379–380.

Mazzi, V. (1969). Biologia della prolattina. *Boll. Zool.*, **36**, 1–60.

Mazzi, V., Peyrot, A., Anzalone, M. R. & Toscano, C. (1966). L'histophysiologie del' adenohypophyse des tritons cretes (*Triturus cristatus carnifex*). *Z. Zellforsch. Mikrosk. Anat.*, **72**, 597–617.

Mazzi, V. & Vellano, C. (1967). Does prolactin activate the hypothalamo-pituitary-thyroid axis in the crested newt? *Ric. Sci.*, **37**, 68–69.

Mazzi, V. & Vellano, C. (1968). The counterbalancing effect of follicle stimulating hormone on the antigonadal activity of prolactin in the male newt. *Triturus cristatus-carnifex* (Laur.). *J. Endocr.*, **40**, 529–530.

Mazzi, V., Vellano, C. & Toscano, C. (1967). Antigonadal effects of prolactin in adult male crested newt. (*Triturus cristatus carnifex* Laur.). *Gen. comp. Endocr.*, 8, 320–324.
Meier, A. H. (1969*a*). Diurnal variations in metabolic responses to prolactin in lower vertebrates. *Gen. comp. Endocr.*, Suppl. 2, 55–62.
Meier, A. H. (1969*b*). Antigonadal effects of prolactin in the white throated sparrow *Zonotrichia albicolis. Gen. comp. Endocr.*, **13**, 222–225.
Meier, A. H. (1970). Thyroxin phases the circadian fattening response to prolactin. *Proc. Soc. exp. Biol. Med.*, **133**, 1113–1116.
Meier, A. H. (1973). Daily hormone rhythms in the white throated sparrow. *Am. Sci.*, **61**, 184–187.
Meier, A. H. (1975). Chronophysiology of prolactin in the lower vertebrates. *Am. Zool.*, **15**, 905–917.
Meier, A. H. & Dusseau, J. W. (1968). Prolactin and the photoperiodic gonadal response in several avian species. *Physiol. Zool.*, **41**, 95–103.
Meier, A. H. & Farner, D. S. (1964). A possible endocrine basis for premigratory fattening in the white-crowned sparrow, *Zonotrichia leucophys gambelii* (Nuttall). *Gen. comp. Endocr.*, **4**, 584–594.
Meier, A. H. & MacGregor, R. (1972*a*). Temporal organisation in avian reproduction. *Am. Zool.*, **12**, 257–271.
Meier, A. H. & MacGregor, R. (1972*b*). Temporal synergism of corticosterone and prolactin controlling fat storage in the white-throated sparrow *Zonotrichia albicollis. Gen. comp. Endocr.*, **17**, 311–318.
Meier, A. H. & Martin, D. D. (1971). Temporal synergism of corticosteroid and prolactin controlling fat storage in the white throated sparrow, *Zonotrichia albicollis. Gen. comp. Endocr.*, **17**, 311–318.
Meier, A. H., Farner, D. S. & King, J. R. (1965). A possible endocrine basis for migratory behaviour in the white-crowned sparrow, *Zonotrichia leucophys gambelii. Anim. Behav.*, **13**, 453–465.
Meier, A. H., Burns, J. T. & Dusseau, J. W. (1969). Seasonal variations in the diurnal rhythm of pituitary prolactin content in the white-throated sparrow *Zonotrichia albicolis. Gen. comp. Endocr.*, **12**, 282–289.
Meier, A. H., John, T. M. & Joseph, M. N. (1971). Corticosterone and the circadian pigeon cropsac response to prolactin. *Comp. Biochem. Physiol.*, **40**, 459–466.
Meier, A. H., Martin, D. D. & MacGregor, R. (1971). Temporal synergism of corticosterone and prolactin controlling gonadal growth in sparrows. *Science*, **173**, 1240–1242.
Meier, A. H., Trobec, T. W., Joseph, M. M. & John, J. M. (1971). Temporal synergism of prolactin and adrenal steroids in the regulation of fat stores. *Proc. Soc. exp. Biol. Med.*, **137**, 408–415.
Meites, J. (1966). Control of mammary growth and lactation. In *Neuroendocrinology*, ed. Martini, L. & Gavong, W. F. p. 667. N.Y.: Academic Press.
Meites, J. (1970). Direct studies of the secretion of the hypothalamic hypophysiotrophic hormones (HHH). In *Hypophysiotrophic Hormones of the Hypothalamus: Assay and Chemistry*, ed. Meites, J. p. 261. Baltimore: Williams & Wilkins.
Meites, J. (1972*a*). In *Breast Cancer Workshop*, ed. Dee, T. Chicago, Illinois: Univ. Chicago Press.
Meites, J. (1972*b*). In *Prolactin and Carcinogenesis*, ed. Boyns, A. R. & Griffiths, K. p. 54. Cardiff: Alpha Omega Alpha.
Meites, J. (1973). In *Human Prolactin*, ed. Pasteels, J. L. & Robyn, C. *Int. Congr. Ser.* **308**, 105. Amsterdam: Excerpta Medica.
Meites, J. & Clemens, T. A. (1971). Hypothalamic control of prolactin secretion. *Vitam. Horm.*, **30**, 166.
Meites, J. M. & Nicoll, C. S. (1966). Adenohypophysis prolactin. *A. Rev. Physiol.*, **28**, 57–88.

Meites, J., Kahn, R. H. & Nicoll, C. S. (1961). Prolactin production by rat pituita *in vitro*. *Proc. Soc. exp. Biol. Med.*, **108**, 440–443.

Meites, J., Nicoll, C. S. and Talwalker, P. K. (1963). In *Advances in Neuroendocrinology*, ed. Nalbondov, A. V. pp. 238–270. Urbana: Univ. Illinois Press.

Meites, J., Lu, K. H., Wuttke, W., Webch, C. W., Nagasawa, H. & Quadri, S. K. (1972). Recent studies on functions and control of prolactin secretion in rats. *Rec. Prog. Horm. Res.*, **28**, 471–526.

Melby, J. C., Dale, S. L., Wilson, J. E. & Nicholas, A. S. (1966). Stimulation of aldosterone secretion by human placental lactogen. *Clin. Res.*, **14**, 283.

Mena, F. & Grosvenor, C. E. (1971). Release of prolactin in rats by exteroceptive stimulation: sensory stimuli involved. *Horm. Behav.*, **2**, 107.

Mena, F., Enjalbert, A., Carbonell, L., Prian, M. & Korda, C. (1976). Effect of suckling on plasma prolactin and hypothalamic monoamine levels in the rat. *Endocrinology*, **99**, 445.

Mendelson, W. B., Jacobs, L. S., Reichnia, J. D., Othner, E., Cryer, P. E., Trivedi, B. & Daughaday, W. H. (1975). Methysergide suppression of sleep related prolactin secretion and enhancement of sleep-related growth-hormone secretion. *J. clin. Invest.*, **56**, 690.

du Mensil, F., du Boisser, F. & du Boisseau, R. (1973). Mechanismes de controle de la fonction luteale chez la truie labrekis et la vache. In *Progress in Endocrinology*, ed. Gual, C. pp. 927–934. Amsterdam: Excepta Medica.

Merckel, C. & Nelson, W. O. (1940). The relation of the estrogenic hormone to the formation and maintenance of corpora lutea in mature and immature rats. *Anat. Rec.*, **76**, 391–409.

Middler, S. A., Kleeman, C. R. & Edwards, E. (1969). Effect of adenohypophysectomy on salt and water balance of the toad *Bufo marinus* with studies on hormonal replacement. *Gen. comp. Endocr.*, **12**, 290–304.

Midgeley, A. R. (1973). Circulating prolactin levels in humans. *Adv. Exp. Biol. Med.*, **36**, 365.

Mikami, S. I., Vitums, A. & Farner, D. S. (1969). Electron microscopic studies on the adenohypophysis of the white crowned sparrow *Zonotrichia leucophrys gambellii*. *Z. Zellforsch. mikrosk. Anat.*, **97**, 1–29.

Mills, E. S. & Topper, Y. J. (1970). Some ultrastructural effects of insulin, hydrocortisone and prolactin on mammary gland explants. *J. Cell Biol.*, **44**, 310–328.

Milne, K. M., Ball, J. H. & Chester Jones, I. (1971). Effects of salinity, hypophysectomy and corticotrophin on branchial Na– and K– activated ATPase in the eel *Anguilla anguilla* L. *J. Endocr.*, **49**, 177–178.

Mira-Moser, F. (1969). Histophysiologie de la fonction thyreotrophe chez le crapaud *Bufo bufo*. L. *Archs. anat. histol. embryol.*, **52**, 87–182.

Miskinsky, J., Khazer, K. & Sulman, F. S. (1968). Prolactin releasing activity of the hypothalamus in post-partum rats. *Endocrinology*, **82**, 611–613.

Mittra, I. (1974). Mammotrophic effect of prolactin enhanced by thyroidectomy. *Nature*, **248**, 525–526.

Mittra, I., Hayward, J. L. V. & McNeilly, A. S. (1974). Hypothalamic-pituitary-prolactin axis in breast cancer. *Lancet*, **1**, 889–891.

Mizuhiva, V., Amakawa, T. & Yamashima, S. (1970. Electron microscopic studies on the localisation of sodium ions and sodium-potassium activated adenosine-triphosphatase in chloride cells of eel gills. *Expl Cell Res.*, **59**, 346–348.

Mizumo, H. & Naito, M. (1956). The effect of locally administered prolactin on the nucleic acid content of the mammary gland in the rabbit. *Endocr., jap.*, **3**, 227–230.

Mizumo, H., Oider, K. & Maita, M. (1955). The role of prolactin in the mammary alveolus formation. *Endocr. jap.*, **2**, 163–167.

Moodbiri, B., Sheth, A. R. & Rao, S. G. (1975). Binding of prolactin by myomas and normal myometrium. *Ind. J. exp. Biol.* **12**, 566.

Moger, W. H. & Geschwind, I. U. (1972). The action of prolactin on the sex accessory glands of the male rat. *Proc. Soc. exp. Biol. Med.*, **141**, 1017–1021.

Moltz, H., Levin, R. & Lean, M. (1969). Prolactin in the post-partum rat synthesis and release in the absence of suckling stimulation. *Science*, **163**, 1083.

Moltz, H., Lubin, M., Lean, M. & Human, M. (1970). Hormonal induction of maternal behaviour in the ovariectomised mulliparous rat. *Physiol. Behav.*, **5**, 1373–1377.

Mori, T. A. & Bern, H. A. (1972). Some early effects of hypophysial isografts in newborn mice. *Jap. Acad.*, **48**, 698–702.

Mori, R., Nagahama, Y. & Bern, H. A. (1974). Ultrastructural changes in vaginal epithelium of mice neonatally treated with estrogen and prolactin. *Anat. Rec.*, **179**, 225–240.

Moriarty, G. C. (1973). Adenohypophysis: ultrastructural cytochemistry. A review. *J. Histochem. Cytochem.*, **21**, 855–894.

Morishige, W. K., Pepe, G. J. & Rothchild, I. (1973). Serum luteinising hormone, prolactin and progesterone levels during pregnancy in the rat. *Endocrinology*, **92**, 1527–1530.

Morishige, W. K. & Rothchild, I. (1974). A paradoxical inhibiting effect of acute stress on the secretion of LH, FSH prolactin and GH in the normal male rat with comments on their statistical evaluation. *Neuroendocrinology*, **16**, 95–107.

Mortimer, C. H., Besser, S. M., McNeilly, A. S., Tunbridge, W. M. G., Gomez-Pan, A. & Hall, R. (1973). Interaction between secretion of the gonadotrophins, prolactin, growth hormone, thyrotropin and corticosteroids in man: the effects of LH/FSH-RH, TRH and hypoglycaemia alone and in combination. *Clin. Endocr.*, **2**, 317.

Morton, M. K. & Mewaldt, L. R. (1962). Some effects of castration on a migratory sparrow (*Zonotrichia atricapilla*). *Physiol. Zool.*, **35**, 237–247.

Motais, R. (1969). Na-K activated adenosine triphosphatase of gills, evidence for two forms of this enzyme in sea water adapted teleost. *Comp. Biochem. Physiol.*, **27**, 604–636.

Mueller, G. P., Chen, H. J. & Meites, J. (1973). *In vivo* stimulation of prolactin release in the rat by synthetic TRH. *Proc. Soc. exp. Biol. Med.*, **144**, 613–615.

Mueller, G. P., Chen, H. J., Dibbet, J. A., Chen, H. J. & Meites, J. (1974). Effects of warm and cold temperatures on release of TSH, GH and prolactin in rats. *Proc. Soc. exp. Biol. Med.*, **147**, 698–700.

Muhlbock, O. & Boot, L. M. (1967). The mode of action of ovarian hormones in the induction of mammary cancer in mice. *Biochem. Pharmacol.*, **16**, 627–630.

Murakawa, S. & Raben, M. S. (1968). Effect of growth hormone and placental lactogen on DNA synthesis in rat costal cartilage and adipose tissue. *Endocrinology*, **83**, 645–650.

Murray, R. M. L., Mozaffarian, G. & Pearson, O. H. (1972). Prolactin levels with L-dopa treatment in metastatic breast cancer. In *Prolactin* and *Carcinogenesis*, ed. Boyns, A. R., Griffiths, K. pp. 158–161. Cardiff: Alpha Omega Alpha.

Murton, R. K., Bagshawe, K. D. & Lofts, B. (1969). The circadia basis of specific gonadotrophin release in relation to avian spermatogenesis. *J. Endocr.*, **45**, 311–312.

Murton, R. K., Lofts, B. & Orr, A. H. (1970). The significance of circadian based photosensitivity in the house sparrow *Passer domesticus*. *Ibis*, **112**, 448–456.

Murton, R. K., Lofts, B. & Orr, A. H. (1970). The significance of circadian photoperiodically controlled spermatogenesis in the greenfinch, *Chloris chloris*. *J. Zool., Lond.*, **161**, 125–136.

Musto, N., Hafiez, A. A. & Bartka, A. (1972). Prolactin increases 17B-hydroxysteroid dehydrogenase activity in the testes. *Endocrinology*, **91**, 1106–1108.

Nader, S., Kjeld, T. M., Blair, C. M., Tooley, M., Gordan, H. & Fraser, T. R. (1975). A study of the effect of bromocriptine on serum oestradiol, prolactin and follicle stimulating hormone levels in pueperal women. *Br. J. Obstet. Gynaec.*, **82**(9), 750–754.

Nagahama, Y., Bern, H. A., Doneen, B. A. & Mishiska, R. S. (1975*a*). Cellular differentiation in the urinary bladder of a euryhaline marine fish *Gillichthys mirabilis* in response to environmental salinity change. *Devl Growth Differ.*

Nagahama, Y., Mishiska, R. S., Bern, H. A. & Gunther, R. L. (1975*b*). Control of prolactin secretion in teleosts with special reference to *Gillichthys mirabilis* and *Tilapia mossambica*. *Gen. comp. Endocr.*, **25**, 166–188.

Nagasawa, H. & Maito, M. (1962). Effects of growth hormone and/or prolactin on the function of the mammary glands of guinea pigs in the declining phase of lactation (I). *Jap. J. zootech. Sci.*, **33**, 165–173.

Nagasawa, H. & Maito, M. (1963). Effects of growth hormone and/or prolactin on the function of the mammary glands of guinea pigs in the declining phase of lactation (II). *Jap. J. zootech. Sci.*, **34**, 174–179.

Nagasawa, H. & Meites, J. (1970). Suppression by ergocornine and ipronazid of carcinogen-induced mammary tumours in rats, effects on serum and pituitary prolactin levels. *Proc. Soc. exp. Biol. Med.*, **135**, 469–472.

Nakajo, S. & Tanaka, K. (1956). Prolactin potency of the cephalic and caudal lobe of the anterior pituitary in relation to broodiness in the domestic fowl. *Poult. Sci.*, **35**, 990–994.

Nakano, R., Mori, A., Kayashima, F., Washio, M. & Tojo, S. (1975). Ovarian response to exogenous gonadotropins in women with elevated serum prolactin. *Am. J. Obstet. Gynec.*, **121**, 187.

Nalbondov, A. V. (1945). A study of the effect of prolactin on broodiness and on the cock testes. *Endocrinology*, **36**, 251–258.

Nandi, S. (1959). Hormonal control of mammogenesis and lactogenesis in the C^3H/He Crgl mouse. *Univ. Calif. Publs Zool.*, **65**, 1–128.

Nandi, S. & Bern, H. A. (1961). The hormones responsible for lactogenesis in BALB/cCrgl mice. *Gen. comp. Endocr.*, **1**, 195–210.

Nasr, H., Mozaffavian, G., Pensky, J. & Pearson, O. H. (1972). Prolactin-secreting pituitary tumours in women. *J. clin. Endocr. Metab.*, **35**, 505–512.

Nasser, B. A., Manku, M. S., Reed, J. D. & Horrobin, D. H. (1974). Actions of prolactin and frusemide on heart rate and rhythm. *Brit. med. J.*, **2**, 27–29.

Negro-Vilar, A., Krulich, L. & McCann, S. (1973). Changes in serum prolactin and gonadotrophins during sexual development of the male rat. *Endocrinology*, **93**, 660–664.

Negro-Vilar, A. & Saad, W. A. (1972). Influence of prolactin secreting pituitary homografts on male sex accessory hormones. *Proc. lv. Int. Congr. Endocr.* Abstr. **184**, 73.

Neill, T. D. (1974). Prolactin; its secretion and control. In *Am. Hdbk. Physiol.* Sect. 7. *Endocrinology* Vol. IV. *The Pituitary Gland.* Pt. 2 pp. 469–488. Am. Physiol. Soc.

Nelson, B. (1968). *Galapagos.* London: Longmans.

Niall, H. D. (1971). Revised primary structure for human growth hormone. *Nature*, **230**, 90–91.

Niall, H. D. (1972). The chemistry of the human lactogenic hormones. In *Prolactin and Carcinogenesis*, ed. Boyns, A. R. & Griffiths, K. pp. 158–161. Cardiff: Alpha Omega Alpha.

Niall, H. D., Hogam, M. L., Sauer, R., Rosenblum, H. & Greenwood, F. C. (1971). Sequences of pituitary and placental lactogenic and growth hormones: evolution from a promordial peptide by gene reduplication. *Proc. natn. Acad. Sci. U.S.A.*, **68**, 866–870.

Nicoll, C. S. (1965). Neural regulation of adenohypophysial prolactin secretion in tetrapods. Identification from *in vitro* studies. *J. exp. Zool.*, **158**, 203–210.

Nicoll, C. S. (1972). Some observations and speculation on the mechanism of 'depletion', 'repletion' and release of adenohypophysical hormones. *Gen. comp. Endocr.* Suppl. 3, 85–96.

Nicoll. C. S. (1974). Physiological actions of prolactin. In *Am. Hdbk. Physiol.* Sect. 7. *Endocrinology* Vol. IV. *The Pituitary Gland*, pp. 253–292. Am. Physiol. Soc.

Nicoll, C. S. & Bern, H. A. (1968). Further analysis of the occurrence of pigeon crop sac – stimulating activity (prolactin) in the vertebrate adenohypophysis. *Gen. comp. Endocr.*, **11**, 5–20.

Nicoll, C. S. & Bern, H. A. (1972). On the actions of prolactin among the vertebrates; is there a common demoninator. In *Lactogenic Hormones*, Ciba Foundn. Symp. pp. 299–319.

Nicoll, C. S. & Fioriado, R. P. (1969). Hypothalamic control of prolactin secretion. *Gen. comp. Endocr.*, Suppl. **2**, 26–31.

Nicoll, C. S. & Licht, P. (1971). Evolutionary biology of prolactin and somatotrophins. II. Electrophoretic comparison of tetrapod somatotrophins. *Gen. comp. Endocr.*, **17**, 490–507.

Nicoll, C. S. & Nicholls, C. W. (1971). Evolutionary biology of prolactins and somatotrophins. I. Electrophoretic comparison of tetrapod prolactins. *Gen. comp. Endocr.*, **17**, 300–310.

Nicoll, C. S., Talwalker, P. K. & Meites, J. M. (1960). Initiation of lactation in rats by non-specific stresses. *Am. J. Physiol.*, **198**, 1103–1106.

Nicoll, C. S., Pfeiffer, E. W. & Fevold, H. R. (1967). Prolactin and nesting behaviour in phalaropes. *Gen. comp. Endocr.*, **17**, 300–310.

Nicoll, C. S., Fioriado, R. P., McKennee, C. T. & Parsons, T. A. (1970). Assay of hypothalamic factors which regulate prolactin secretion. In *Hypophysiotrophic Hormones of the Hypothalamus. Assay and Chemistry*, ed. Meites, J. pp. 115–150. Baltimore: Williams & Wilkins.

Nichols, C. W. (1973). Somatotrophic effects of prolactin and growth hormone in juvenile snapping turtles (*Chelydra serpentina*). *Gen. comp. Endocr.*, **21**, 219–224.

Nieber, R. & Tomlinson, R. W. (1970). The effect of amiloride on sodium transport in the normal and moulting frog skin. *Acta. physiol. scand.*, **77**, 85–94.

Niell, J. D. (1970). Effect of stress on serum prolactin and luteinising hormone levels during the oestrus cycle in the rat. *Endocrinology*, **87**, 1192.

Niell, J. D. & Reichart, L. E. (1971). Development of a radioimmunoassay for rat prolactin and evaluation of the NIAMD rat prolactin radioimmunoassay. *Endocrinology*, **88**, 548–555.

Niell, J. D., Freeman, M. E. & Fillser, S. A. (1971). Control of the proestrus surge of prolactin and LH secretion by oestrogens in the rat. *Endocrinology*, **89**, 1448–1453.

Nikolaisky, G. V. (1963). The ecology of fishes. New York: Academic Press.

Nilsson, K., Wide, L. & Holfelt, B. (1975). The effect of apomorphine on basal and TRH stimulated release of thyrotrophin and prolactin in man. *Acta Endocr., Copnh.*, **80**(2), 220–229.

Niswender, G. D. (1972). The effect of ergocornine on reproduction in sheep. *Biol. Reprod.*, **7**, 138.

Niswender, G. D. (1974). Influence of 2-Br.-alpha-ergocryptine on serum levels of prolactin and the estrous cycle in sheep. *Endocrinology*, **94**, 612–615.

Niswender, G. D., Midgeley, A. R., Monroe, S. E. & Reichart, L. E. (1968). Radioimmunoassay for rat luteinising hormone with anti-ovine LH serum and ovine LH ^{131}I. *Proc. Soc. exp. Biol. Med.*, **128**, 807–811.

Niswender, G. D., Chen, C. L., Midgeley, A. R., Meites, J. & Ellis, S. (1969). Radioimmunoassay for rat prolactin. *Proc. Soc. exp. Biol. Med.*, **130**, 793–797.

Niwelenski, J. (1958). The effect of prolactin and sommatotrophin on the regeneration of the forelimb in the newt. *Triturus alpestris. Folia biol., Praha*, **6**, 9–36.

Noble, S. K. (1931). *The biology of the amphibia*. New York: Dover Publications.

Noble, G. K., Kumpf, K. F. & Billings, V. H. (1936). The induction of brooding behaviour in the jewel fish. *Anat. Rec.*, **67**, 50–51.

Noble, G. K., Kumpf, K. F. & Billings, V. H. (1938). The induction of brooding behaviour in the jewel fish. *Endocrinology*, **23**, 353–359.

Noel, G. L., Suh, H. K. & Frantz, A. G. (1973) Stimulation of prolactin release by stress in humans. *J. clin. Endocr.*, **38**, 1255.

Noel, G. L., Suh, H. K., Stone, J. G. & Frantz, A. G. (1972). Human prolactin and growth hormone release during surgery and other conditions of stress. *J. clin. Endocr. Metab.*, **35**, 840–851.

Noel, G. L., Dimond, R. C., Wartofsky, L., Earll, J. M. & Frantz, A. G. (1974). Studies of prolactin and TSH secretion by continuous infusion of small amounts of thyrotropin-releasing hormone (TRH). *J. clin. Endocr. Metab.*, **39**, 6–17.

Nokin, J., Vekemans, M., L'Hermite, M. & Robyn, C. (1972). Circadian periodicity of serum prolactin concentration in man. *Brit. med. J.*, **ii**, 561–562.

Norgren, A. (1966). Effects of different doses of oestrone and progesterone on mammary glands of gonadectomised rabbits. *Acta. Univ. lund.* Section II, No. 31, p. 24.

Norgren, A. (1968). Modifications of mammary development of rabbits injected with ovarian hormones. *Acta. Univ. lund.* Section II, No. 4, p. 41.

Nozaki, M., Tatsumi, Y. & Ichikawa, T. (1974). Histological changes in the prolactin cells of the rainbow trout *Salmo gairdnerii uridens* at the time of hatching. *Annotnes. zool. jap.*, **47**, 15–21.

Oduleye, S. O. (1976). The effects of calcium on water balance of the brown trout (*Salmo trutta*). *J. exp. Biol.*, **63**, 343–356.

Odum, E. P. (1960). Premigratory hyperphagia in birds. *Am. J. clin. Nutr.*, **8**, 621–629.

Ogawa, M. (1974). The effects of bovine prolactin sea water and environmental calcium on water influx in isolated gills of Japanese eel, *Anguilla japonica Comp. Biochem. Physiol.*, **52A**, 539–543.

Ogawa, M. (1975). The effects of prolactin, cortisol and calcium-free environment on water influx in isolated gills of Japanese eel, *Anguilla japonica. Comp. Biochem. Physiol.*, **49A**, 545–533.

Ogawa, M. & Johanson, P. H. (1967). A note on the effect of hypophysectomy on the mucous cells of the goldfish *Carassius auratus. Can. J. Zool.*, **45**, 885–886.

Ogawa, M., Yagasaki, M. & Yamazaki, F. (1973). The effect of prolactin on water influx in isolated gills of the goldfish, *Carassius auratus. J. comp. Biochem. Physiol.*, **44A**, 1177–1183.

Ogle, C. (1934). Adaptation of sexual activity to environmental stimulation. *Am. J. Physiol.*, **107**, 628.

Ojeda, S. R., Harms, P. G. & McCann, S. M. (1974*a*). Effect of blockade of dopaminergic receptors on prolactin and LH release. Median eminence and pituitary sites of action. *Endocrinology*, **94**, 1650–1657.

Ojeda, S. R., Harms, P. G. & McCann, S. M. (1974*b*). Central effects of prostaglandin E_1(PGE_1) on prolactin release. *Endocrinology*, **95**, 613.

Oka, T. (1974). Spermidine in hormone dependant differentiation of mammary gland *in vitro. Science*, **184**, 78–80.

Oka, T. & Perry, J. W. (1974*a*). Arginasa affects lactogenesis through its influence on the biosynthesis of spermidine. *Nature*, **250**, 660–661.

Oka, T. & Perry, J. W. (1974*b*). Studies on the function of glucocorticoid in mouse mammary gland epithelial cell differentiation *in vitro*. Stimulation of glucose-6-phosphate dehydrogenase. *J. biol. Chem.*, **249**, 3586–3591.

Oka, T. & Perry, J. W. (1974*c*). Spermidine as a possible mediator of glucocorticoid effect on milk protein synthesis in mouse mammary epithelium *in vitro. J. biol. Chem.*, **249**, 7647–7652.

Oka, T. & Perry, J. (1976). Studies on regulatory factors of ornithuce decarboxylase activity during development of the mouse mammary tissue *in vitro*. *J. biol. Chem.*, **251**, 1738–1744.

Oka, T. & Topper, Y. J. (1972). Is prolactin mitogenic for mammary epithelium. *Proc. natn. Sci. U.S.A.*, **69**, 1693–1696.

Olivereau, M. (1966). Influence d'un sejour en eau deminiralisée sur la système hypophyse-serenalien de l'anguille. *Annls Endocr.*, **27**, 665–678.

Olivereau, M. (1968). Etude cytologique de l'hypophyse du muge, en particulier en relation avec la salinité exterieure. *Z. Zellforsch. mikrosk. Anat.*, **87**, 545–561.

Olivereau, M. (1970). Structure histologique du rein et electrolytes plasmatiques chez l'anguille après autotransplantation de l'hypophyse. *Z. vergl. Physiol.*, **71**, 350–364.

Olivereau, M. (1971*a*). Effets de l'adaptation au milieu privé des electrolytes sur les cellules a chlorures de la branchie de l'anguille. *C.r. Soc. Biol.*, **165**, 1009–1013.

Olivereau, M. (1971*b*). Action de la reserpine chez l'Angulle. I. Cellules à prolactine de l'hypophyse du male. *Z. Zellforsth. mikrosk. Anat.*, **121**, 232–243.

Olivereau, M. (1973). Dopaminergic control of prolactin secretion in eels. In *International Symposium on Neurosecretion*, London. 1973, VI, p. 60.

Olivereau, M. (1975). Dopamine, prolactin control and osmoregulation in eels. *Gen. comp. Endocr.*, **26**, 550–561.

Olivereau, M. & Ball, J. N. (1966). Histological study of functional ectopic pituitary transplants in a teleost fish (*Poecilia formosa*). *Proc. R. Soc. Lond.*, Ser. B., **164**, 106–129.

Olivereau, M. & Ball, J. N. (1970). Pituitary influences on osmoregulation in teleosts. *Mem. Soc. Endocr.*, **18**, 57–85.

Olivereau, M. & Chester-Baraduc, M. M. (1965). Action de la prolactin chez l'Anguille intact et hypophysectomisée. II. Effects sur les electrolytes plasmatiques (sodium, potassium et calcium). *Gen. comp. Endocr.*, **7**, 27–36.

Olivereau, M. & Lemoine, A. M. (1968). Action de la prolactine chez l'anguille intacte. III. Effect sur la structure histologique du rein. *Z. Zellforsch. mikrosk. Anat.*, **88**, 576–590.

Olivereau, M. & Lemoine, A. M. (1970). Presence d'acide N-acetyl-neuraminique dans la peau d'*Anguilla anguilla*. I. Interet de son dosage dans l'étûde del'osmoregulation. *Z. vergl. Physiol.*, **73**, 22–33.

Olivereau, M. & Lemoine, A. M. (1971). Effect of prolactin in intact and hypophysectomised eels. VII. Effect on the concentration of sialic acid (N-acetyl-neuraminic acid) in the skin. *Z. vergl. Physiol.*, **73**, 34–43.

Olivereau, M. & Lemoine, A. M. (1972). Effets de variations de la salinité extrême sur la teneur en acide-N-acetyl-neuraminique (ANAN) de la peau chez l'anguille. Modifications simultanées des cellules a prolactine de l'hypophysis. *J. comp. Physiol.*, **79**, 411–422.

Olivereau, M. & Lemoine, A. M. (1973). Action de la prolactine chez l'Anguille intacte et hypophysectomisee. VIII. Effets sur les electrolytes plasmatiques. *J. comp. Physiol.*, **86**, 65–75.

Olivereau, M. & Olivereau, J. (1971). Influence de l'hypophysectomie et d'un traitement prolactinique sur les cellules à mucus de la branchie chez l'anguille. *C.r. Soc. Biol.*, **165**, 2267.

Olivereau, M., Lemoine, A. M. & Dimovska, A. (1971). Denaers sur la contrôle de la fonction prolactinique chez l'Anguille. *Annls Endocr.*, **32**, 271.

De Olmos, J. G. (1972). The amygdaloid projection field in the rat as studied with the ouprie-silver method. In *The Neurobiology of the Amygdala*, ed. Eleftheriou, B. E. p. 295. New York & London: Plenum Press.

Ong, H. C. (1976). Human placental lactogen; physiological role in pregnancy. *Med. J. Malaysia*, **30**, 165–167.

Ortmann, R. & Etkin, W. (1963). The cytology of the pars distalis of metamorphosing and immature *Rana pipiens*. *Gen. comp. Endocr.*, **11**, 139–150.

Osewood, T. & Fiedler, K. (1968). Die Wirkung von Sauger-Prolactin auf die Schilddrusse des Segelflossers *Pterophyllum sealare* (Cichlidae, Teleosteii). *Z. Zellforsch. mikrosk. Anat.*, **91**, 617–632.

Oxender, W. D., Hafs, H. D. & Ingalls, W. G. (1972). Serum growth hormone LH and prolactin in the bovine fetus and neonate. *J. Anim. Sci.*, **35**, 51–55.

Ozegovic, B. & Milkovic, S. (1972). Effects of adrenocorticotrophic hormone, growth hormone, prolactin, adrenalectomy and corticoids upon the weight, protein and nucleic acid content of the female rat preputial glands. *Endocrinology*, **90**, 903–908.

Palmiter, R. D. (1969). Hormonal induction and regulation of lactose synthetase in mouse mammary gland. *Biochem. J.*, **113**, 409–417.

Paluden, K. (1951). Contributions to the breeding biology of *Larus argentatus* and *Larus fuscus*. *Vidensk. Meddr. dansk. naturh. Foren.*, **114**, 1–128.

Pang, P. K. T. (1973). Endocrine control of calcium metabolism in teleosts. *Am. Zool.*, **13**.

Pang, P. K. T. & Sawyer, W. H. (1974). Effects of prolactin on hypophysectomised mud puppies, *Necturus maculosus*. *Am. J. Physiol.*, **226**, 458–462.

Pang, P. K. T., Griffith, R. W. & Pickford, G. E. (1971). Hypocalcemia and tetonic seizures in hypophysectomised biltifish *Fundulus heteroclitus*. *Proc. Soc. exp. Biol. Med.*, **136**, 85–87.

Paris, A. L. & Romaley, J. A. (1972). Effect of short-term stress upon fertility. I. Before puberty. *Fert. Steril.*, **24**, 540–545.

Paris, A. L., Kelly, P. & Romaley, J. A. (1973). Effects of short-term stress upon Fertility. II. Puberty and after. *Fert. Steril.*, **24**, 546–552.

Parsons, T. A. (1970). Effect of cations on prolactin and growth hormone secretion by the rat adenohypophysis *in vitro*. *J. Physiol.*, **228**, 1221–1222.

Pasqualini, R. Q. (1953). La funcion de la gonadotrofina C (Luteotrofina, prolactina) en el mache. *Prensa. Med. Argent.*, **40**, 2658–2660.

Pasteels, J. L. (1961). Premiers resultats de culture combinée *in vitro* d'hypophyse et d'hypothalamus dans le but d'en apprecier la secretion de prolactine. *C.r. Acad. Sci., Paris*, **253**, 3074–3075.

Pasteels, J. L. (1962). Elaboration par l'hypophyse humaine en culture de tissus, d'une substance stimulant le jabot de pigeon. *C.r. Acad. Sci., Paris*, D, **254**, 4083–4085.

Pasteels, J. L. (1967). Hormone de croissance et prolactine dan l'hypophyse humaine. *Annls Endocr.*, D, **260**, 4381–4384.

Pasteels, J. L. (1971). Morphology of prolactin secretion. In *Lactogenic Hormones*, ed. Wolstenholme, G. E. & Knight, J. pp. 241–257. London: Ciba Foundation.

Pasteels, J. L. (1976). *Human Prolactin*. Amsterdam: Excerpta Medica.

Pasteels, J. L., Robyn, C. & Hubinst, P. O. (1965). Un unimuserum neutralisant la prolactine humaine. *C.r. Acad. Sci., Paris*, D, **260**, 4381–4384.

Pasteels, J. L., Donguy, A., Frenolte, M. & Ectors, F. (1971). Inhibition de la secretion de prolactine par l'ergocornine et la 2 Br-α-ergocryptine: Action direct sur l'hypophyse en culture. *Annls Endocr.*, **32**, 188–192.

Patel, M. D. (1936). The physiology of the formation of 'pigeons milk'. *Physiol Zool.*, **9**, 129–152.

Payan, P. & Maetz, J. (1970). *Bull. Inform. Sci. Tech. Commiss. Energ. Atom.*, **146**, 77.

Payan, P. & Maetz, J. (1971). Balance hydrique chez les elasmobranches: arguments en favour d'un contrôle endocrinien. *Gen. comp. Endocr.*, **16**, 535–554.

Payne, F. (1955). Acidophillic granules in the gonadotrophin secretory basophils of laying hens. *Anat. Rec.*, **100**, 49–55.

Payne, R. B. (1966). Absence of brood patch in cassin auklets. *Condors*, **68**, 209–210.

Peake, G. T., McKeel, D. W., Jarett, L. & Daughaday, W. H. (1969). Ultrastructural histologic and hormonal characterization of a prolactin-rich human pituitary tumour. *J. clin. Endocr. Metab.*, **29**, 1383–1393.

Peaker, M. (1971). Avian salt glands. *Phil. Trans. R. Soc., B*, **262**, 289–300.

Peaker, M. & Linzell, J. L. (1972). *Salt glands in birds and reptiles.* (Physiological Society Monograph), London: Cambridge University Press.

Peaker, M., Phillips, J. G. & Wright, A. (1970). The effect of prolactin on the secretory activity of the nasal gland of the domestic duck (*Anas platyrhynchus*). *J. Endocr.*, **47**, 123–127.

Peaker, M., Peaker, S. J., Phillips, J. G. & Wright, A. (1971). The effects of corticotrophin, glucose and potassium chloride on secretion by the nasal gland of the duck *Anas platyrhynchus*. *J. Endocr.*, **50**, 293–301.

Pearse, A. G. E. (1951). The application of cytochemistry to the localisation of gonadotrophic hormone in the pituitary. *J. Endocr.*, **7**, xiviii–1.

Pearson, O. H., Herena, O., Herena, L., Molina, A. & Butler, T. (1969*b*). Prolactin-dependent rat mammary cancer: a model for man? *Trans. Assoc. Am. Phys.*, **82**, 225–238.

Pelletier, T. (1973). Evidence for photoperiodic control of prolactin release in rams. *J. Reprod. Fert.*, **35**, 143–147.

Pennycuik, R. R. (1966). Factors affecting the survival and growth of young mice born and reared at 36°C. *Aust. J. exp. Biol. med. Sci.*, **44**, 405.

Perek, M., Eckstein, B. & Sobel, H. (1957). Histological observations on the anterior lobe of the pituitary gland of laying and moulting hens. *Poult. Sci.*, **36**, 954–958.

Peter, R. E. (1972). Feedback effects of thyroxine in goldfish *Carassius auratus* with an autotransplanted pituitary. *Neuroendocrinology*, **10**, 273–281.

Peter, R. E. & McKeown, B. A. (1973). Control of prolactin secretion in the goldish *Carassius auratus*. In *Sixth International Symposium on Neurosecretion*, ed. Knowles, F. Heidlberg: Springer-Verlag.

Peyne, A., Rowault, J. P. & Laporte, P. (1968). Effet potentialisateur de la prolactin endogine sur les effects sexuels males soumis à la testosterone. *C.r. Soc. Biol.*, **162**, 1592–1595.

Peyrot, A. (1969). La fonction thyreotrophe de l'adenohypophyse chez la triton crète après lesion permanente de l'eminence mediale. *Gen. comp. Endocr.*, **13**, 525–526.

Pfaff, D. W. & Keiner, M. (1972). Estradiol concentrating cells in the rat amygdala as part of a limbic system. In *The Neurobiology of the Amygdala*. ed. Eleftheriou. p. 775. New York and London: Plenum Press.

Pharris, B. B. & Hunter, K. K. (1971). Interrelationships of prostaglandin F_2a and gonadotrophins in immature female rats. *Proc. Soc. exp. Biol. Med.*, **136**, 503–506.

Piaceck, B. E. & Meites, J. (1967). Stimulation by light of gonadotrophin release from transplanted pituitaries of hypophysectomised rats. *Neuroendocrinology*, **2**, 129–137.

Pickford, G. E. (1956). Melanogenesis in *Fundulus heteroclitus*. *Anat. Rec.*, **125**, 603–604.

Pickford, G. E. (1973). Introductory remarks. *Am. Zool.*, **13**, 711–717.

Pickford, G. E. & Phillips, J. G. (1959). Prolactin, a factor in promoting survival of hypophysectomised billifish in freshwater. *Science*, **130**, 454–455.

Pickford, G. E., Robertson, E. E. & Sawyer, W. H. (1965). Hypophysectomy, replacement therapy and the tolerance of the euryhaline billifish *Fundulus heteroclitus* to hypotonic media. *Gen. comp. Endocr.*, **5**, 169–180.

Pickford, G. E., Pang, P. R. T. & Sawyer, W. H. (1966). Prolactin and serum osmolality of hypophysectomised billifish *Fundulus heteroclitus* in freshwater. *Nature, Lond.*, **209**, 1040–1041.

Pickford, G. E., Griffith, R. W., Tarretti, J., Hendler, E. & Epstein, F. (1970). Branchial reduction and renal stimulation of (Na^+, K^+) ATPase by prolactin in hypophysectomised billifish in freshwater. *Nature, Lond.*, **228**, 378–379.

Pittendrigh, C. S. & Minois, D. H. (1964). The entrainment of circadian oscillations by light and their role as photoperiodic clocks. *Am. Nat.*, **98**, 261–294.
Popovici, D. G. (1963). Recherches neurophysiologiques sur la reflexe d'evacuation du lait. *Rev. Biol., Bucharest*, **8**, 75–81.
Posner, B. I. (1976). Regulation of lactogenic specific binding sites in rat liver, studies on the role of lactogens and estrogens. *Endocrinology*, **99**, 1168–1177.
Posner, B. I., Kelly, P. A. & Friesen, H. G. (1975). Prolactin receptors in rat liver, possible induction by prolactin. *Science*, **188**, 57–59.
Posner, B. I., Kelly, P. A., Shiu, R. & Friesen, H. G. (1974). Studies on insulin, growth hormone and prolactin binding: tissue distribution, species variation and characterisation. *Endocrinology*, **95**, 521–531.
Potts, W. T. W. & Fleming, W. R. (1970). The effects of prolactin and divalent ions on the permeability to water of *Fundulus kansae*. *J. exp. Biol.*, **53**,
Potts, W. R. & Parry, G. W. (1960). *Osmotic and ionic regulation in animals*. Oxford: Pergamon Press.
del Pozo, E. & Ohrihaus, E. E. (1976). Lack of effect of acute prolactin suppression on renal water, sodium and potassium excretion during sleep. *Horm. Res.*, **7**, 11–15.
del Pozo, E., Brundelhe, R., Varga, L., & Friesen, H. G. (1972). The inhibition of prolactin secretion in man by CB-154 (2 Br-α-ergocryptine). *J. clin. Endocr. Metab.*, **35**, 768.
Purves, H. D. (1966). Cytology of the adenohypophysis. In *The pituitary gland*, ed. Harris, G. W. & Donovan, B. T. Vol. 1. pp. 147–232. London: Butterwick.

Quadri, S. K. & Meites, J. (1971). Regression of spontaneous mammary tumours in rats by ergot drugs. *Proc. Soc. exp. Biol. Med.*, **138**, 999–1001.
Quijada, M., Illner, P., Krulich, L. & McCann, S. M. (1973). The effect of catecholamines on hormone release from anterior pituitaries and ventral hypothalamic incubated *in vitro*. *Neuroendocrinology*, B, 151–163.
Quinn, D. L. & Everett, J. W. (1967). Delayed pseudopregnancy induced by selective hypothalamic stimulation. *Endocrinology*, **80**, 155–162.

Rabbin, R. M., Swann, M., Shapiro, D. J. & Isaacson, L. (1974). Effect of growth hormone on sodium transport and osmotic water flow across toad skin. *Horm. Metab. Res.*, **6**, 129–132.
Radacot, J. (1963). Contributions à l'etude des types cellulaire du lobe anterieur de l'hypophyse chez quelques mammifères. In *Cytology de l'adenohypophyse*, C.N.R.S. Paris.
Rakoff, J. S., Siler, T. M., Sinha, Y. N. & Yen, S. S. C. (1973). Prolactin and growth hormone release in response to sequential stimulation by argine and synthetic TRF. *J. clin. Endocr. Metab.*, **37**, 641.
Ramirez, V. D. & McCann, S. M. (1964). Induction of prolactin secretion by implants of estrogen into the hypothalamo hypophysial region of female rats. *Endocrinology*, **75**, 206–214.
Ramsey, D. H. & Bern, H. A. (1972). Stimulation by ovine prolactin of fluid transfer in everted sacs of rat small intestine. *J. Endocr.*, **53**, 453–459.
Rankin, C. J. & Maetz, J. (1971). A perfused teleostean gill preparation; vascular actions of neurohypophysial hormones and catecholamines. *J. Endocr.*, **51**, 621–635.
Ratner, A. & Meites, J. (1964). Depletion of prolactin-inhibiting activity of rat hypothalamus by estradiol or suckling stimulus. *Endocrinology*, **75**, 377.
Ratner, A. & Peake, G. T. (1974). Maintenance of hyperprolactinemia by gonadal steroids in androgen – sterilised and spontaneously constant oestrus rats. *Proc. Soc. exp. Biol. Med.*, **146**, 680–683.

Ratner, A., Talwalker, P. K. & Meites, J. (1963). Effect of oestrogen administration *in vivo* on prolactin release by rat pituitary *in vitro*. *Proc. Soc. exp. Biol. Med.*, **112**, 12–15.
Ratner, A., Talwalker, P. K. & Meites, J. (1965). Effect of reserpine on prolactin-inhibiting activity of rat hypothalamus. *Endocrinology*, **77**, 315–319.
Raud, H. R., Kiddy, C. A. & Odell, W. D. (1971). The effect of stress upon the determination of prolactin by radioimmunoassay. *Proc. Soc. exp. Biol. Med.*, **136**, 689–693.
Ravault, J. P. & Courot, M. (1975). Blood prolactin in the male lamb from birth to puberty. *J. Reprod. Fert.*, **42**, 563–566.
Ray, E. W., Averill, S. C., Lyons, W. R. & Johnson, R. E. (1965). Rat placental lactogen activities corresponding to those of pituitary mammotrophin. *Endocrinology*, **56**, 359–373.
Re, P. H., Kourides, E. C., Ridgeway, E. C., Weintraub, B. D. & Maleof, F. (1976). The effect of glucocorticoid administration on human pituitary secretion of thyrotropin and prolactin. *J. clin. Endocr. Metab.*, **431**(2), 838–846.
del Re, R., Del Pozo, E., de Grandi, O., Friesen, H. G., Hinselmann, M. & Wyss, H. (1973). Prolactin inhibition and suppression of puerperal lactation by a Br.-ergocryptine (CB-154). A comparison with osetrogen. *Obstet. Gynaec.*, **41**, 834.
Reddi, A. H. (1969). Role of prolactin in the growth and secretory activity of the prostate and other accessory glands of mammals. *Gen. comp. Endocr.*, Suppl. 2, 81–96.
Reece, R. P. & Turner, J. (1937). Effect of stimulus of suckling upon galactin content of the rat pituitary. *Proc. Soc. exp. Biol. Med.*, **35**, 621.
Rees, E. D. & Huggins, C. (1960). Steroid influences on respiration, glycolysis and levels of pyridine nucleotide-linked dehydrogenase of experimental mammary cancer. *Cancer Res.*, **20**, 963–971.
Reier, P. J., Merishige, W. K. & Rothchild, I. (1974). The effect of ether and laparotomy on serum prolactin levels in progesterone-treated intact and ovariectomised rats. *Neuroendocrinology*, **16**, 43–51.
Reinke, H. E. & Chadwick, C. S. (1940). Inducing land stage of *Triturus vividescens* to assume water habitat by pituitary implants. *Proc. Soc. exp. Biol. Med.*, **40**, 691–693.
Reiter, R. J. (1974). Pituitary and plasma prolactin levels in male rats as influenced by the pineal gland. *Endocr. Res. Comm.*, **1**, 169–180.
Reiter, R. J., Black, D. E. & Vaughan, M. K. (1976). A counter antigonadotrophic action of melatonin in male rats. *Neuroendocrinology*, **19**, 72–80.
Relkin, R. (1967). Neurological pathways involved in lactation. *Dis. nerv. Syst.*, **28**, 94–97.
Relkin, R. (1972). Effects of variations in environmental lighting on pituitary and plasma prolactin levels in the rat. *Neuroendocrinology*, **9**, 278.
Relkin, R. & Adachi, M. (1973). Prolactin secretion and sodium deprivation. *Fedn. Proc.*, **32**, 1031.
Relkin, R., Adachi, M. & Kahen, S. A. (1972). Effects of pinealectomy and constant light and darkness on prolactin levels in the pituitary and plasma and on pituitary ultrastructure of the rat. *J. Endocr.*, **54**, 263.
Reyes, F. I., Winter, J. S. D. & Faiman, C. (1972). Pituitary ovarian interrelationships during the puerperium. *Am. J. Obstet. Gynec.*, **114**, 589–594.
Reyes, F. I., Winter, J. S. D., Faiman, C. & Hobson, W. C. (1975). Serial serum levels of gonadotrophins, prolactin and sex steroids in the non-pregnant and pregnant chimpanzee. *Endocrinology*, **96**, 1447–1455.
Richards, R. C. & Benson, S. K. (1970). Ultrastructural changes associated with hormonally induced inhibition of mammary involution. *Acta. endocr., Copnh.*, Suppl. 138, 257.
Richardson, B. P. (1973). Evidence for a physiological role of prolactin in osmoregulation after its inhibition by 2 bromo-α-ergocryptine. *Br. J. Pharmac.*, **47**, 623P–624P.
Richardson, D. (1945). Thyroid and pituitary hormones in relation to regeneration. II. Regeneration of the hind limb of the newt *Triturus viridescens* with different combinations of thyroid and pituitary hormones. *J. exp. Zool.*, **100**, 417–429.

Richardson, K. C. (1935). The secretory phenomena in the oviduct of the fowl including the process of shell formation examined by the microincumeration technique. *Phil. Trans. R. Soc., B*, **225**, 149–195.

Riddle, O. (1963*a*). Prolactin or progesterone as key to parental behaviour: a review. *Anim. Behav.*, **11**, 419–432.

Riddle, O. (1963*b*). Prolactin in vertebrate function and organisation. *J. Natn. Cancer Inst.*, **31**, 1039–1110.

Riddle, O. & Braucher, P. F. (1931). Studies on the physiology of reproduction in birds: control of special secretion of the crop gland in pigeons by anterior pituitary hormone. *Am. J. Physiol.*, **97**, 617–625.

Riddle, O. & Dykshorn, S. W. (1932). Secretion of crop milk in the castrate male pigeon. *Proc. Soc. exp. Biol. Med.*, **29**, 1213–1215.

Riddle, O. & Lahr, E. L. (1944). On broodiness of Ring Doves following implants of certain steroid hormones. *Endocrinology*, **35**, 255–260.

Riddle, O., Bates, R. W. & Dykshorn, S. W. (1933). The preparation, identification and assay of prolactin – a hormone of the anterior pituitary. *Am. J. Physiol.*, **105**, 191–216.

Riddle, O., Lahr, E. L. & Bates, R. W. (1934). Maternal behaviour induced in virgin rats by prolactin. *Proc. Soc. exp. Biol. Med.*, **32**, 730–734.

Riddle, O., Bates, R. W. & Lahr, E. L. (1935). Prolactin induces broodiness in fowl. *Am. J. Physiol.*, **111**, 352–360.

Ridley, A. (1964). Effects of osmotic stress and hypophysectomy on ion distribution in bullfrogs. *Gen. comp. Endocr.*, **4**, 486–491.

Rimoin, D. L., Holzmann, G. B., Merinee, T. J., Rabinowitz, D., Barnes, A. C., Tyson, J. E. A. & McKusick, V. A. (1968). Lactation in the absence of human growth hormone. *J. clin. Endocr. Metab.*, **28**, 1183–1188.

Rillema, J. A. (1974). Possible role of cyclic nucleotides in mediating the effects of prolactin in the mammary gland. *Fedn Proc.*, **33**, 214.

Rillema, J. A. (1975). Possible role of prostaglandin $F_{2\alpha}$ in mediating effects of prolactin on RNA synthesis in mammary gland explants of mice. *Nature*, **253**, 466–467.

Rivier, C. & Vale, W. (1974). *In vivo* stimulation of prolactin secretion in the rat by thyrotropin releasing factor, related peptides and hypothalamic extracts. *Endocrinology*, **95**, 978–983.

Riviera, E. M. (1964). Differential responsiveness to hormones of C3H and A mouse mammary tissues in organ culture. *Endocrinology*, **74**, 853–864.

Robinson, D. S. (1963). Changes in the lipolytic activity of the guinea pig mammary gland at parturition. *J. Lipid Res.*, **4**, 21–23.

Robyn, C. & Vekemans, M. (1976). Influence of low dose oestrogen on circulating prolactin, LH and FSH levels in post-menopausal women. *Acta. endocr., Copnh.*, **83**(1), 9–14.

Robyn, C., Delvaye, P., Nobin, J., Vekemous, M., Badawi, M., Perez-Lopez, F. R. & L'Hermite, M. (1973). Prolactin and human reproduction. In *Human Prolactin*, ed. Pasteels, J. L. & Robyn, C. p. 167. Amsterdam: Excerpta Medica.

Robertson, M. C. & Friesen, H. G. (1975). The purification and characterisation of rat prolactin. *Endocrinology*, **97**, 621–629.

Rocenziveig, A., L'Hermite, M., Bila, S., Henson, J. C. & Robyn, C. (1973). Comparative study of L-dopa and 2 Br-α-ergocryptine as inhibitors of prolactin secretion in post menopausal women. *Eur. J. Cancer.*, **41**, 237.

Rogol, A. D. & Rosen, S. W. (1974). Prolactin of apparent large molecular size: the major immunoreactive prolactin component in plasma of a patient with a pituitary tumour. *J. clin. Endocr. Metab.*, **38**, 714–717.

Rolland, R. & Schellebus, L. (1973). A new approach to the inhibition of puerperal lactation. *J. Obstet. Gynaec. Br. Commonw.*, **80**, 945.

Rolland, R., Grusalus, G. L. & Hammond, J. M. (1976). Demonstration of specific binding of prolactin by porcine corpora lutea. *Endocrinology*, 98(5), 1083–1091.

Rolland, R., Lequin, R. M., Schellobus, L. A. & de Jong, F. H. (1975*a*). The role of prolactin in the restoration of ovarian function during the early postpartum period in the human female. 1. A study during physiological lactation. *Clin. Endocr.*, **4**, 15–25.

Rolland, R., de Jong, F. H., Schellobus, L. A. & Lequin, R. M. (1975*b*). The role of prolactin in the restoration of ovarian function during the early postpartum period in the human female. II. A study during inhibition of lactation by bromoergocryptine. *Clin. Endocr.*, **4**, 27–38.

Rosenblatt, J. S. (1967). Mean-hormonal basis of maternal behaviour in the rat. *Science*, **156**, 1512–1514.

Rosenfield, A. G. (1974). Injectable long acting progestogen contraception, a neglected modality. *Am. J. Obstet. Gynec.*, **120**, 537.

Ross, F. & Mussynowitz, M. L. (1968). A syndrome of primary hypothyroidism, amenorrhea and galactorrhea. *J. clin. Endocr. Metab.*, **28**, 591.

Rothchild, I. (1960). The corpus luteum – pituitary relationship: the lack of an inhibiting effect of progesterone on the secretion of pituitary luteotrophin. *Endocrinology*, **67**, 54–61.

Rothchild, I. (1965). Interrelations between progesterone and the ovary, pituitary and central nervous system in the control of ovulation and the regulation of progesterone secretion. *Vitam. Horm.*, **23**, 209.

Rothchild, I. & Schubert, R. (1963). The corpus luteum – pituitary relationship: the induction of pseudopregnancy in the rat by progesterone. *Endocrinology*, **72**, 969–972.

Rowell, T. E. (1961). Maternal behaviour in non-maternal golden hamsters (*Mesocricatus auratus*). *Anim. Behav.*, **9**, 11–15.

Rubin, R. T., Gowin, P. R., Lubin, A., Poland, R. E. & Pirke, K. M. (1975). Nocturnal increase of testosterone in men. Relation to gonadotrophins and prolactin. *J. clin. Endocr. Metab.*, **40**, 1027.

Russell, D. H. & McVicker, T. A. (1972). Polyamine biogenesis in the rat mammary gland during pregnancy and lactation. *Biochem. J.*, **1**, 71–76.

Saaman, N., Yen, S. C. C., Friesen, H. & Pearson, O. H. (1966). Serum placental lactogen levels during pregnancy and in trophoblastic disease. *J. clin. Endocr. Metab.*, **26**, 1303.

Saaman, N., Yen, S. C. C., Gonsalez, D. & Pearson, O. H. (1968). Metabolic effects of placental lactogen (HPL) in man. *J. clin. Endocr. Metab.*, **28**, 485–491.

Saaman, N. A., Leavens, M. E. & Jksse, J. H. Jr. (1977). Serum prolactin in patients with 'functionless' chromophobe adenomes before and after therapy. *Acta endocr.*, **84**(3), 449–460.

Sacki, Y. & Tanake, Y. (1955). Changes in prolactin content of fowl pituitary during broody periods and some experiments on the induction of broodiness. *Poult. Sci.*, **34**, 909–919.

Sage, M. (1970). Control of prolactin release and its role in colour change in the teleost *Gillichthys mirabilis*. *J. exp. Zool.*, **173**, 121–128.

Sage, M. (1973). The relationship between the pituitary content of prolactin and blood sodium levels in mullet (*Mugil cephalus*) transferred from sea water to freshwater. *Contr. mar. Sci.*, **17**, 163–167.

Sage, M. & Bern, H. A. (1972). Assay of prolactin in vertebrate pituitaries by its dispersion of xanthophae pigment in the teleost *Gillichthys mirabilis*. *J. exp. Zool.*, **180**, 169–174.

Sage, M. & de Vlaming, V. L. (1973). A diurnal rhythm in hormone effectiveness and in pituitary content of prolactin in *Fundulus similis*. *Tex. Rep. Biol. Med.*, **31**, 101–102.

Saint-Giron, H. (1963). Histologie comparee de l'hypophyse chez les Reptiles. *CNRS, Colloq. Intern. Paris*, **128**, 275–285.
Saint-Giron, H. (1965). Dennees histologiques sur le lobe anterieur de l'hypophyse chez *Sphenodon punctalis*. *Archs Anat. microsc. Morph. exp.*, **54**, 633–634.
Saint-Giron, H. (1967). Morphologie comparee de l'hypophyse chez les Squamata. *Annls Sci. nat. (Zool.)*, 12 Ser., **9**, 229–308.
Saito, T. & Saxena, B. B. (1975). Specific receptors for prolactin in the ovary. *Acta Endocr., Copnh.*, **80**, 126–137.
Saji, M. A. (1966). An immunological estimation for prolactin in sheep blood. In *Reproduction in the Female Mammal*, ed. Lamming, C. E. & Amoroso, E. C. London: Butterworths.
Salil, H., Flax, H., Brander, W. & Hobbs, J. R. (1972). Prolactin dependence in human breast cancers. *Lancet*, **2**, 1103.
Sampietro, A. & Vericelli, P. (1968). Effetti della prolattina sul tasso ematieo del sodio nel Tritone cristato normale ed ipofisectomizzato. *Boll. Zool.*, **35**, 419.
Sapag-Hagar, M. & Greenbaum, A. L. (1974*a*). The role of cyclic nucleotides in the development and function of rat mammary tissue. *FEBS Lett.*, **46**, 180–183.
Sapag-Hagar, M. & Greenbaum, A. L. (1974*b*). Adenosine 3′, 5′-monophosphational hormone interrelationships in the mammary gland of the rat during pregnancy and lactation. *Europ. J. Biochem.*, **47**, 303–312.
Sapag-Hagar, M., Greenbaum, A. L. & Lewis, D. J. (1974). The effects of cyclic AMP on enzymatic and metabolic changes in explants of rabbit mammary tissue. *Biochem. biophys. Res. Commun.*, **59**, 261–268.
Sar, M. & Meites, J. (1968). Effects of suckling on pituitary release of prolactin, GH and TSH in postpartum lactating rats. *Neuroendocrinology*, **4**, 25.
Sar, M. & Meites, J. (1969). Effects of suckling on pituitary release of prolactin, growth hormone and TSH in postpartum lactating rats. *Neuroendocrinology*, **4**, 25.
Sasaki, G. H. & Leung, B. S. (1975). On the mechanism of hormone action in 7, 12 dimethylbenz (a) anthracene-induced mammary tumour. 1. Prolactin and progesterone effects on oestrogen receptor *in vitro*. *Cancer*, **35**, 645–651.
Sasame, H. A., Perez-Cruet, J., di Chiara, G. (1972). Evidence that methadone blocks dopamine receptors in the brain. *J. Neurochem.*, **19**, 1953–1957.
Sassin, J. F., Frantz, A. G., Wettgman, E. D. & Kapen, S. (1972). Human prolactin: 24 hour pattern with increased release during sleep. *Science*, **177**, 1205–1207.
Sate, T., Junjo, Y., Iesaka, T., Ishikawa, T. & Igarashi, M. (1974). Follicle stimulating hormone and prolactin release induced by prostaglandins in the rat. *Prostaglandins*, **5**. 483.
Sawin, P. B., Dononberg, V. H., Ross, S., Hafter, E. & Zarrow, M. X. (1960). Materna behaviour in the rabbit: Hair loosening during gestation. *Am. J. Physiol.*, **198**, 1099–1102.
Sawyer, C. H. (1967). Some endocrine aspects of forebrain inhibition. *Brain Res.*, **6**, 48–59.
Saxena, B. N., Emerson, K. & Selenkow, H. A. (1969). Serum placental lactogen (HPL) levels as an index of placental function. *New Engl. J. Med.*, **281**, 225.
Scanes, C. G., Bolton, N. J. & Chadwick, A. (1976). Purification and properties of an avian prolactin. *Gen. comp. Endocr.*, **27**, 371–379.
Scaraimuzzi, R. J., Blake, C. A., Norman, R. L., Hilliard, J. & Sawyer, C. H. (1972). Testosterone and gonadotrophin levels in serum of male rats following hypothalamic deafferentiation. *J. Endocr.*, **55**, xvi.
Schamble, M. K. & Mentwig, M. R. (1974). Temperature and prolactin as control factors in newt forelimb regeneration. *J. exp. Zool.*, **187**, 335–344.
Schams, D. (1973). *Untersuchungen über Prolaktin beim Rind.* Habilitationsschrift Technische Universitat Munchen-Weidenstephen.

Schams, D. & Karg, H. (1970). Untersuchungen über Prolaktin in Rinderblut mit einer radioimmunologischen Bestimnungsmethode. *Zentbl. vet. Med.*, **A17**, 193.

Schemberg, D. W. (1969). The concept of a uterine luteolytic hormone. In *The Gonads*, ed. McKerns, K. W. pp. 383–414. New York: Appleton.

Schenber, J. G., Ben-David, M. & Polishuk, W. Z. (1975). Prolactin in normal pregnancy; relationship of foetal, maternal and amniotic fluid levels. *Am. J. Obstet. Gynaec.*, **123**, 834–838.

Schenber, J. G., Ben-David, M. & Polishuk, W. Z. (1976). The role of prolactin in reproduction. *Harefuah*, **89**, 333–335.

Schooley, J. P. (1937). Pituitary cytology in pigeons. *Cold Spring Harb. Symp. quant. Biol.*, **5**, 165–179.

Schooley, J. P. & Riddle, O. (1938). The morphological basis of pituitary function in pigeons. *Am. J. Anat.*, **62**, 313–350.

Schooley, J. P., Riddle, O. & Bates, R. W. (1938). Effective stimulation of crop sacs by prolactin in hypophysectomised and adrenalectomised pigeons. *Proc. Soc. exp. Biol. Med.*, **136**, 408.

Schotté, O. E. & Talon, A. (1960). The importance of autotransplanted pituitaries for survival and regeneration of adult Triturus. *Experientia*, **16**, 72–74.

Schreibman, M. P. & Kallman, K. D. (1966). Endocrine control of freshwater tolerance in teleosts. *Gen. comp. Endocr.*, **6**, 144–155.

Schultheiss, H., Hanke, W. & Maetz, J. (1972). Hormonal regulation of the skin diffusional permeability to water during development and metamorphosis of *Xenopus laevis* Daudin. *Gen. comp. Endocr.*, **18**, 400–404.

Schultz, R. B. & Blizzard, R. M. (1966). A comparison of human placental lactogen (HPL) and human growth hormone (HGH) in hypopituitary patients. *J. clin. Endocr. Metab.*, **26**, 921–924.

Schulz, K. D., Czygon, P. J., del Pozo, E. & Friesen, H. G. (1973). Varying response of human metastasing breast cancer to the treatment with 2 Br-α-ergocryptine (CB 154). In *Human Prolactin*, ed. Pasteels, J. L. & Robyn, C. Amsterdam: Excerpta Medica.

Schulze, A. B. & Turner, C. W. (1933). Experimental initiation of milk secretion in the albino rat. *J. Dairy Sci.*, **16**, 129–133.

Schwarts, M. B. (1964). Acute effects of avariectomy on pituitary LH, uterine weight and vaginal cornification. *Am. J. Physiol.*, **207**, 1251–1259.

Sciarra, J. J., Kaplan, S. L. & Grumbach, M. M. (1963). Localisation of antihuman growth hormone serum within the human placenta: evidence for a human chorionic growth hormone. *Nature*, **199**, 1005–1006.

Selander, R. K. (1964). The problem of timing of the development of the incubation patch in male birds. *Condor*, **66**, 75–76.

Selander, R. K. & Kuich, A. (1963). Hormonal control and development of the incubation patch in icterids with notes on the behaviour of cowbirds. *Condor*, **62**, 65.

Selander, R. K. & Young, S. Y. (1966). The incubation patch of the house sparrow, Passer domesticus Limnaeus. *Gen. comp. Endocr.*, **6**, 325–333.

Selye, H. (1951). General adaptation syndrome. *A. Rev. Med.*, **2**, 327–342.

Selye, H., Collys, J. B. & Thomson, D. L. (1935). Effect of oestrin on ovaries and adrenals. *Proc. Soc. exp. Biol. Med.*, **32**, 1377–1381.

Seppala, M. & Hirronen, E. (1975). Raised serum prolactin levels associated with hirsutism and amenorrhea. *Br. med. J.*, **4**, 144.

Seppala, M. & Ruoslahti, E. (1970). Serum concentration of human placental lactogenic hormone (HPL) in pregnancy complications. *Acta obstet. gynaec. scand.*, **49**, 143–147.

Seppala, M., Hirronen, E., Unnerus, H. A., Ranta, T. & Laatikainen, T. (1976). Prolactin and testosterone, independent circulating levels in hyperprolactinemic and normoprolactinemic amenorrhea. The effect of prolactin suppression by Bromocriptine. *J. clin. Endocr. Metab.*, **43**, 198.

Shaar, C. J. & Clemens, J. A. (1974). The role of catecholamines in the release of anterior pituitary prolactin *in vitro*. *Endocrinology*, **95**, 1202–1212.

Shaar, C. J., Smalstig, E. B. & Clemens, J. A. (1973). The effect of catecholamines, apomorphine and monoamine oxidase on rat anterior pituitary prolactin release *in vitro*. *Pharmacologist*, **15**, 256.

Shani, J., Furr, B. J. A. & Forsyth, I. A. (1973). A re-examination of the antigonadal effect of prolactin in the male fowl *Gallus domesticus*. *J. Reprod. Fert.*, **34**, 167–170.

Shani, J., Sulunovici, S., Goldhaber, G., Givant, Y., Sulman, F. G., & Lumenfield, B. (1976). Cyclic AMP and the effect of prolactin on the crop sac. *J. Endocr.*, **69**, 169–173.

Sherry, W. E. & Nicoll, C. S. (1967). RNA and protein synthesis in the response of the pigeon crop to prolactin. *Proc. Soc. exp. Biol. Med.*, **126**, 824–829.

Sheth, N. A., Ranadive, K. J., Suranja, J. N. & Sheth, A. R. (1975). Circulating levels of prolactin in human breast cancer. *Br. J. Cancer*, **32**(2), 160–168.

Sherman, L. (1976). Prolactin and breast cancer. *CA*, **25**, 258–263.

Sherwood, L. M. (1967). Similarities in the chemical structure of human placental lactogen and pituitary growth hormone. *Proc. natn. Acad. Sci. U.S.A.*, **58**, 2307–2314.

Sherwood, L. M. (1969).

Sherwood, L. M. & Hardwerger, S. (1969). Correlations between structure and function of human placental lactogen and human growth hormone. In *Fifth Rochester Trophoblast Conf.*, ed. Lund, C. & Choate, J. W. pp. 230–255. Rochester: Rochester Univ. Press.

Sherwood, L. M., Handwerger, S. & McLaurin, W. D. (1971). Amino acid sequence of human placental lactogen. *Nature*, **233**, 59–61.

Sherwood, L. M., Handwerger, S. & McLaurin, W. D. (1972). The structure and function of human placental lactogen. In *Lactogenic Hormones*. ed. Wolstenholme, G. E. & Knight, J. E. pp. 27–45. Ciba Foundation.

Shikado, Y. (1966). Studies on the effect of chronic electrical stimulation of the hypothalamus on the artificial corpora lutea and mammary gland in rabbits. *J. Jap. obstet. gynaec. Soc.*, **13**, 168.

Shinino, M., Arimura, A. & Remels, E. G. (1974). Effects of blinding, olfactory bulbectomy, and pinealectomy on prolactin and growth hormone cells of the rat, with special reference to ultrastructure. *Am. J. Anat.*, **139**, 191–207.

Shiu, R. P. & Friesen, H. G. (1974*a*). Solubilization and purification of a prolactin receptor from the rabbit mammary gland. *J. biol. Chem.*, **249**, 7902–7911.

Shiu, R. P. & Friesen, H. G. (1974*b*). Properties of a prolactin receptor from the rabbit mammary gland. *Biochem. J.*, **140**, 301.

Shiu, R. P., Kelly, P. A. & Friesen, H. G. (1973). Radioreceptor assay for prolactin and other lactogenic hormones. *Science*, **180**, 968–970.

Shoemaker, V. H. (1969). Excretion and osmoregulation in birds. In *Avian Biology*, ed. Farner, D. S. & King, J. K. Vol. 1. pp. 271–294. London & New York: Academic Press.

Shome, B. & Friesen, H. G. (1971). Purification and characterisation of monkey placental lactogen. *Endocrinology*, **89**, 631–641.

Shulkes, A. A., Cavelli, M. D. & Denton, D. A. (1972). Hormonal factors influencing salt appetite in lactation. *Aust. J. exp. Biol. med. Sci.*, **50**, 819–826.

Shymala, G., Meri, T. & Bern, H. A. (1974). Nuclear and cytoplasmic oestrogen receptors in vaginal and uterine tissue of mice treated neonatally with steroids and prolactin. *J. Endocr.*, **63**, 275–284.

Simpson, A. A., Simpson, M. H. W. & Kulkarni, P. M. (1973*a*). Effect of perphenazine during late pregnancy on prolactin production and lactogenesis in the rat. *J. Endocr.*, **57**, 431–436.

Simpson, A. A., Simpson, M. H. W., Sinha, Y. N. & Schmidt, G. H. (1973*b*). Changes in concentration of prolactin and adrenal corticosteroids is not plasma during pregnancy and lactation. *J. Endocr.*, **58**, 675–676.

Simpson, H. W., Cole, E. H., Hume, P. D., Fleming, K. A. & Ravadia, H. B. (1975). Is prolactin involved in circadian rhythm of sodium excretion? *Lancet*, **i**, 342.

Singer, W., Desjardins, P. & Friesen, H. G. (1970). Human placental lactogen. An index of placental function. *Obstet. Gynaec.*, **36**, 222–232.

Singhas, C. A. & Dent, J. M. (1975). Hormonal control of the tail fin and of the nuptial pads in the male red-spotted newt. *Gen. comp. Endocr.*, **26**, 382–393.

Sinha, Y. N. & Tucker, A. H. (1969). Relationship of pituitary prolactin and LH to mammary and uterine growth of prepubertal rats during the oestrus cycle. *Proc. Soc. exp. Biol. Med.*, **131**, 908.

Sinha, Y. M., Selby, F. W. & Vanderhaan, W. P. (1974). Effects of ergot drugs on prolactin and growth hormone secretion and a mammary nucleic acid content in C3H/Bi mice. *J. natn. Cancer Inst.*, **52**, 189–191.

Sinha, Y. M., Selby, F. W., Lewis, U. & Vanderhaan, W. P. (1973). A homologous radioimmunoassay for human prolactin. *J. clin. Endocr. Metab.*, **20**, 1095–1106.

Skadhauge, E. (1974). Renal concentrating ability in xesophilic birds. *Symp. zool. Soc. Lond.*, **31**, 113–131.

Smalstig, E. B., Sawyer, B. D. & Clemens, J. A. (1974). Inhibition of rat prolactin release by apomorphine *in vivo* and *in vitro*. *Endocrinology*, **95**, 123–129.

Smith, M. S., Freeman, M. E. & Neill, J. D. (1974). The control of progesterone secretion during the estrous cycle and early pseudopregnancy in the rat. Prolactin, Gonadotrophin and steroid levels associated with rescue of the corpus luteum of pseudopregnancy. *Endocrinology*, **96**, 219–226.

Smith, M. S. & Niell, J. D. (1975*a*). A 'critical period' for cervically-stimulated prolactin release. *Endocrinology*, **98**, 324.

Smith, M. S. & Niell, J. D. (1975*b*). Termination at midpregnancy of the true daily surges of plasma prolactin initiated by mating in the rat. *Endocrinology*, **96**, 696–699.

Smith, R. J. F. & Hoar, W. S. (1967). The effects of prolactin and testosterone on the parental behaviour of the male stickleback, *Gasterosteus aculeatus*. *Anim. Behav.*, **15**, 342–352.

Smith, V. G. & Convey, E. M. (1974). TRH-stimulation of prolactin release from bovine pituitary cells. *Proc. Soc. exp. Biol. Med.*, **149**, 70–74.

Smith, V. G., Beek, T. W., Conroy, E. M. & Tucker, H. A. (1974). Bovine serum prolactin, growth hormone cortisol and milk yield after ergocryptine. *Neuroendocrinology*, **15**, 172–181.

Snart, R. S. & Dalton, T. R. (1972). Response of toad bladder to prolactin. *Comp. Biochem. Physiol.*, **45A**, 307–311.

Solyom, J., Ludwig, E. & Vajda, A. (1971). Aldosteronotrophic effect of pituitary incubation medium in the rat. *Acta physiol. hung.*, **39**, 343–349.

de Sombre, E. R., Kledzik, G. & Marshall, S. (1976). Estrogen and prolactin receptor concentrations in rat mammary tumours and response to endocrine ablation. *Cancer Res.*, **36**, 3830–3833.

Soulairac, A. (1958). Les regulations psychophssiologiques de la faim. *J. Physiol., Paris*, **50**, 663–783.

Spark, R. F., Palleta, J., Naftolic, F. & Clemens, R. (1976). Galactorrhea-amenorrhea syndromes: etology and treatment. *Ann. intern. Med.*, **84**, 532.

Spellacy, W. H. & Buhi, W. C. (1969). Pituitary growth hormone and placental lactogen levels measured in normal term pregnancy and at early and late postpartum periods. *Am. J. Obstet. Gynec.*, **105**, 888–896.

Spellacy, W. H., Buhi, W. C. & Birk, S. A. (1970*a*). Human growth hormone and placental lactogen levels in midpregnancy and late postpartum. *Obstet. Gynaec.*, **56**, 238–243.

Spellacy, W. H., Buhi, W. C. & Birk, S. A. (1970*b*). Normal lactation and blood growth hormone studies. *Am. J. Obstet. Gynec.*, **107**, 244–249.

Spellacy, W. H., Buhi, W. C. & Birk, S. A. (1975). Stimulated plasma prolactin levels in women using medoxyprogesterone acetate or an intruterine device for contraception. *Fert. Sterl.*, **26**(10), 970–981.
Spellacy, W. H., Buhi, W. C., Schram, J. D., Birk, S. A. & McCreary, S. A. (1971). Value of human chorionic sommatotrophin in managing high risk pregnancies. *Obstet. Gynec.*, **37**, 567–573.
Spies, H. G., & Niswender, G. D. (1971). Levels of prolactin, LH and FSH in the serum of intact and pelvic-neurectomised rats. *Endocrinology*, **88**, 937.
Spies, H. G., Hilliard, J. & Sawyer, C. H. (1968). Pituitary and uterine factors controlling regression of corpora lutea in intact and hypophysectomised rabbits. *Endocrinology*, **83**, 354–367.
Stanley, J. G. & Fleming, W. R. (1967). Effect of prolactin and ACTH on the serum and urine sodium levels of *Fundulus kansae. Comp. Biochem. Physiol.*, **20**, 199–208.
Steele, E. & Hinde, R. A. (1963). Hormonal control of brood patch and oviduct development in domesticated canaries. *J. Endocr.*, **26**, 11–24.
Steele, E. & Hinde, R. A. (1964). Effect of exogenous oestrogen on brood patch development of intact and ovariectomised canaries. *Nature, Lond.*, **202**, 718–719.
Stefano, F. J. & Denoso, A. O. (1967). Nerepinephrine levels in the rat hypothalamus during the estrous cycle. *Endocrinology*, **81**, 1405–1406.
Sterental, A., Dominiquez, M., Weisman, C. & Pearson, O. H. (1963). *Cancer Res.*, **23**, 481.
Stetson, M. H., Lewis, R. A. & Farner, D. S. (1973). Some effects of exogenous gonadotrophins and prolactin on photostimulated and photorefractory white crowned sparrows. *Gen. comp. Endocr.*, **21**, 424–430.
Strong, C. R. & Dils, R. (1972). Fatty acid biosynthesis in rabbit mammary gland during pregnancy and early lactation. *Biochem. J.*, **128**, 1303–1309.
Stroobants, W. L., Van Zanten, A. K. & de Bruijn, H. W. (1975). Serial human placental lactogen estimations in serum and placental weight for dates. *Br. J. Obstet. Gynaec.*, **82**, 899–902.
Stumpf, W. E. & Sar, M. (1971). Estradiol concentrating neurons in the amygdala. *Proc. Soc. exp. Biol. Med.*, **136**, 102.
Subramanian, M. G. & Gala, R. R. (1976). The influence of cholinergic serotoninergic and adrenergic drugs on the afternoon surge of plasma prolactin in ovariectomised estrogen treated rats. *Endocrinology*, **98**, 842–848.
Sud, S. C., Clemens, J. A. & Meites, J. (1970). *Indian J. exp. Biol.*, **8**, 81.
Sundararaj, B. I. & Goswani, S. V. (1966). Effects of mammalian hypophysical hormones, placental gonadotrophins, gonadal hormones and adrenal corticosteroids on ovulation and spawning in hypophysectomised catfish *Heteropreustes fossilis* Bloch. *J. exp. Zool.*, **161**, 287–296.
Sundararaj, B. I. & Nayyar, S. K. (1969). Effect of prolactin on the 'seminal vesicles' and neural regulation of prolactin secretion in the catfish *Heteropreustes fossilis* Bloch. *Gen. Endocr.*, Suppl. 2., 69–80.
Suwa, S. & Friesen, H. G. (1969*a*). Biosynthesis of human placental proteins and human placental lactogen (HPL) *in vitro*. I. Identification of ^{3}H-labelled HPL. *Endocrinology*, **85**, 1028–1036.
Suwa, S. & Friesen, H. G. (1969*b*). Biosynthesis of human placental proteins and human placental lactogen (HPL) *in vitro*. II. Dynamic studies of normal term placentas. *Endocrinology*, **85**, 1037–1045.
Swanson, L. V. & Hafs, H. D. (1971). LH and prolactin in blood serum from estrus to ovulation in Hobtein heifers. *J. Anim. Sci.*, **33**, 1038.
Swingle, W. W., Seay, P., Perlmutt, J., Collins, E. J., Barlow, G. & Feder, E. J. (1951*a*). An experimental study of pseudopregnancy in the rat. *Am. J. Physiol.*, **167**, 586–593.
Swingle, W. W., Feder, E. J., Barlow, G., Collins, E. J. & Perlmutt, J. (1951*b*). Induction of pseudopregnancy in rat following adrenal removal. *Am. J. Physiol.*, **167**, 593–598.

Szako, M. & Frohman, L. A. (1976). Dissociation of prolactin-releasing activity from thyrotropin releasing hormone in porcine stalk media ominence. *Endocrinology*, **98**, 1451.

Tai, S. W. & Chadwick, A. (1976). The effect of salt loading on the prolactin cells of the chicken pituitary gland. *I.R.C.S. med. Sci.*, **4**, 509–513.

Takahava, J., Arimura, A. & Sahally, A. V. (1974*a*). Stimulation of prolactin and growth hormone release by TRH infused into a hypophysial portal vessel. *Proc. Soc. exp. Biol. Med.*, **146**, 831–835.

Takahava, J., Arimura, A. & Sahally, A. V. (1974*b*). Effect of catecholamines on the TRH-stimulated release of prolactin and growth hormone from sheep pituitaries *in vitro*. *Endocrinology*, **95**, 1490–1494.

Takanati, O., Kumaoka, S., Sakauchi, N., Kuwahara, T. & Matsushima, S. (1967). A case report of pituitary tumour presenting as Forkes-Albright syndrome: determination of pituitary prolactin content. *Endocr. jap.*, **14**, 95–100.

Talamantes, F. (1973). Mammotrophic activity of chinchilla, hamster and rat placental. *Endocrinology*, **92**, Suppl. A, 275.

Talamantes, F. (1975). Comparative study of the occurrence of placental prolactin among mammals. *Gen. comp. Endocr.*, **27**, 115–121.

Talbot, J. A. & Reiter, R. J. (1974). Influence of metalanic 5-Methoxytryptopol and pinealectomy on pituitary and plasma gonadotrophin and prolactin levels in castrated adult male rats. *Neuroendocrinology*, **13**, 164–172.

Talwalker, P. K. & Meites, J. (1961). Mammary lobulo-alveolar growth induced by anterior pituitary hormones in adreno-ovariectomised and adreno-ovariectomised hypophysectomised rats. *Proc. Soc. exp. Biol. Med.*, **107**, 880–883.

Talwalker, P. K. & Meites, J. (1964). Mammary lobulo-alveolar growth in adreno-ovariectomised rats following transplant of 'mammotropic" pituitary tumour. *Proc. Soc. exp. Biol. Med.*, **117**, 121–124.

Talwalker, P. K., Meites, J. & Nicoll, C. S. (1960). Effects of hydrocortisone, prolactin and oxytocin on lactational performance of rats. *Am. J. Physiol.*, **199**, 1070–1072.

Talwalker, P. K., Ratner, A. & Meites, J. (1963). *In vitro* inhibition of pituitary prolactin synthesis and release by hypothalamic extract. *Am. J. Physiol.*, **205**, 213–218.

Tashjian, A. H. (1972). *Proc. 4th Int. Congr. Endocr.* Washington D.C. ed. Scow, R. Amsterdam: Excerpta Medica.

Tashjian, A. H. & Hoyt, R. F. (1972). Transient controls of organ specific functions in pituitary cells in culture. In *Molecular genetics and developmental biology*, ed. Sussman, M. pp. 353–387. Englewood Cliffs, New Jersey: Prentice-Hall Inc.

Tashjian, A. H., Levine, L. & Wilhelmi, A. E. (1965). Immunochemical studies with antisera to fractions of human growth hormone which are high or low in pigeon crop gland-stimulating activity. *Endocrinology*, **77**, 1023–1036.

Tashjian, A. H., Bancroft, F. C. & Levine, L. (1970). Production of both prolactin and growth hormone by clonal strains of rat pituitary tumour cells. Differential effects of hydrocortisone and tissue extracts. *J. Cell Biol.*, **47**, 61–70.

Tassawa, R. A. (1969). Hormonal and nutritional requirements for limb regeneration and survival of adult newts. *J. exp. Zool.*, **170**, 33–53.

Tavolga, W. N. (1955). Effects of gonadectomy and hypophysectomy on pre-spawning behaviour in males of the gobiid fish *Bothygobius soporator*. *Physiol. Zool.*,

Taylor, J. C., Reaker, M. & Linzell, J. L. (1975). Effect of prolactin on ion movements across the mammary epithelium of the rabbit. *J. Endocr.*, **65**, 26–27P.

Terenius, L. (1971). Anti-oestrogen and breast cancer. *Europ. J. Cancer*, **7**, 57–74.

Terkel, J. (1970). *Aspects of maternal behaviour in the rat with special reference to humoural factors underlying maternal behaviour at parturition.* Ph.D. Thesis. Rutgers Univ. Newark N.J.

Terkel, J. & Rosenblatt, R. S. (1968). Maternal behaviour induced by maternal blood plasma injected into virgin rats. *J. comp. Physiol. Psychol.*, **65**, 479–482.
Terkel, J., Blake, C. A. & Sawyer, C. H. (1972). Serum prolactin levels in lactating rats after suckling or exposure to ether. *Endocrinology*, **91**, 49.
Terranova, P. F. & Kent, G. C. (1974). Uterine trauma; its anobolic effect and influence to uterine sensitivity to prolactin in ovariectomised hamsters. *Endocrinology*, **94**, 1484–1486.
Thaphlijal, J. P. & Saxena, R. H. (1964). The effect of prolactin on the hypophyses and testes of an Indian weaver bird (*Ploceus phillippinus*). *Gen. comp. Endocr.*, **4**, 119–123.
Thatcher, W. W. & Tucker, H. A. (1970). Lactational performance of rats injected with oxytocin, cortisol acetate, prolactin and growth hormone during prolonged lactation. *Endocrinology*, **86**, 237–240.
Thoman, E. B., Sproul, M., Seder, B. & Larne, S. (1970). Influence of adrenatectomy in female rats on reproductive processes including effects on the foetus and offspring. *J. Endocr.*, **46**, 297–303.
Thomas, J. A. & Manandhar, M. (1974). Effect of prolactin and/or testosterone on cyclic AMP in the rat prostate gland. *Horm. Metab. Res.*, **6**, 529–530.
Thomas, J. A. & Manandhar, M. (1975). Effects of prolactin and/or testosterone on nucleic acid levels in prostate glands of normal and castrated rats. *J. Endocr.*, **65**, 149–150.
Thorburn, G. D. & Schneider, W. (1972). The progesterone concentration in the plasma of the goat during the oestrus cycle and pregnancy. *J. Endocr.*, **52**, 23.
Thorner, M. O., McNeilly, A. S., Hagan, C. & Besser, G. M. (1974). Long term treatment of galactorrhea and hypogonadism with bromocriptine. *Brit. med. J.*, **2**, 419.
Von Tienhoven, E. (1961). Endocrinology of reproduction in birds. In *Sex and Internal Secretions*, **2**, ed. Young, W. C. 3rd ed. pp. 1088–1169. Baltimore: Williams & Wilkins.
Tindall, J. S. (1974). Hypothalamic control of secretion and release of prolactin. *J. Reprod. Fert.*, **39**, 437–461.
Tindall, T. S. & Knaggs, G. S. (1969). An ascending pathway for release of prolactin in the brain of the rabbit. *J. Endocr.*, **45**, 111.
Tindall, J. S. & Knaggs, G. S. (1970). Environmental stimuli and the mammary gland. *Mem. Soc. Endocr.*, **18**, 239.
Tindall, J. S. & Knaggs, G. S. (1972). Pathways in the forebrain of the rabbit concerned with the release of prolactin. *J. Endocr.*, **52**, 253.
Tindall, D. J. & Means, A. H. (1976). Concerning the hormonal regulation of androgen binding protein in rat testis. Endocrinology, **99**, 809–818.
Tindall, J. S., Knaggs, G. S. & Turney, A. (1967). Central nervous control of prolactin secretion in the rabbit: effect of local oestrogen implants in the amygdaloid complex. *J. Endocr.*, **37**, 279.
Tixier-Vidal, A. (1965). Caracteres ultrastructuraux des types cellulaires de l'adenohypophyse du Canard mule. *Archs Anat. microsc. Morph. exp.*, **54**, 719–780.
Tixier-Vidal, A. (1967). Modifications cytologiques ultrastructurals de la prehypophyse après divers procédés de deconnexion hypothalamo-hypophysaires chez les Oiseaux. *Biol. Med.*, **56**, 318–331.
Tixier-Vidal, A. (1969). *Biol. Bull.*, **53**, 495–505.
Tixier-Vidal, A. (1970). *Cytologie hypophysaire et photoregulation de la reproduction.* Cell Inten. C.N.R.S. Photoregulation de la reproduction, Montpelier.
Tixier-Vidal, A. & Assemacher, I. (1966). Etude cytologique de la prehypophyse du Pigeon pendant la couvaison et la lactation. *Z. Zellforsch. mikrosk. Anat.*, **69**, 489–519.
Tixier-Vidal, A. & Gourdji, D. (1965). Evolution cytologique de l'hypophyse du Canard en culture organotypique. Elaboration autonome de prolactine par les explants. *C.r. Acad. Sci.*, D, **261**, 805–808.

Tixier-Vidal, A. & Gourdji, D. (1972). Cellular aspects of the control of prolactin secretion in birds. *Gen. comp. Endocr.*, Suppl. 3, 51–64.

Tixier-Vidal, A. & Picart, R. (1967). Etude quantitative par radioautographie au microscope electronique de l'utilisation de la DL-leucine-^{3}H par les cellules de l'hypophyse du canard en culture organotypique. *J. Cell Biol.*, **35**, 501–509.

Tixier-Vidal, A., Follett, B. K. & Farner, D. S. (1968). The anterior pituitary of the Japanese quail, *Cotumix coturnix japonica*. The cytological effects of photoperiodic stimulation. *Z. Zellforsch. mikrosk. Anat.*, **92**, 610–635.

Tokon, H., Josimovich, J. B. & Salazar, H. (1972). The ultrastructure of the mammary gland during prolactin induced lactogenesis in the rabbit. *Endocrinology*, **90**, 1569–1577.

Topper, Y. J. (1969). Multiple hormone interactions related to the growth and differentiation of mammary gland *in vitro*. In *Progress in Endocrinology*, ed. Gual, C. pp. 973–978, Amsterdam: Excerpta Medica.

Topper, Y. J. (1970). Multiple hormone interactions in the development of mammary gland *in vitro*. *Rec. Prog. Horm. Res.*, **26**, 287–302.

Topper, Y. J. & Oka, T. (1971). In *Symposium on effects of drugs on cellular control mechanisms*, ed. Rabin, B. & Freedman, R. B. pp. 131–150. London: Macmillan.

Trobec, T. N. (1974). Daily rhythms in the hormonal control of fat storage in lizards. In *Chrone Biology*, ed. Scheving, L. E., Halberg, F. & Pauly, J. E. pp. 147–151. Tokyo: Igabu Shoir Ltd.

Tuchmann-Duplessis, H. (1949). Action de l'hormone gonadotrope et lactogène sur le compartement et les caractères sexual secondaires du Triton normal et castre. *Archs Anat. microsc. Morph. exp.*, **38**, 302–317.

Turkington, R. W. (1968). Hormone synthesis of DNA by mammary gland *in vitro*. *Endocrinology*, **82**, 540–546.

Turkington, R. W. (1970). Changes in hybridizable nuclear RNA during differentiation of mammary cells. *Biochem. biophys. Acta*, **213**, 484–494.

Turkington, R. W. (1971*a*). In *Developmental aspects of the cell cycle*. ed. Cameron, I. L. pp. 315–355. New York: Academic Press.

Turkington, R. W. (1971*b*). Measurement of prolactin activity in human serum by the induction of specific milk proteins, *in vitro*. *J. clin. Endocr. Metab.*, **33**, 210–216.

Turkington, R. W. (1971*c*). Ectopic production of prolactin. *New Engl. J. Med.*, **285**, 1455–1458.

Turkington, R. W. (1972*a*). Measurement of prolactin activity in human serum by the induction of specific milk proteins *in vitro*. Results in various clinical disorders. In *Lactogenic Hormones*, ed. Wolstenholme, G. E. and Knight, J. pp. 169–183. Ciba Foundation.

Turkington, R. W. (1972*b*). Molecular biological aspects of prolactin. In *Lactogenic Hormones*, ed. Wolstenholme, G. W. & Knight, J. pp. 111–127. Ciba Foundation.

Turkington, R. W. (1972*c*). Serum prolactin levels in patients with gynaecomastia. *J. clin. Endocr. Metab.*, **34**, 62–66.

Turkington, R. W. & Frant, W. L. (1972). In *Prolactin and Carcinogenesis*, 4th Tenovus workshop, ed. Boyns, A. R. & Griffiths, K. pp. 39–47. Cardiff: Alpha Omega Alpha.

Turkington, R. W. & Hill, R. L. (1969). *Science*, **163**, 1458–1460.

Turkington, R. W. & Topper, Y. J. (1966). Stimulation of casein synthesis and histological development of mammary gland by human placental lactogen *in vitro*. *Endocrinology*, **79**, 175–181.

Turkington, R. W. & Riddle, M. (1969). Hormone-dependent phosphorylation of nuclear proteins during mammary gland differentiation *in vitro*. *J. biol. Chem.*, **244**, 6040–6047.

Turkington, R. W. & Riddle, R. M. (1970). Hormone-dependent formation of polysomes in mammary cells *in vitro*. *J. biol. Chem.*, **245**, 5145–5152.

Turkington, R. W. & Ward, O. T. (1969). Hormonal stimulation of RNA polymerase in mammary gland *in vitro*. *Biochem. biophys. Acta*, **174**, 291–301.
Turkington, R. W., Juergens, W. G. & Topper, Y. J. (1967*a*). Steroid structural requirements for mammary gland differentiation *in vitro*. *Endocrinology*, **80**, 1139–1142.
Turkington, R. W., Lockwood, D. H. & Topper, Y. J. (1967*b*). The induction of milk protein synthesis in post-mitotic mammary epithelia cells exposed to prolactin. *Biochim. biophys. Acta*, **148**, 475–480.
Turkington, R. W., Underwood, L. E. & Van Wyk, J. J. (1971). Elevated serum prolactin after pituitary stalk section in man. *New Engl. J. Med.*, **285**, 707–710.
Turkington, R. W. (1973). Human prolactin, an ancient molecule provides new insight for clinical medicine. *Am. J. Med.*, **53**, 389–394.
Turkington, R. W., Brew, K., Vanaman, T. C. & Hill, R. L. (1968). The hormonal control of lactose synthetise in the developing mouse mammary gland. *J. biol. Chem.*, **243**, 3382–3387.
Turkington, R. W., Majunder, G. C., Kadohama, N., MacIndoe, J. H. & Frantz, W. L. (1973). Hormonal regulation of gene expression in mammary cells. *Rec. Prog. Horm. Res.*, **29**, 417–449.
Turpen, C., Johnson, D. C. & Dunn, J. S. (1975). Stress induced prolactin and gonadotrophic secretion patterns. *Neuroendocrinology*, **20**, 339–351.
Tyson, J. E. & Friesen, H. (1973). Factors influencing the secretion of human prolactin and growth hormone in menstrual and gestational women. *Am. J. Obstet. Gynec.*, **116**, 377.
Tyson, J. E., Austin, K. L. & Farrinholt, J. W. (1971). Prolonged nutritional deprivation in pregnancy: changes in human chorionic sematomammotrophin and growth hormone secretion. *Am. J. Obstet. Gynec.*, **109**, 1080–1082.
Tyson, J. E., Friesen, H. G. & Anderson, M. S. (1972*a*). Human lactational and ovarian response to endogenous prolactin release. *Science*, **177**, 897.
Tyson, J. E., Hurang, P., Gudya, H. & Friesen, H. G. (1972*b*). Studies of prolactin secretion in human pregnancy. *Am. J. Obstet. Gynec.*, **113**, 14–20.

Utian, W. H., Begg, G., Vinik, A. I., Paul, M. & Shumann, L. (1975). Effect of bromocriptine and chlorotrianisene on inhibition of lactation and serum prolactin. A comparative double blind study. *Br. J. Obstet. Gynaec.*, **82**, 755–9.
Utida, S., Kanija, M. & Shirai, N. (1971). Relationship between the activity of Na^+, K^+ activated adenosine triphosphatase and the number of chloride cells in eel gills with special reference to sea water adaptation. *Comp. biochem. Physiol.*, **38**, 443–447.
Utida, S., Matai, S., Hirano, T. & Kamemoto, T. I. (1971). Effect of prolactin on survival and plasma sodium levels in hypophysectomised medaka *Oryzias* latipes. *Gen. comp. Endocr.*, **16**, 566–573.
Utida, S., Hirano, J., Oida, H., Ando, M., Tohusen, D. W. & Bern, H. A. (1972). Hormonal control of the intestine and urinary bladder in teleost osmoregulation. *Gen. comp. Endocr.*, Suppl. 3, 317–327.

Vale, W., Blackwell, R., Grant, G. & Guillema, R. (1973). TRF and thyroid hormones on prolactin secretion by rat anterior pituitary cells *in vitro*. *Endocrinology*, **93**, 26–33.
Valverde, R. C., Chieffo, V. & Reichlin, S. (1972). Prolactin releasing factor in porcine and rat hypothalamic tissue. *Endocrinology*, **91**, 982–993.
Van der Laar, W. P. (1973). Changing concepts of prolactin in man. *Calif. Med.*, **118**, 28–37.
Van der Laar, W. P., Owens, D. & Topper, Y. J. (1973). Superactive forms of placental lactogen and prolactin. *Biochim. biophys. Res. Comm.*, **60**, 323–330.
Vanhaebt, L., Golstein, J., Van Cauter, E., L'Hermite, M. & Robyn, C. (1973). Etude simultanee des variations circadiennes des taux sanguins de la thyreotropine (TSH)

et de la prolactine hypophysaires chez l'homme. *C.r. Acad. Sci.*, D, **276**, 1875–1877.
Van Oordt, P. G. W. J. (1965). Nomenclature of the hormone producing cells in the adenohypophysis. A report of the International Committee for Nomenclature of the Adenohypophysis. *Gen. comp. Endocr.*, **5**, 131–134.
Van Oordt, P. G. W. J. (1966). Changes in the pituitary of the common toad *Bufo bufo* during metamorphosis, and the identification of the throtropic cells. *Z. Zellforsch. mikrosk. Anat.*, **75**, 47–56.
Varga, L., Lutterbeck, P. M., Pryor, J. S., Wenner, R. & Erb, H. (1972). Suppression of puerperal lactation by an ergot alkaloid. A double blind study. *Br. med. J.*, **2**, 743–744.
Varma, S. K., Sonbeen, P. H., Varma, K., Stuart-Soeldner, J., Sedenkow, H. A. & Emerson, K. (1971). Measurement of human growth hormone in pregnancy and correlation with human placental lactogen. *J. clin. Endocr. Metab.*, **32**, 328–332.
Vebemans, M., Delvaye, P., L'Hermite, M. & Robyn, C. (1972). Evolution des taux seriques de prolactine au cours du cycle menstruel. *C.r. Acad. Sci.*, D, **275**, 2247–2250.
Vellano, C., Peyrot, A., Mazzi, V. (1967). Effects of prolactin on the pituitary thyroid axis, integument and behaviour of the adult male crested newt. *Monit. Zool. ital.*, **1**, 207–227.
Vellano, C., Mazzi, V. & Sacerdote, M. (1970). Tail height, a prolactin dependent ambisexual character in the newt (*Trituturus cristatus carnifex* Laur.). *Gen. comp. Endocr.*, **14**, 535–541.
Vellano, C., Peyrot, A. & Crasta, G. P. (1973). Effect of prolactin on hydrotropism in the radiothyroidectomised crested newt. *Atti Accad. Sci. Torino*, **107**, 437–445.
Vellano, C., Lodi, G., Benn, M., Sacerdote, M. & Mazzi, V. (1970). Analysis of the integumentary effect of prolactin in the hypophysectomised crested newt. *Monit. Zool. ital.* (M.S.), **4**, 115–146.
Vermouth, N. L. & Deis, R. P. (1974). Prolactin release and lactogenesis after ovariectomy in pregnant rats: effect of ovarian hormones. *J. Endocr.*, **63**, 13–20.
Vietti, M., Ciancia Perone, A. & Guardabassi, A. (1973). Pronephric degeneration in normal and prolactin treated larvae from *Rana temporaria* in correlation with metamorphosis. *Boll. Zool.*, **40**, 401–404.
Vivien, J. H. (1938). Sur les effets de l'hypophysectomie chez un teleosteon marin, *Gobius paganellus*. *C.r. Acad. Sci.*, D, **207**, 1452–1455.
Vivien, J. H. (1941). Contribution a l'étude de la physiologie hypophysaire dans ses relations avec l'appareil genital, la thyroide et les corps supravenaux, chez les poissons selaciens et teleosteons. *Bull. biol.* Fr. Belg., **75**, 257–309.
Vizeolyi, E. & Perks, A. M. (1974). The effect of arginine vasotocin on the isolated amniotic membrane of the guinea pig. *Can. J. Zool.*, **52**, 371–386.
de Vlaming, V. L. & Pardo, R. J. (1974). *In vitro* effects of prolactin on liver lipid metabolism in the teleost fish *Notemigonus chrysoleucas*. *Am. Zool.*, **14**, 230 (Abstr.).
de Vlaming, V. L. & Sundararaj, B. I. (1972). Endocrine influences on seminal vesicles in the estuarine gobiid fish, *Gillichthys mirabilis*. *Biol Bull.*,
de Vlaming, V. L., Sage, M. & Beitz, B. (1975). Aspects of endocrine control of osmoregulation in the euryhaline elasmobranch *Dasyatis sakina*. *Comp. Biochem. Physiol.*, **52A**, 75–76.
de Vlaming, V. L., Sage, M. & Charlton, C. B. (1973). The effects of melatonin and prolactin on body lipids in teleosts and amphibians and an effect of melatonin on pituitary prolactin in a teleost. *Am. Zool.*, **13**, 125 (Abstr.).
de Vlaming, V. L., Sage, M., Charlton, C. B. & Tiegs, R. (1974). The effects of melatonin on lipid deposition in cyprinodontid fishes and on pituitary prolactin activity in *Fundulus similis*. *J. comp. Physiol.*,
Voci, V. E. & Carbon, N. R. (1971). Maternal behaviour induced by maternal blood plasma injected into virgin rats. Quoted in Zarrow, Gondelman & Denenberg (1971).

Voci, V. E. & Carbon, N. R. (1973). Enhancement of maternal behaviour and nest building following systemic and dioneophalic administration of prolactin and progesterone in the mouse. *J. comp. Physiol. Psychol.*, **81**, 388–393.

Voogt, J. L., Chen, C. L. & Meites, J. (1970). Serum and pituitary prolactin levels before, during and after puberty in female rats. *Am. J. Physiol.*, **212**, 396–399.

Wagner, H. O. (1956). Die Bedeutung von Umweltfaktoren und Geschlecht-hormonen fur den Jahresthythms der Zugvogel. *Z. vergl. Physiol.*, **38**, 355–369.

Wagner, H. O. (1957). Vogelzug Umweltrauze und Hormone. In *Verhandlungen der Deutschen Zoologischen Gesellschaft in Graz.* Akad. Verlagagesellschaft. K-G, Leipzig: Geest & Portig.

Wagner, H. O. & Thomas, I. (1957). Die hormonale Blockeurung des Zuginpulses der vogel wahrend der Fortpflanzungszeit. *Z. vergl. Physiol.*, **40**, 73–84.

Wakabaghshi, H., Arimura, A. & Schally, A. V. (1974). Effect of pentobarbital and ether on serum prolactin levels in rats. *Proc. Soc. exp. Biol. Med.*, **37**, 1181–1193.

Wany, D. Y., Hallowes, R. C., Bealing, J., Strong, C. & Dib, R. (1971). Effect of prolactin and growth hormone on fatty acid synthesis in mouse mammary gland explants in ergen culture. *J. Endocr.*, **51**, xxx.

Wartofsky, L., Dinard, R. C. & Noel, G. L. (1975). Failure of propanolol to alter thyroid iodine release, thyroxine turnover, or the TSH and PRL responses to thyrotropin-releasing hormone in patients with thyroxicosis. *J. clin. Endocr. Metab.*, **41**(3), 485–490.

Wartofsky, L., Dinard, R. C., Noel, G. L., Frantz, A. G. & Earll, J. M. (1976). Effect of acute increases in serum triiodothyronine on TSH and prolactin responses to TRH and prolactin in normal subjects and in patients with primary hypothyroidism. *J. clin. Endocr. Metab.*, **42**(3), 443–458.

Watson, J., Krulich, L. & McCann, S. M. (1971). Effect of crude rat hypothalamic extract on serum gonadotrophin and prolactin levels in normal and orchidectomised male rats. *Endocrinology*, **89**, 1412–1418.

Weber, W. (1962). Zur Histologie und Cytologie der Kropfmilchbildung der Taube. *Z. Zellforsch. mikrosk. Anat.*, **56**, 247–276.

Weichert, C. K., Boyd, R. W. & Cohen, R. S. (1934). A study of certain endocrine effects on the mammary glands of female rats. *Am. J. Physiol.*, **218**, 396–399.

Weidmann, U. (1956). Observations and experiments on egg laying in the black headed gull (*Larus ridibundus*). *Br. J. Anim. Behav.*, **4**, 150–161.

Weiner, R. A., Blake, C. A. & Sawyer, C. H. (1972). Integrated levels of plasma LH and prolactin following hypothalamic deafferentation in the rat. *Neuroendocrinology*, **10**, 349.

Weintraub, B. D. & Rosen, S. V. (1971). Ectopic production of human chorionic somatomammotropic by non-trophoblastic cancers. *J. clin. Endocr. Metab.*, **32**, 94–107.

Weiss, G., Dierscke, D. J., Karoch, F. J., Hotchbiss, J., Butter, W. R. & Knobil, E. (1973). The influence of lactation upon luteal function in the rhesus monkey. *Endocrinology*, **93**, 954.

Weiss, J. & Kabisch, K. (1973). Jahreszyklische untersuchungen der catecholamine im hypothalamus von *Triturus vulgaris* (L.). *Acta Histochem.* (*Jena*), **46**, 319–321.

Weist, W. G. (1970). Progesterone and 20 ahydroxypregnenolone in plasma ovaries and uteri during pregnancy in the rat. *Endocrinology*, **87**, 43–48.

Weist, W. G. & Kidwall, W. R. (1969). The regulation of progesterone secretion by ovarian dehydrogenases. In *The Gonads*, ed., McKeno, K. W. pp. 295–326. New York: Appleton.

Wellings, S. R. (1969). Ultrastructural basis of lactogenesis. In *Lactogenesis: The Initiation of Milk Secretion at Parturition*, ed. Reynolds, M. & Folley, S. J. pp. 5–25. Philadelphia: University of Pennsylvania Press.

Welsch, C. W., Nagasawa, H. & Meites, J. (1971). Increased incidence of spontaneous mammary tumours in female rats with induced hypothalamic lesions. *Cancer Res.*, **30**, 2310–2313.

Welsch, C. W., Squires, M. D., Cassell, E., Chen, C. L. & Meites, J. (1971). Median emminence lesions and serum prolactin: influence of ovariectomy and ergocornine. *Am. J. Physiol.*, **221**, 1714.

Wendelaar Bonga, S. E. & Veerhuis, M. (1974). The effects of prolactin on the kidney cells of the stickleback, *Gasterosteus aculeatus*, after transfer from the sea to freshwater, studied by ultra thin sectioning and freeze etching. *Gen. comp. Endocr.*, **22**, 382 (Abstr.).

Wents, A. C., Jones, G. S., Rocco, L. & Matthews, R. R. (1975). Gonadotrophin response to luteinising hormone releasing hormone administration in secondary amenorrhea and galactorrhea syndrones. *Obstet. Gynaec.*, **45**, 256.

von Werder, K. & Clemm, C. (1974). Evidence for 'big' and 'little' components of circulating immunoreactive prolactin in humans. *FEBS Lett.*, **47**, 181–184.

Wilberson, J. A. (1963). The role of growth hormone in regeneration of the forelimb of the hypophysectomised newt. *J. exp. Zool.*, **154**, 223–230.

Wigham, T. R. & Ball, J. N. (1974). Evidence for dopaminergic inhibition of prolactin secretion in the teleost, *Poecilia latipinna. J. Endocr.*, **63**, 46–47P.

Wigham, T. R. & Ball, J. N. (1976). *Neuroendocrinology*, **21**, 47–56.

Wigham, T. R., Ball, J. N. & Ingleton, P. M. (1975). Secretion of prolactin and growth hormone by teleost pituitaries *in vitro*. III. Effect of dopamine. *J. comp. Physiol.*, B, **104**, 87–96.

Wiley, M. L. & Collette, B. B. (1970). Breeding tubercles and contact organs in fishes, their occurrence structure and significance. *Am. Mus. Natur. Hist. Bull.*, **143**, 147–153.

Wilhelmi, A. E. (1961). Fractionation of human pituitary glands. *Can. J. Biochem.*, **39**, 1659–1668.

Wilson, G. D. & Ensor, D. M. (1977). Dopaminergic and cyclic nucleotide interactions in the control of prolactin secretion. *J. Endocr.* (in press).

Wingstrand, K. G. (1966). Comparative anatomy and evolution of the hypophysis. In *The pituitary gland*, ed. Harris, G. W. & Donovan, B. T. Vol. 1, pp. 58–126. London: Butterworths.

Winters, J. S., Faiman, C. & Hobson, W. C. (1975). Pituitary-gonadal relations in infancy: I. Patterns of serum gonadotrophic concentrations from birth to four years of age in man and chimpanzee. *J. clin. Endocr. Metab.*, **40**, 545–551.

Witorsch, R. J. & Kitay, J. I. (1972). Pituitary hormones affecting adrenal 5- -reductase activity: ACTH, growth hormone and prolactin. *Endocrinology*, **91**, 264–769.

Witochi, E. & Fugo, N. W. (1940). Response of sex characters of adult female starling to synthetic hormones. *Proc. Soc. exp. Biol. Med.*, **45**, 10–14.

Wittouck, P. (1972*b*). Intensification par la prolactine de l'absorption d'ions sodium au riveau des branchies isolées de larves d'*Ambystoma mexicanum. Arch. int. Physiol Biochem.*, **80**, 825–827.

Wolfe, J. M. (1935). Reaction of ovaries of mature female rats to injection of oestrin. *Proc. Soc. exp. Biol. Med.*, **32**, 757–759.

Wolfson, A. (1959). Role of light and darkness in regulation of refractory period in gonadal and fat cycles of migratory birds. *Physiol. Zool.*, **32**, 160–176.

Woods, M. C. & Simpson, M. E. (1961). Pituitary control of the testes of the hypophysectomised rat. *Endocrinology*, **69**, 91–125.

Wookey, J. T. & Linton, J. R. (1976). Isolation and characterisation of prolactin from the grey mullett *Mugilcephalus. Comp. Biochem. Physiol.*, **53B**, 133–137.

Wright, D. W. & Snart, R. S. (1971). Simultaneous measurement of the effect of vasopression on sodium and water transport across toad bladder. *Life Sci.*, **10**, 301–308.

Wurtman, R. J., Chu, E. W. & Axelrod, J. (1963). Relation between the oestrus cycle and the binding of catecholamines in the rat uterus. *Nature, Lond.*, **198**, 547–548.

Wuttke, W. & Meites, J. (1970). Effects of ether and pentobarbital on serum prolactin and LH levels in proestrus rats. *Proc. Soc. exp. Biol. Med.*, **135**, 648–652.

Wuttke, W. & Meites, J. (1972*a*). Luteolytic role of prolactin during the oestrus cycle of the rat. *Proc. Soc. exp. Biol. Med.*, **137**, 988–995.

Wuttke, W. & Meites, J. (1972*b*). Effects of electrochemical stimulation of the hypothalamus on serum prolactin and LH in rats. *Pflugers. Arch. Europ. J. Physiol.*, **337**, 71.

Wuttke, W., Cassall, E. & Meites, J. (1971). Effects of ergocornine on serum prolactin and LH and a hypothalamic content of PIF and LRF. *Endocrinology*, **88**, 737.

Yanai, R. & Nagasawa, H. (1970). Inhibition of mammary tumourigenesis by ergot alkaloids and promotion of mammary tumourigenesis by pituitary isografts in adreno-ovariectomised mice. *J. natn. Cancer Inst.*, **48**, 715–719.

Yen, S. S. C., Ehara, Y. & Siler, T. M. (1974). Augmentation of prolactin secretion by estrogen in hypogonadal women. *J. clin. Invest.*, **53**, 652–655.

Yochim, J. M. & Defeo, V. J. (1963). Hormonal control of the onset magnitude and duration of uterine sensitivity in the rat by steroid hormones of the ovary. *Endocrinology*, **72**, 317–326.

Yokoyama, A., Shinde, Y. V. & Ota, K. (1969). Endocrine control of changes in lactase content of the mammary gland in rats shortly before and after parturition. In *Lactogenesis: The Initiation of Milk Secretion at Parturition.* ed. Reynolds, M. & Folley, S. J. pp. 65–71. Philadelphia: University of Pennsylvania Press.

Yoshinga, K., Hawkins, R. A. & Stocker, J. F. (1969). Estrogen secretion by the rat ovary *in vivo* during the estrous cycle and pregnancy. *Endocrinology*, **85**, 103–112.

Yoshizato, K. & Yasumasu, I. (1970). Effect of prolactin on the tadpole tail fin. I. Stimulatory effect of prolactin on the collagen synthesis of the tadpole tail fin. *Devl. Growth Diff.*, **11**, 305–317.

Yoshizato, K. & Yasumasu, I. (1971). Effect of prolactin on the tadpole tail fin. II. Stimulatory effect of prolactin on the synthesis of acid mucopolysaccharide of the tadpole tail fin. *Devl. Growth Diff.*, **12**, 265–272.

Yoshizato, K. & Yasumasu, I. (1972*a*). Effect of prolactin on the tadpole tail fin. III. Effect of prolactin on the hydroxylation of protocellagen-proline on the tail fin. *Devl. Growth Diff.*, **12**, 265–272.

Yoshizato, K. & Yasumasu, I. (1972*b*). Effect of prolactin on the tadpole tail fin. IV. Effect of prolactin on the metabolic fate of hyaluronic acid, collagen and RNA with special reference to catabolic process. *Devl. Growth Diff.*, **14**, 119–127.

Yoshizato, K. & Yasumasu, I. (1972*c*). Effect of prolactin on the tadpole tail fin. V. Stimulatory effect of prolactin on the incorporation of 3H-thymidine into DNA of the tadpole tail fin. *Devl. Growth Diff.*, **14**, 129–132.

Zambrano, D. (1971). The nucleus lateralis tuberalis system of the gobiid fish *Gillichthys mirabilis.* III. Functional modifications of the neurons and the gonadotrophic cells. *Gen. comp. Endocr.*, **17**, 164–182.

Zambrano, D., Nishicka, R. S. & Bern, H.A. (1972). The innervation of the pituitary gland of teleost fishes. Its origin, nature and significance. In *Brain-endocrine interactions. Median Emminence: Structure and Function*, (International Symposium, Munich, 1971). pp. 50–66. Basel: Karger.

Zambrano, D., Clarke, W. C., Hawkins, E. F., Sage, M. & Bern, H. A. (1974). Influence of 6-hydroxydopamine on hypothalamine control of prolactin and ACTH secretion in the teleost fish, *Tilapia mossambica. Neuroendocrinology*, **13**, 284–298.

Zarate, A., Canales, E. S., Soria, J., Ruiz, F. & MacGregor, C. (1972). Ovarian refractoriness during lactation in women: effect of gonadotrophin stimulation. *Am. J. Obstet. Gynec.*, **112**, 1130.

Zarate, A., Canales, E. S., Jacobs, L. S., Moneiro, P. J., Soria, J. & Daughaday, W. H., (1973). Restoration of ovarian function in patients with the amenorrhea-galactorrhea syndrome after long term therapy with L. dopa. *Fert. Steril.*, **24**, 340.

Zarate, A., Canales, E. S., Soria, J., Lean, C., Garrido, J. & Fonesca, E. (1974*a*). Refractory postpartum ovarian response to gonadal stimulation in non-lactating women. *Obstet. Gynaec.*, **44**, 819.

Zarate, A., Canales, E. S., Soria, J., Garrido, J., Jacobs, L. S. & Schally, A. V. (1974*b*). Pituitary secretory reserve in patients with amenorrhea associated with galactorrhea. *Annls Endocr.*, **35**, 535.

Zarate, A., Schally, A. V., Soria, J., Jacobs, L. S. & Canales, E. S. (1974*c*). Effect of thyrotropin releasing hormone (TRH) on the menstrual cycle in women. *Obstet. Gynaecol.*, **43**, 487.

Zarate, A. *et al.* (1974). Gonadotrophin and prolactin secretion in human pseudocyosis, effect of synthetic luteinising hormone-releasing hormone (LH-RH) and thyrotropin releasing hormone (TRH). *Annls Endocr.*, **35**(4), 445–50.

Zarrow, M. X., Donenberg, V. H. & Kalberor, W. D. (1965). Stain differences in the endocrine basis of maternal nest building in the rabbit. *J. Reprod. Fert.*, **10**, 397–407.

Zarrow, M. X., Gondelman, R. V. & Donenberg, V. H. (1971). Prolactin: is it an essential hormone for maternal behaviour in the mammal. *Horm. Behav.*, **2**, 343–354.

Zarrow, M. X., Johnson, N. P., Donenberg, V. H. & Bryant, L. P. (1973). Maintenance of lactational diestrum in the postpartum rat through tactile stimulation in the absence of suckling. *Neuroendocrinology*, **11**, 150.

Zarrow, M. X., Faroog, A., Donenberg, V. H., Sawin, P. B. & Ross, S. (1963). Maternal behaviour in the rabbit: Endocrine control of maternal nest building. *J. Reprod. Fert.*, **6**, 375–383.

Zeilmaker, G. H. (1965). Normal and delayed pseudopregnancy in the rat. *Acta endocr. Copnh.*, **49**, 558–565.

Zeilmaker, G. H. (1968). Prolonged lactation in mice and its effect on mammary tumourigenesis. *Int. J. Cancer*, **3**, 291–303.

Zepp, E. A., Thomas, J. A., Mawhinney, M. G. & Lloyd, J. W. (1973). Differential response to either bovine or ovine prolactin on testosterone –^{3}H (T–H^{3}) metabolism of various lobes of the rat prostate gland. *Pharmacologist*, **15**, 256.

Zinder, O., Hamosh, M., & Fleck, T. R. (1974). Effect of prolactin on lipoprotein lipase in mammary glands and adipone tissue of rats. *Am. J. Physiol.*, **226**, 742–748.

Zipser, R. D., Licht, P. & Bern, H. A. (1969). Comparative effects of mammalian prolactin and growth hormone on growth in toads *Bufo boreas* & *Bufo marinus*. *Gen. comp. Endocr.*, **13**, 382–391.

Zmigrod, A., Linden, H. R. & Lamprecht, S. A. (1972). Reductive pathways of progesterone metabolism in the rat ovary. *Acta endocr., Copnh.*, **69**, 141–152.

Index